Higher MATHS

For SQA 2019 and beyond

Revision + Practice
2 Books in 1

001/02102019

10 9 8 7 6 5 4 3 2 1

ISBN 9780008365233

Published by
Leckie & Leckie Ltd
An imprint of HarperCollins*Publishers*
Westerhill Road, Bishopbriggs, Glasgow, G64 2QT
T: 0844 576 8126 F: 0844 576 8131
leckieandleckie@harpercollins.co.uk www.leckieandleckie.co.uk

Publisher: Sarah Mitchell
Project Manager: Gillian Bowman

Special thanks to
Jouve (layout and illustration); Louise Robb (proof editing)

A CIP Catalogue record for this book is available from the British Library.

Acknowledgements
Whilst every effort has been made to trace the copyright holders, in cases where this has been unsuccessful, or if any have inadvertently been overlooked, the Publishers would gladly receive any information enabling them to rectify any error or omission at the first opportunity.
Printed and bound by Grafica Veneta S.p.A., Italy

To access the ebook version of this Success Guide visit
www.collins.co.uk/ebooks
and follow the step-by-step instructions.

Contents

Contents

This book, your course and your exam

Complete Revision and Practice

This **Complete two-in-one Revision and Practice** book is designed to support you as a student of Higher Mathematics. It can be used either in the classroom, for regular study and homework, or for exam revision. By combining **a revision guide and two full sets of practice exam papers**, this book includes everything you need to be fully familiar with the Higher Mathematics exam. As well as including ALL the core course content with practice opportunities, there is comprehensive assignment and exam preparation advice, a Topic Index to help with targeted topic practice and easy reference with an Index and Quick Test answers.

Using the revision guide

TOP TIP

Always revisit questions you failed to solve. Wait a few days then try to solve them again. Use this guide to help.

About this book

This revision guide was written to help you understand and revise the Higher Mathematics course.

However, you don't become good at mathematics just by reading a guide like this one, although it will give you the knowledge and skills that you need. To become good at mathematics you will need to practise problems – the more you solve problems, the better you will become. So use this book as a starting point – come back to it for the essential knowledge and skills that you need to tackle questions and solve problems. If you get stuck there will be a similar question in this guide to help you.

Top Tips

Throughout this guide you will fnd Top Tips about the mathematics you are learning and about exam technique. You should study these carefully.

Quick Tests

You will find the Quick Tests after each topic. If you have diffculty with the questions in a test, then you need to continue to revise that topic. The answers to the tests are at the end of the revision guide.

Sample assessment questions

At the end of each chapter you will find a selection of check-up questions and also typical questions from the end-of-course exam. Detailed solutions to these questions are to be found on pages 122 to 136. You should spend time attempting these questions and then compare your solutions carefully with the given solutions.

TOP TIP

Use the Quick Tests and assessment questions to identify your strengths and weaknesses.

Using the practice exam papers

This book contains practice exam papers, which mirror the actual SQA exam as much as possible. The layout, paper colour and question level are all similar to the actual exam that you will sit, so that you are familiar with what the exam paper will look like.

The full worked solution is given to each question so that you can see how the right answer has been arrived at. The solutions are accompanied by a commentary that includes further explanations and advice. There is also an indication of how the marks are allocated and, where relevant, what the examiners will be looking for. Reference is made at times to the relevant sections in the revision guide.

The practice exam papers can be used in two main ways:

1. You can complete an entire practice paper as preparation for the final exam. If you would like to use the book in this way, you can either complete the practice paper under exam-style conditions by setting yourself a time for each paper, and answering it as well as possible without using any references or notes. Alternatively, you can answer the practice paper questions as a revision exercise, using your notes to produce a model answer. Your teacher may mark these for you.
2. You can use the Topic Index, provided before the first practice paper, to find all the questions within the papers that deal with a specific topic. This allows you to focus specifically on areas that you particularly want to revise or, if you are mid-way through your course, it lets you practise answering exam-style questions for just those topics that you have studied.

The assessment structure

To gain a course award you have to pass the end-of-course exam.

The end-of-course exam will consist of two papers as follows:

Paper 1 (non-calculator)	1 hour 30 minutes	worth 70 marks
Paper 2 (calculator allowed)	1 hour 45 minutes	worth 80 marks

Both papers will consist of some short response and some extended response questions.

Depending on how well you performed in this end-of-course exam you will be awarded a grade A,B,C or D.

The formulae list

The following formulae list will be available to you during your end-of-course exam:

FORMULAE LIST

Circle:

The equation $x^2 + y^2 + 2gx + 2fy + c = 0$ represents a circle centre $(-g, -f)$ and radius $\sqrt{g^2 + f^2 - c}$.

The equation $(x - a)^2 + (y - b)^2 = r^2$ represents a circle centre (a, b) and radius r.

Scalar Product : $\mathbf{a.b} = |\mathbf{a}||\mathbf{b}| \cos\theta$, where θ is the angle between $\mathbf{a}$ and $\mathbf{b}$

or $\mathbf{a.b} = a_1b_1 + a_2b_2 + a_3b_3$ where $\mathbf{a} = \begin{pmatrix} a_1 \\ a_2 \\ a_3 \end{pmatrix}$ and $\mathbf{b} = \begin{pmatrix} b_1 \\ b_2 \\ b_3 \end{pmatrix}$

Trigonometric formulae:

$$\sin(A \pm B) = \sin A \cos B \pm \cos A \sin B$$
$$\cos(A \pm B) = \cos A \cos B \mp \sin A \sin B$$
$$\sin 2A = 2\sin A \cos A$$
$$\cos 2A = \cos^2 A - \sin^2 A$$
$$= 2\cos^2 A - 1$$
$$= 1 - 2\sin^2 A$$

Table of standard derivatives:

$f(x)$	$f'(x)$
$\sin ax$	$a \cos ax$
$\cos ax$	$-a \sin ax$

Table of standard integrals:

$f(x)$	$\int f(x)dx$
$\sin ax$	$-\frac{1}{a} \cos ax + C$
$\cos ax$	$\frac{1}{a} \sin ax + C$

Higher MATHS

For SQA 2019 and beyond

Revision Guide

Ken Nisbet

Chapter 1

Some preliminary notation

Set notation

Sets are collections of things, usually numbers. The **members** or **elements** of a set are **listed** or **described** inside 'curly brackets' **{ }**.

The **Empty Set** is the set with no members.

A collection of only some of the members of a given set is called a **subset** of that set.

$\in$ means 'is a member of'.

$\notin$ means 'is not a member of'.

Sets of numbers

You will need to recognise the following sets of numbers:

N = **{Natural Numbers}** = {1, 2, 3, 4, ...}

W = **{Whole Numbers}** = {0, 1, 2, 3, ...}

Z = **{Integers}** = {..., −3, −2, −1, 0, 1, 2, 3, ...}

(..., −3, −2, −1: Negative Integers; 1, 2, 3, ...: Positive Integers)

Q = **{Rational Numbers}**

These are numbers that can be written as a 'Ratio' of two integers, eg $\frac{2}{3}$, $-4 = \frac{-4}{1}$, $1{\cdot}25 = \frac{5}{4}$

R = **{Real Numbers}**

These are numbers that can be represented by **all** the points on the Real Number line, e.g.

−3, $-\sqrt{3}$, 0, $\frac{1}{3}$, $\sqrt{2}$, 2, π

Note:

N is a subset of **W**, **W** is a subset of **Z**, **Z** is a subset of **Q**, **Q** is a subset of **R**.

Examples

2$\in$**N**. This means '2 is a Natural Number'.

The Even Numbers and the Odd Numbers are subsets of the Natural Numbers.

1$\notin$ {Primes}. This means '1 is not a Prime Number'.

The set of solutions from **R** of $x^2 = -4$ is the Empty Set. In other words the equation has no Real solutions.

Functions and their graphs

What is a function?

A function, f, consists of:

1. a **formula**, $f(x)$, which tells you what to do with a given value of x.
2. a **domain**, which describes the values of x you are allowed to use in the formula.

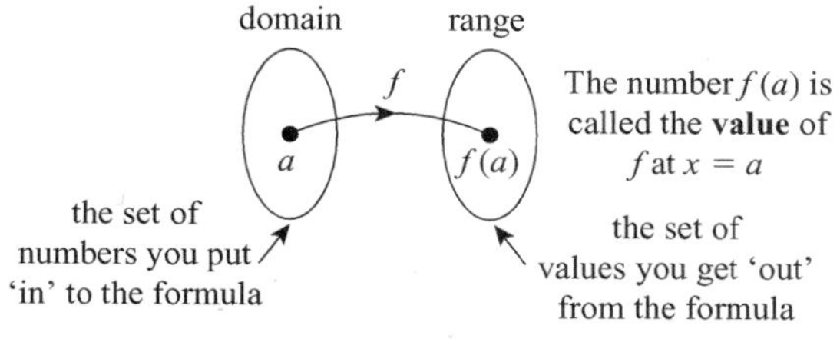

Note 1: When considering the domain of a given function, **avoid** numbers that will cause:

- Division by zero
- Square-rooting a negative number.

Note 2: The **range** is the set of all possible values of the function.

Examples

Describe a suitable domain for the functions defined by:

a) $f(x) = \frac{x+1}{x^2+x-6}$ b) $g(x) = \sqrt{x-3}$

Solution

a) Avoid $x^2 + x - 6 = 0$

$(x - 2)(x + 3) = 0$

$x = 2$ or $x = -3$

A suitable domain is: all real numbers apart from -3 and 2

b) Avoid $x - 3 < 0$ so avoid $x < 3$

A suitable domain is: all real numbers $x \geq 3$

Function graphs

A typical **graph** of a function f shows the points $(a, f(a))$ for all values $x = a$ in the domain of f.

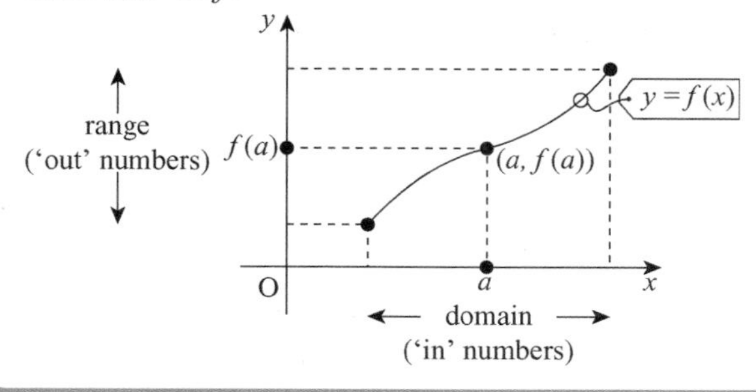

Example

The point $(2,k)$ lies on the graph $y = f(x)$ where $f(x) = 3x - 1$

Find the value of k.

Solution:

$f(2) = 3 \times 2 - 1 = 5$ so $k = 5$ ((2,5) lies on the graph)

TOP TIP

For a point on the graph: the value of the x-coordinate is put into the formula of the function $f(x)$ to get the value of the y-coordinate.

Quick Test 1

1. State a suitable domain for the function f. a) $f(x) = \frac{x+1}{x+2}$ b) $f(x) = \sqrt{10-x}$

2. In each case the given point lies on the graph $y = f(x)$. Find the value of b.

   a) $(-4, b)$, $f(x) = 2x + 3$ b) $(-4, b)$ $f(x) = x^2 - x - 1$

Composite and inverse functions

Composite functions

Two functions f and g can be combined 'one after the other':

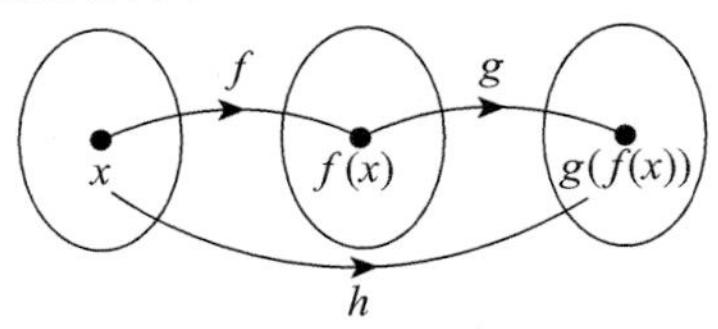

This gives a new function h defined by

$$h(x) = g(f(x))$$

Example

If $f(x) = 2x - 1$ and $g(x) = 3x^2$, find $f(g(x))$ and $g(f(x))$ and show $g(f(x)) - 2f(g(x)) = 5 - 12x$

Solution

$f(g(x)) = f(3x^2) = 2(3x^2) - 1 = \mathbf{6x^2 - 1}$

$g(f(x)) = g(2x - 1) = 3(2x - 1)^2$
$= \mathbf{12x^2 - 12x + 3}$

So $g(f(x)) - 2f(g(x))$
$= 12x^2 - 12x + 3 - 2(6x^2 - 1)$
$= 12x^2 - 12x + 3 - 12x^2 + 2 = \mathbf{5 - 12x}$

Inverse functions

A function f can have an inverse $\boldsymbol{f^{-1}}$ which 'undoes' f:

$$f^{-1}(f(a)) = a$$

for all values $x = a$ in the domain of f.

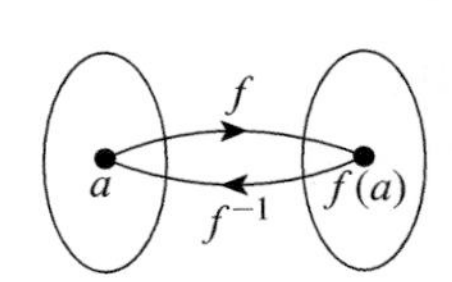

TOP TIP

Some simple inverses:

$f(x) = x + k \Rightarrow f^{-1}(x) = x - k$

$f(x) = x - k \Rightarrow f^{-1}(x) = x + k$

$f(x) = kx \Rightarrow f^{-1}(x) = \frac{x}{k}$

$f(x) = \frac{x}{k} \Rightarrow f^{-1}(x) = kx$

where k is a non-zero constant.

Inverse of a linear function

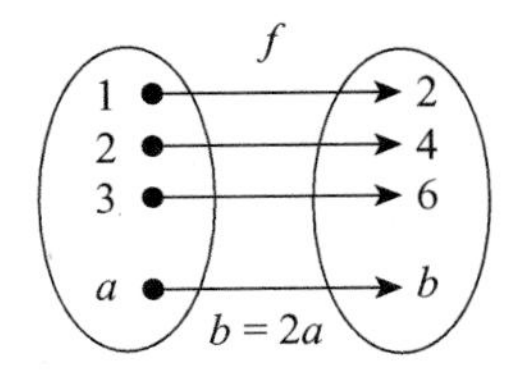

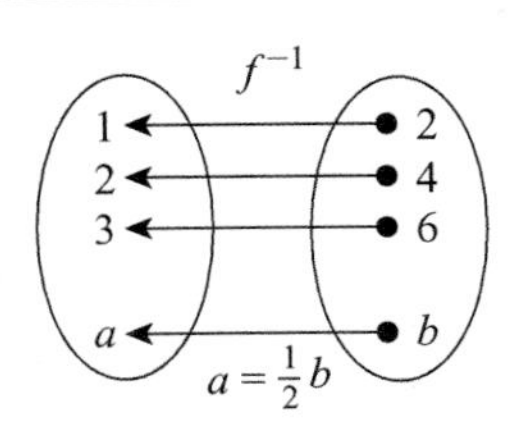

The inverse of the *doubling* function is the *halving* function.

You change the subject of $b = 2a$ from b to a to get $a = \frac{1}{2}b$

So when $f(x) = 2x$ then $f^{-1}(x) = \frac{1}{2}x$

To find the inverse of function f:

Step 1 Replace x by a in the formula for f. In other words find $f(a)$

Step 2 Let $b = f(a)$

Example

In each case find the inverse function f^{-1}

a) $f(x) = \frac{2x}{3}$ b) $f(x) = 5 - 3x$

Solution

a) $f(a) = \frac{2a}{3}$ so $b = \frac{2a}{3} \Longrightarrow 3b = 2a$

$\Longrightarrow \frac{3b}{2} = a$ so $f^{-1}(x) = \frac{3x}{2}$

b) $f(a) = 5 - 3a$ so $b = 5 - 3a$

$\Longrightarrow 3a = 5 - b$

$\Longrightarrow a = \frac{5-b}{3}$ so $f^{-1}(x) = \frac{5-x}{3}$

Step 3 Change the subject of the formula in Step 2 from b to a

Step 4 Change b to x to get the inverse formula

Graphs and inverse functions

Function Graph $y = f(x)$

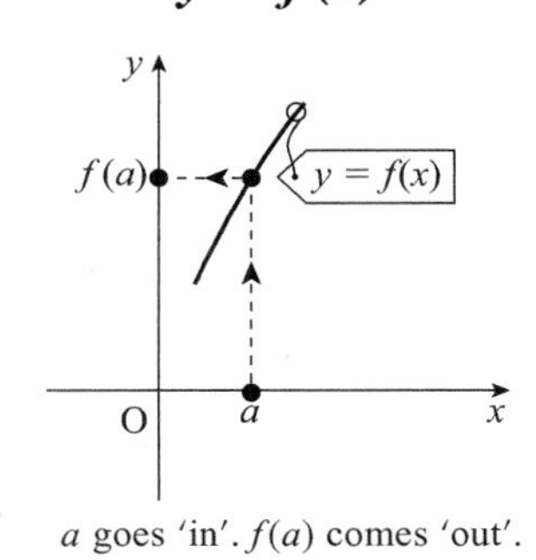

a goes 'in'. f(a) comes 'out'.

reverse the process ⟹ 'flip' in the line $y = x$

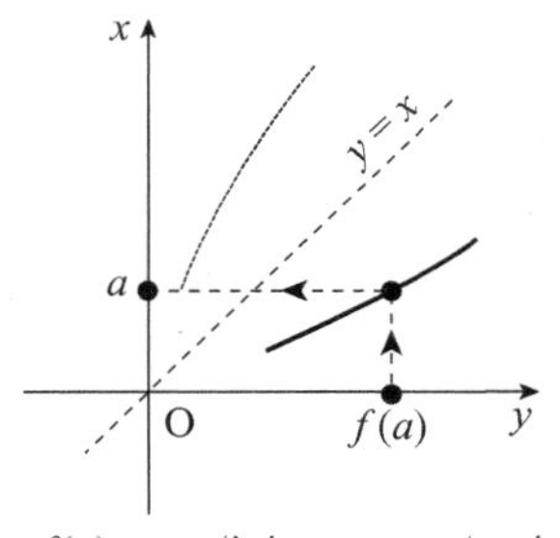

f(a) goes 'in'. a comes 'out'.

relabel ⟹ the axes

Inverse Function Graph $y = f^{-1}(x)$

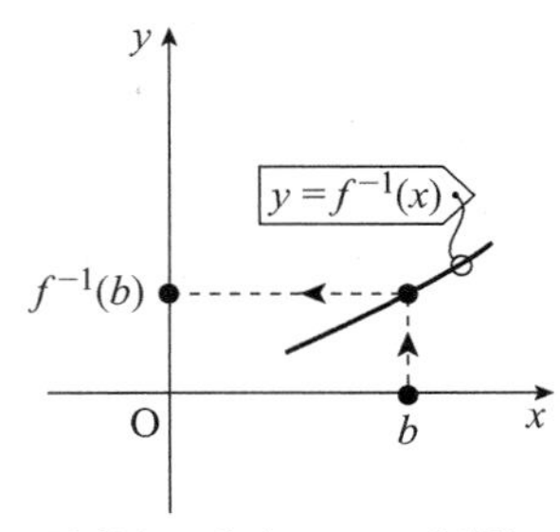

If $f(a) = b$ then $a = f^{-1}(b)$.

Here are three examples of inverse functions and their graphs:

Doubling $f(x) = 2x$

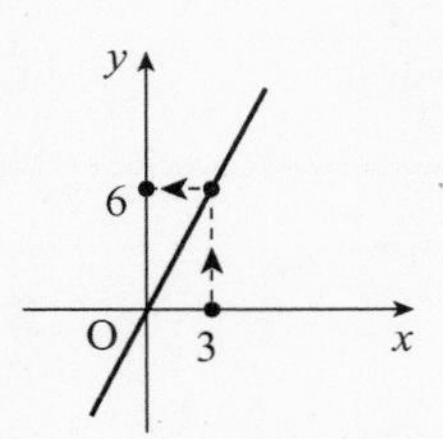

⇓

Halving $f^{-1}(x) = \frac{1}{2}x$

Squaring $f(x) = x^2\ (x \geq 0)$

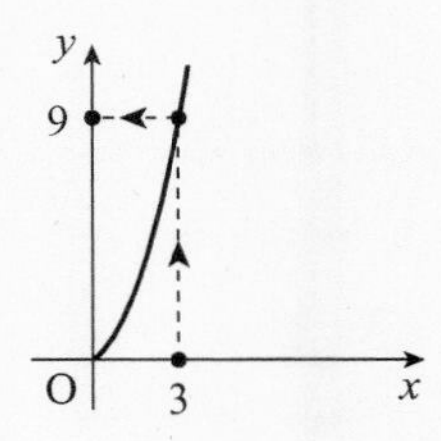

⇓

Square-rooting $f^{-1}(x) = \sqrt{x}\ (x \geq 0)$

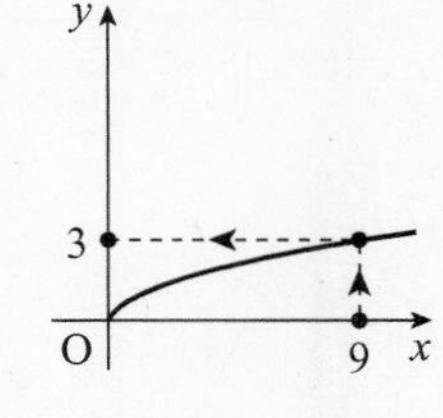

Inverting $f(x) = \frac{1}{x}\ (x \neq 0)$

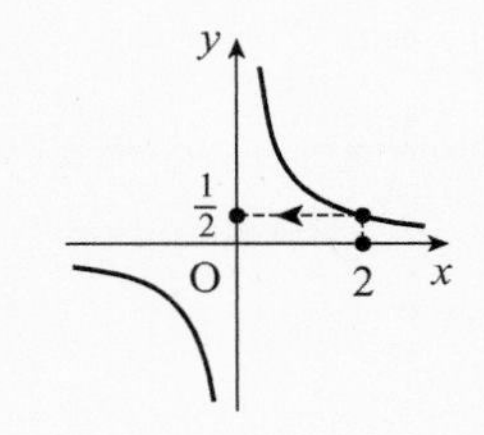

⇓

Inverting $f^{-1}(x) = \frac{1}{x}\ (x \neq 0)$

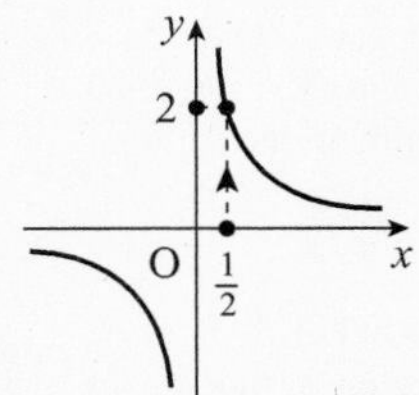

f is its own inverse!

TOP TIP

Flip the graph $y = f(x)$ in the line $y = x$ to get the graph $y = f^{-1}(x)$

Quick Test 2

1. $f(x) = 3x - 1$ and $g(x) = x^2$, find:

   a) $g(f(x))$ b) $f(g(x))$ c) $f(f(x))$ d) $g(g(x))$

2. In each case find the inverse function f^{-1}

   a) $f(x) = \frac{1}{2}x + 1$ b) $f(x) = 7 - x$ c) $f(x) = \sqrt{x - 1}$

Chapter 1

Quadratic functions

Completing the square

The Method

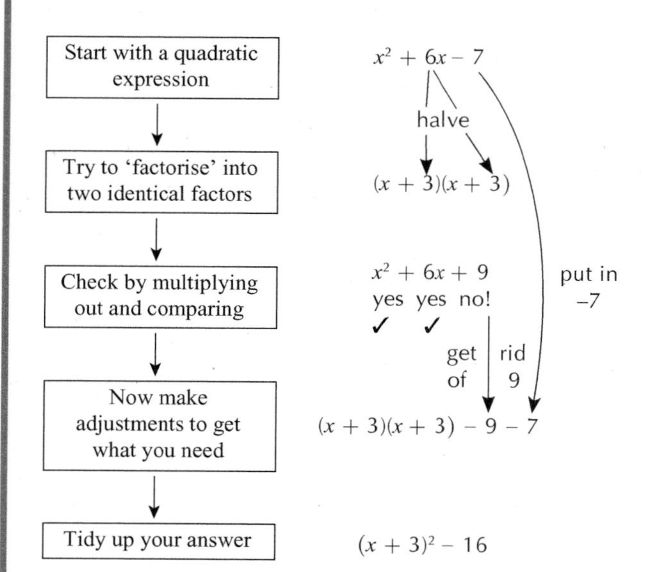

Example

Express $3 + 8x - 2x^2$ in the form $a + b(x + c)^2$ where a, b and c are constants.

Solution

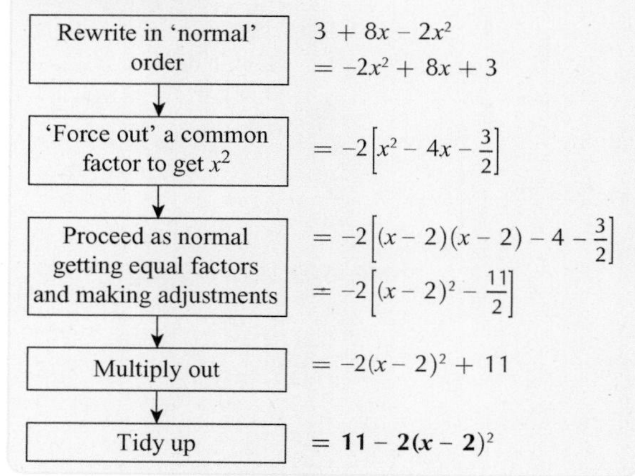

Graphs related to $y = x^2$

TOP TIP

In these examples $y = f(x) = x^2$ but the effects work for any graph, e.g. $y = \sin x$, $y = \sqrt{x}$ etc.

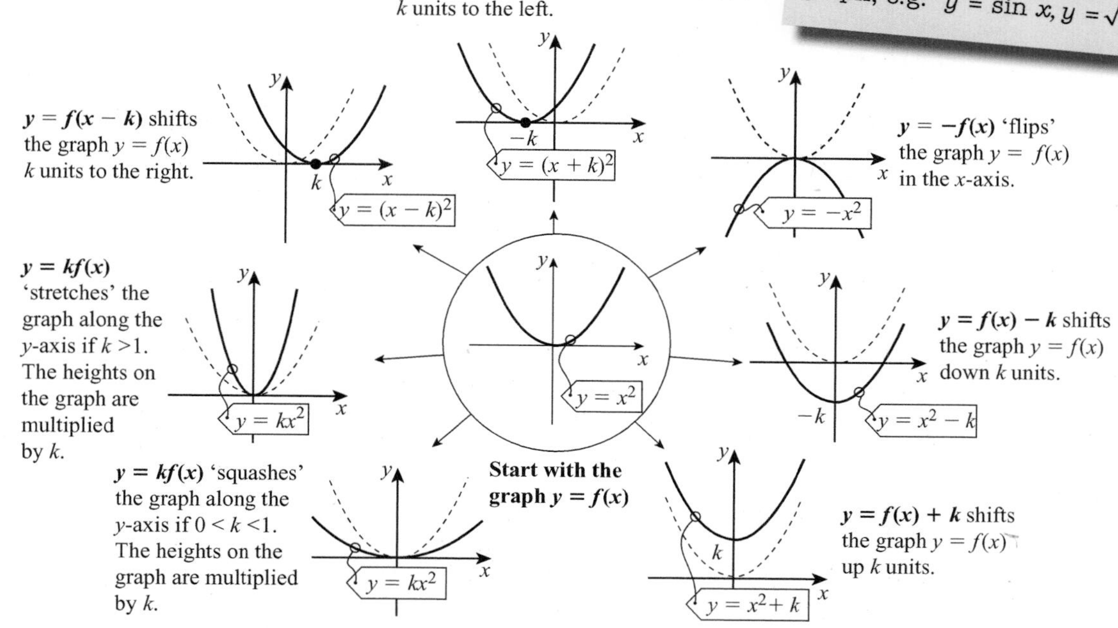

Sketching quadratic graphs

Some hints: $y = ax^2 + bx + c$

- 

  If $a > 0$ then the parabola is concave upwards.

  If $a < 0$ then the parabola is concave downwards.

- Where does it cross the y-axis? → Set $x = 0$ to find y

- Complete the square: $y = (x + d)^2 + e$

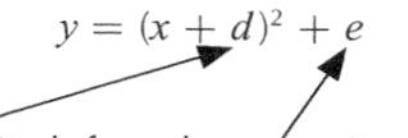

So move the graph $y = x^2$ d units left and up e units.

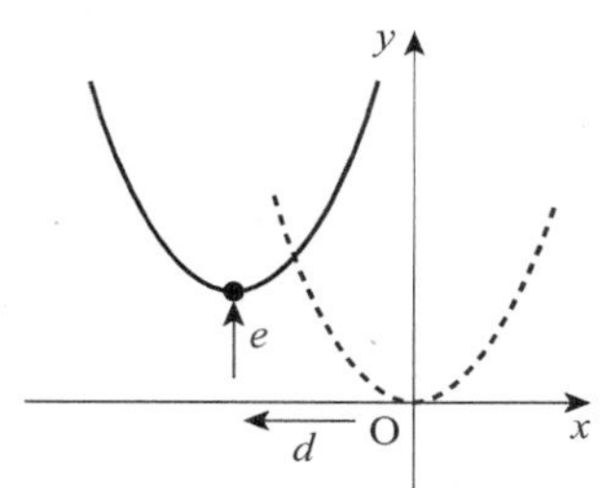

Note the turning point is $(-d,e)$.

Example 1

Sketch $y = x^2 - 6x + 10$ and give the coordinates of the minimum turning point.

Solution When $x = 0$, $y = 10$ so the y-intercept is $(0, 10)$

Also $y = x^2 - 6x + 10 = (x - 3)^2 + 1$

So $y = x^2$ is moved 3 right and 1 up.

The minimum turning point is **(3, 1)**

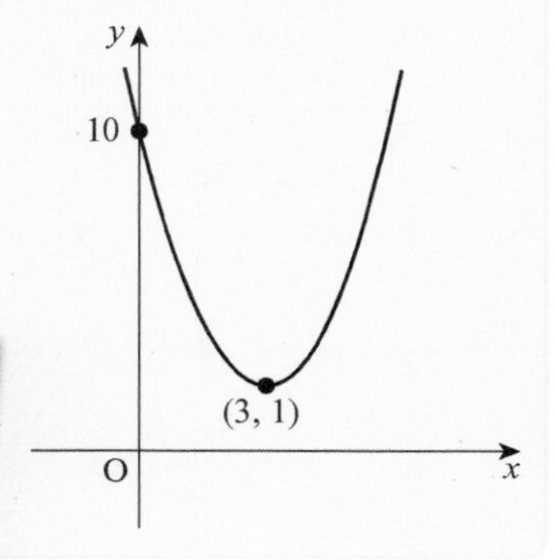

TOP TIP

Not all quadratic graphs $y = ax^2 + bx + c$ cross the x-axis. So solving $y = 0$ to find the x-axis intercepts may not work!

Example 2

The sketch shows the graph $y = a(x + b)^2 + c$

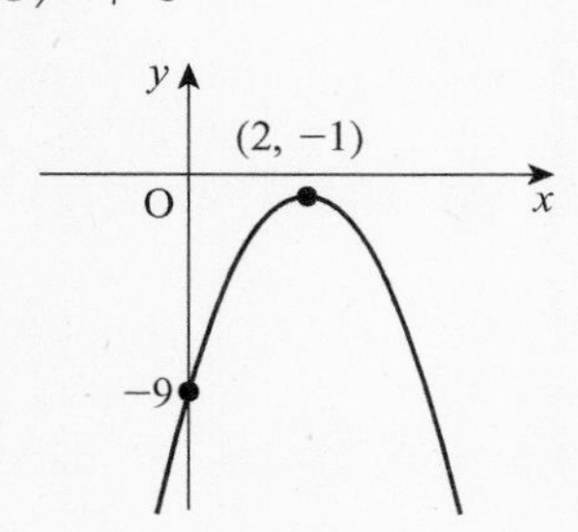

Find the values of a, b and c.

Solution

Start with $y = x^2$. 'Flip' in the x-axis, so value of a is negative.

Then move graph 2 units right, $\boldsymbol{b} = \mathbf{-2}$, and 1 unit down, $\boldsymbol{c} = \mathbf{-1}$

This gives $y = a(x - 2)^2 - 1$.

Since $x = 0$ gives $y = -9$ (y-intercept) then $-9 = a(0 - 2)^2 - 1$ so $-9 = 4a - 1$ so $4a = -8$ giving $\boldsymbol{a} = \mathbf{-2}$

Quick Test 3

1. Express in the form $a(x + b)^2 + c$ and state the values of a, b and c
   a) $2x^2 - 4x + 6$
   b) $3x^2 - 6x + 2$
2. The graph shows $y = k(x + m)^2 + n$
   Find the values of k, m and n

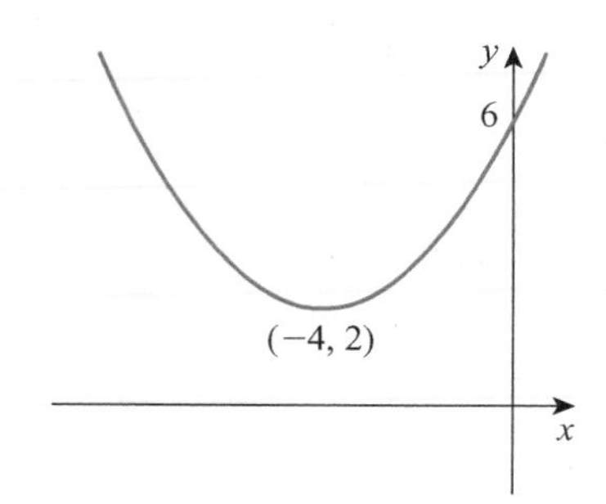

Chapter 1

Related graphs

> **TOP TIP**
>
> These effects on the graph $y = f(x)$ are called transformations. The left/right up/down effects are called translations. The flips are called reflections.

Summary of effects

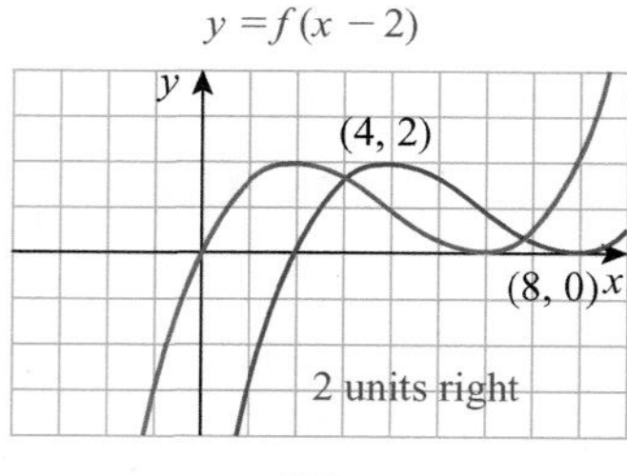

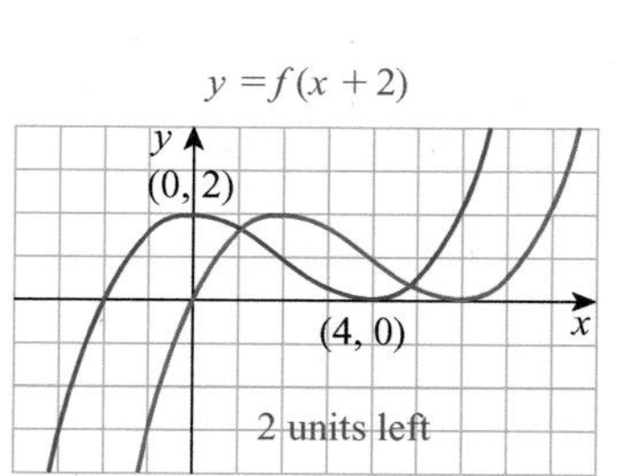

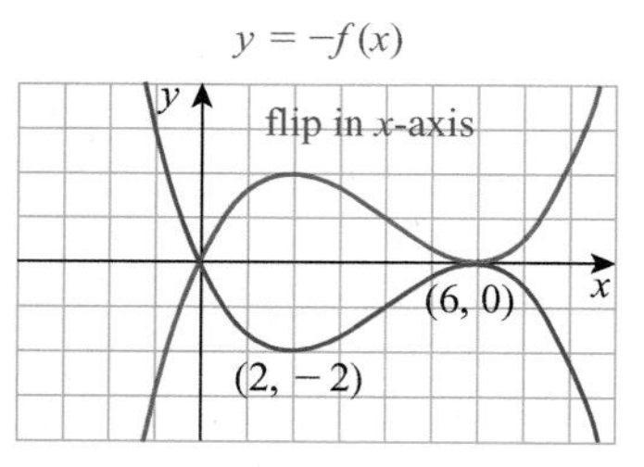

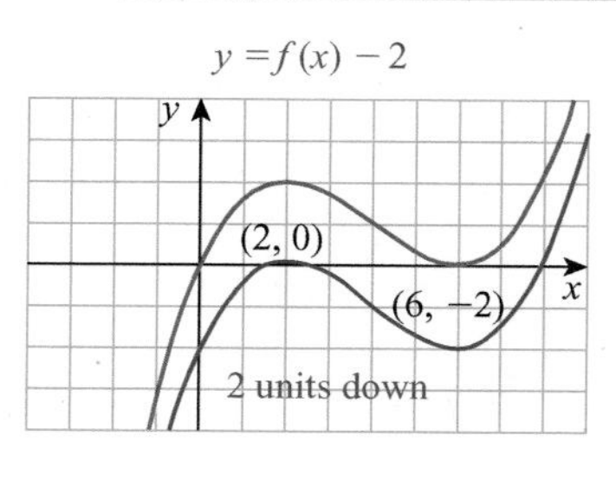

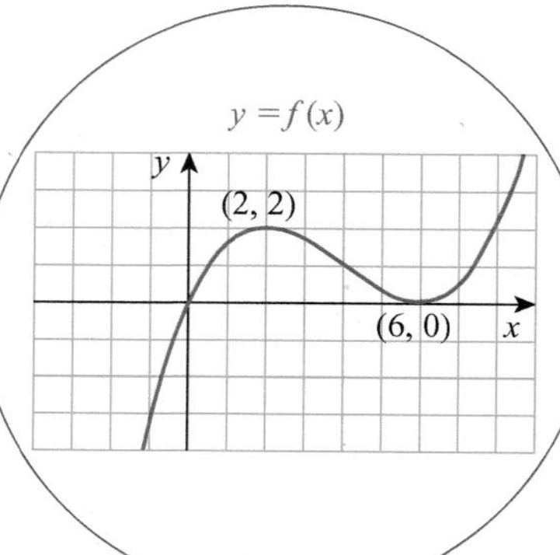

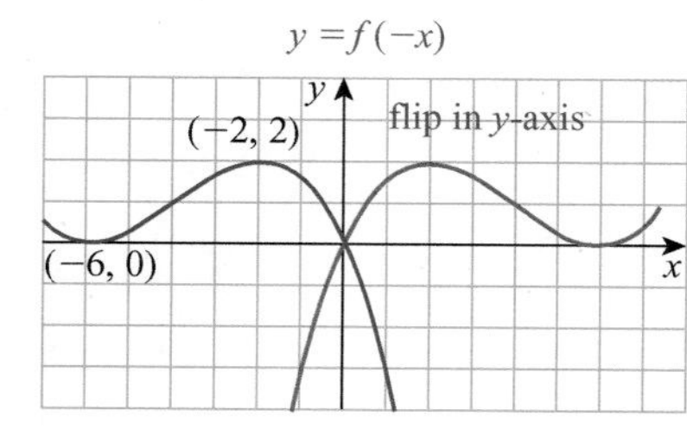

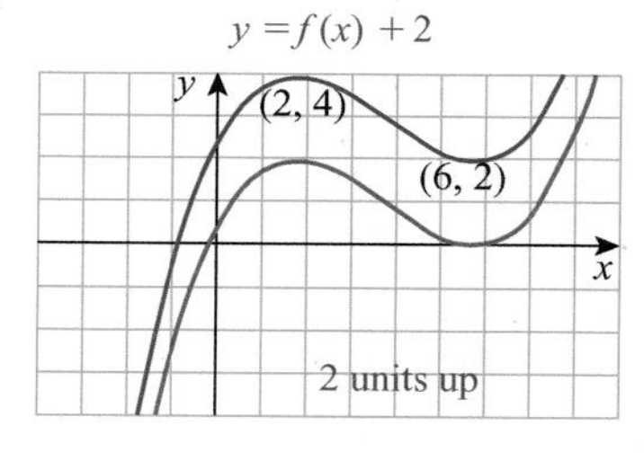

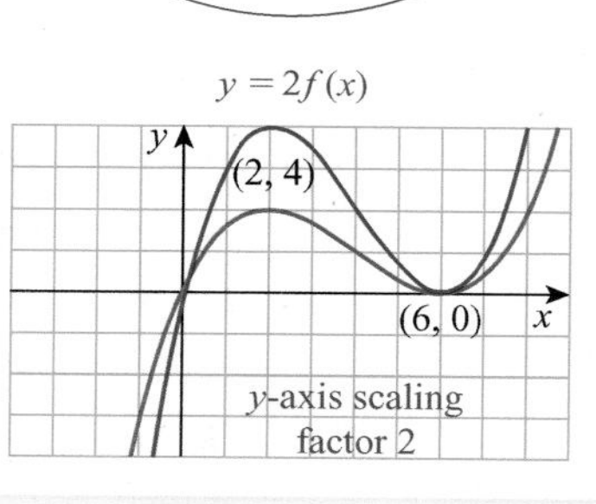

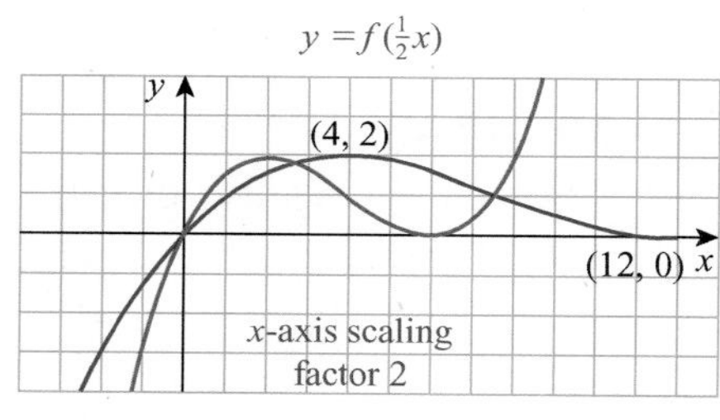

Example

The graph of $y = f(x)$ is shown. Sketch the indicated graphs showing clearly the images of the named points.

a) $y = f(x + 2)$

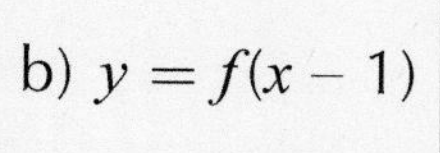

b) $y = f(x - 1)$

c) $y = -f(x)$

d) $y = f(x) - 1$

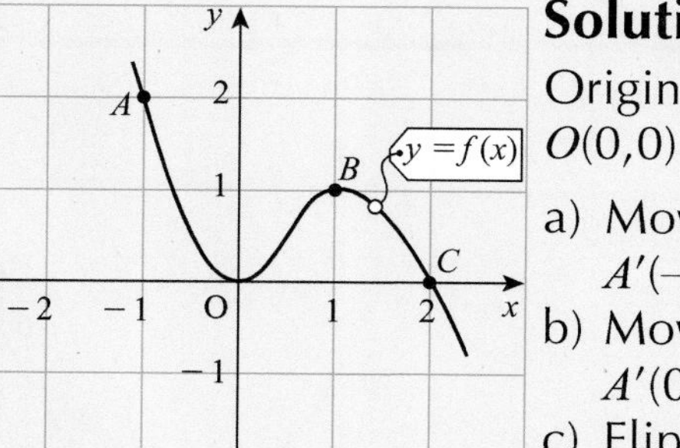

Solution

Original points: $A(-1,2)$, $B(1,1)$, $C(2,0)$ and $O(0,0)$

a) Move graph 2 units left
$A'(-3, 2)$, $B'(-1, 1)$, $C'(0, 0)$ and $O'(-2, 0)$

b) Move graph 1 unit right
$A'(0, 2)$, $B'(2, 1)$, $C'(3, 0)$ and $O'(1, 0)$

c) Flip graph in x-axis
$A'(-1, -2)$, $B'(1, -1)$, $C'(2, 0)$ and $O'(0, 0)$

d) Move graph down 1 unit
$A'(-1, 1)$, $B'(1, 0)$, $C'(2, -1)$ and $O'(0, -1)$

Combining effects

When several effects are combined you will need to split up this combined effect into individual effects. Some typical splits you might encounter are explained below. Usually you are given the graph $y = f(x)$ and your aim is to draw a related graph. You will also know a few points on the graph $y = f(x)$ and you should be able to find where these points are on the related graph.

Note: A and B are on $y = f(x)$ and A' and B' are on the related graph.

Aim:	$y = 3f(x - 5)$		
Start:	$y = f(x)$	$A(4, 2)$	$B(-4, -1)$
5 units right:	$y = f(x - 5)$	$(9, 2)$	$(1, -1)$
y-axis scaling (factor 3):	$y = 3f(x - 5)$	$A'(9, 6)$	$B'(1, -3)$

Aim:	$y = -\frac{1}{2}f(x + 1) - 3$		
Start:	$y = f(x)$	$A(3, 6)$	$B(0, -1)$
1 unit left:	$y = f(x + 1)$	$(2, 6)$	$(-1, -1)$
y-axis scaling $\left(\text{factor } \frac{1}{2}\right)$:	$y = \frac{1}{2}f(x + 1)$	$(2, 3)$	$\left(-1, -\frac{1}{2}\right)$
flip in x-axis:	$y = -\frac{1}{2}f(x + 1)$	$(2, -3)$	$\left(-1, \frac{1}{2}\right)$
3 units down:	$y = -\frac{1}{2}f(x + 1) - 3$	$A'(2, -6)$	$B'\left(-1, -\frac{5}{2}\right)$

TOP TIP

If you know the coordinates of a point on the graph $y = f(x)$ then a corresponding point should appear on your sketch of the related graph.

Example

The graph of $y = f(x)$ is shown:

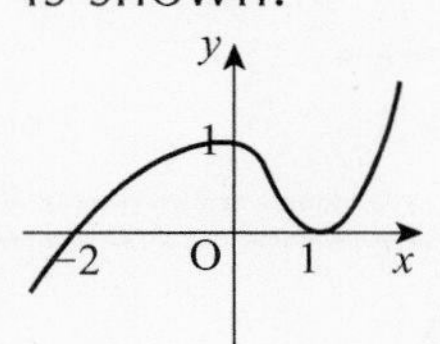

Sketch the graph $y = 2 - f(x)$

Solution

Split into two steps:

Step 1 graph of $y = -f(x)$

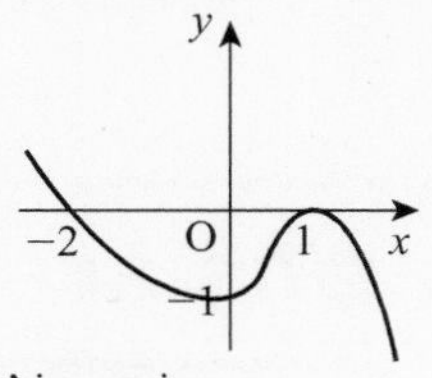

'flip' in x-axis

Step 2 graph of $y = 2 - f(x)$

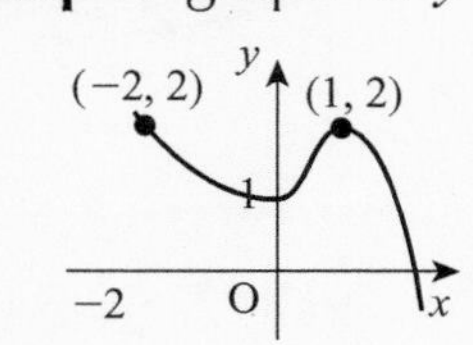

shift up 2 units

Note: $2 - f(x)$ can be rewritten as $-f(x) + 2$ resulting in moving the graph $y = -f(x)$ up 2 units

Quick Test 4

The graph of $y = f(x)$ is shown. Sketch the indicated graphs showing clearly the images of the named points.

1. $y = -f(x) + 2$
2. $y = f(x + 1) - 1$

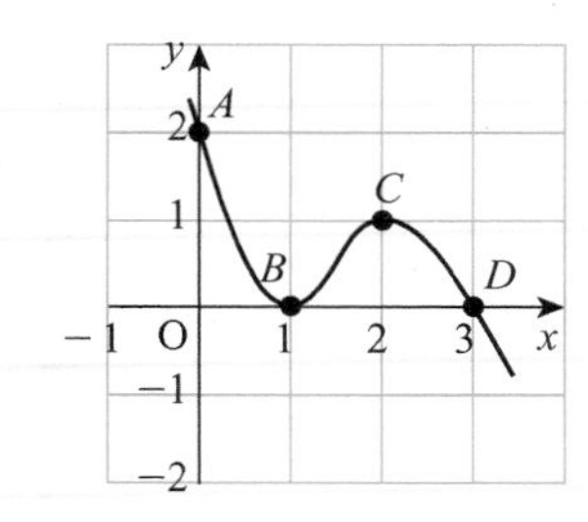

Exponential functions

Growth functions ($a > 1$)

The functions defined by $f(x) = a^x$ are called **exponential functions**. The number a is the **base** and x is the **exponent**. When $a > 1$ they are **growth functions**.

'The powers of 2' $y = 2^x$

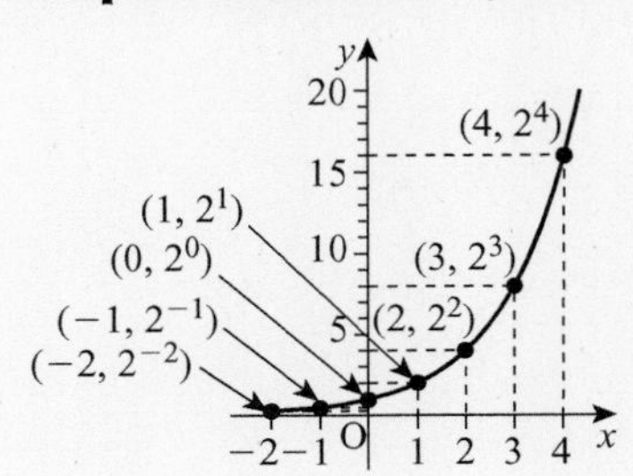

Other examples

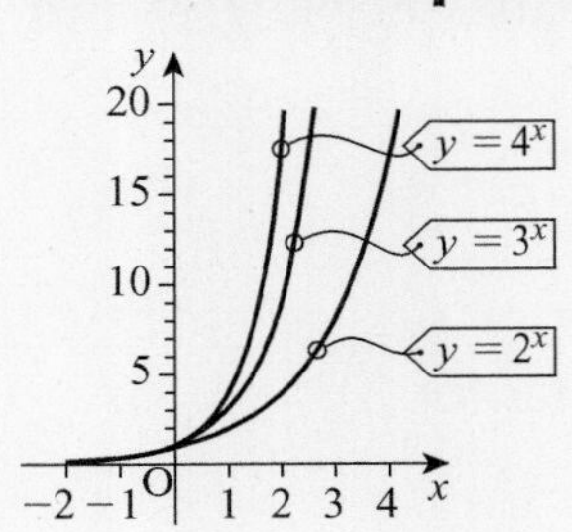

Decay functions ($0 < a < 1$)

When $0 < a < 1$ they are **decay functions**.

'The powers of $\frac{1}{2}$' $y = \left(\frac{1}{2}\right)^x$

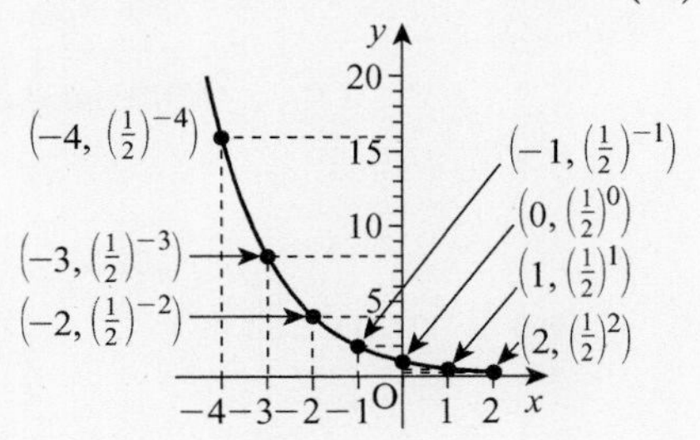

Other examples

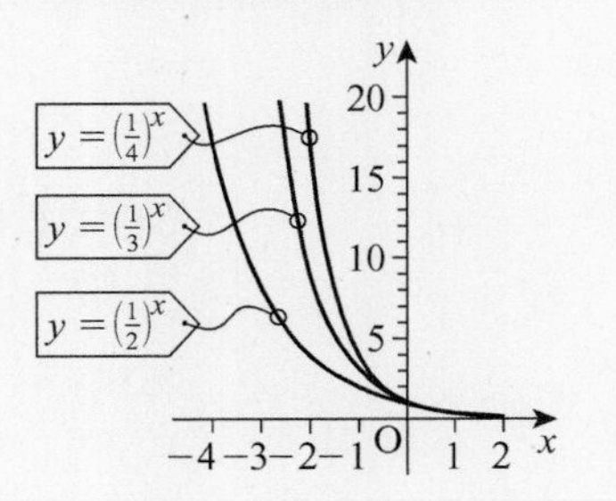

Growth and decay graphs are related!

A growth curve ($a > 1$)

A decay curve ($0 < a < 1$)

Related graphs

Start with $f(x) = a^x$
Flip in y-axis to get

$$f(-x) = a^{-x} = \frac{1}{a^x} = \left(\frac{1}{a}\right)^x$$

If $a > 1$ (growth function)

then $0 < \frac{1}{a} < 1$ (decay function)

Note: $y = a^x$ passes through $(0, 1)$. The graphs approach but do not touch the x-axis.

Reminder: the rules of indices

Rule	Examples
$x^m \times x^n = x^{m+n}$	$a^2 \times a^3 = a^{2+3} = a^5$
$\frac{x^m}{x^n} = x^{m-n}$	$\frac{c^7}{c^3} = c^{7-3} = c^4$
$(x^m)^n = x^{mn}$	$(y^3)^4 = y^{3\times 4} = y^{12}$
$x^0 = 1$	$2^0 = 1$ $\left(\frac{1}{2}\right)^0 = 1$ $(a+b)^0 = 1$
$x^{-n} = \frac{1}{x^n}$	$a^{-1} = \frac{1}{a^1} = \frac{1}{a}$ $a^{-3} = \frac{1}{a^3}$
$x^{\frac{m}{n}} = (\sqrt[n]{x})^m$	$a^{\frac{3}{2}}$ (power 3, square root) $= (\sqrt{a})^3$ $a^{\frac{2}{3}}$ (power 2, cube root) $= (\sqrt[3]{a})^2$

Example

$f(x) = a^x$ and $g(x) = \left(\frac{1}{a}\right)^x$

a) If $a = 2^{-1}$ find formulae for $f(x)$ and $g(x)$ in simplest form with positive indices.

b) Hence, evaluate $f(-\frac{1}{2}) + g(\frac{3}{2})$ for $a = \frac{1}{2}$

Solution

a) $f(x) = (2^{-1})^x = 2^{-1\times x} = 2^{-x} = \frac{1}{2^x}$

$g(x) = \left(\frac{1}{2^{-1}}\right)^x = (2^1)^x = 2^{1\times x} = 2^x$

b) So $f\left(-\frac{1}{2}\right) + g\left(\frac{3}{2}\right) = \frac{1}{2^{-\frac{1}{2}}} + 2^{\frac{3}{2}} = 2^{\frac{1}{2}} + 2^{\frac{3}{2}}$

$= \sqrt{2} + (\sqrt{2})^3 = \sqrt{2} + \sqrt{2} \times \sqrt{2} \times \sqrt{2}$

$= \sqrt{2} + 2\sqrt{2} = 3\sqrt{2}$

Exponential models

Scotland in 2011 had a population growth rate of 0·6% per year and a population of 5·3 million.

Population in 2012: $5{\cdot}3 \times 1{\cdot}006$
Population in 2013: $5{\cdot}3 \times 1{\cdot}006 \times 1{\cdot}006$
$= 5{\cdot}3 \times 1{\cdot}006^2$
Population in 2014: $5{\cdot}3 \times 1{\cdot}006^2 \times 1{\cdot}006$
$= 5{\cdot}3 \times 1{\cdot}006^3$

So the population after x years is given by $5{\cdot}3 \times 1{\cdot}006^x$

This is an exponential function that describes mathematically how the population behaves. It is called a mathematical model.

Example

Estimate the population of Scotland in 2020 using the same mathematical model as opposite.

Solution

Use $x = 9$ as 2020 is 9 years after 2011

Population $= 5{\cdot}3 \times 1{\cdot}006^9$
$= 5{\cdot}59316...$

Estimate is 5·6 million (to 1 decimal place)

Quick Test 5

1. a) For this graph write down an inequality that a satisfies.

b) The point (–1, 3) lies on the curve. Calculate the value of the base a.

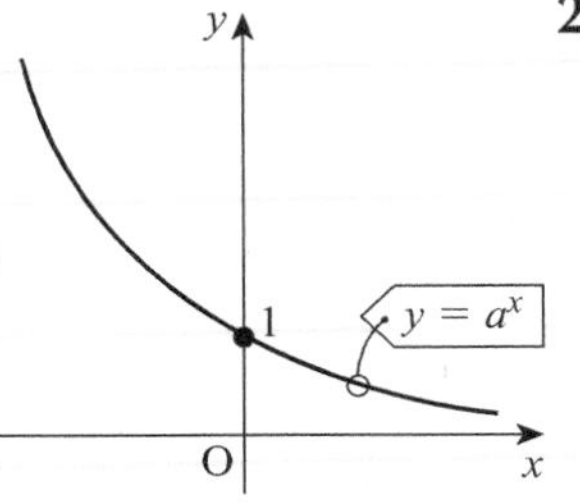

2. The price £P_m of a house in Edinburgh after m months is given by the exponential model:

$$P_m = P_0\,(1{\cdot}0035)^m$$

where P_0 is the price at the start of 2014. If a house in Edinburgh is worth £280,000 at the start of 2014, estimate its value by mid-2016.

Chapter 1

Logarithmic functions

The inverse of an exponential function

Let's look at the inverse of the base 2 exponential function $f(x) = 2^x$

Function graph
(exponential)
$f(x) = 2^x$

Inverse function graph
(logarithmic)
$f^{-1}(x) = \log_2 x$

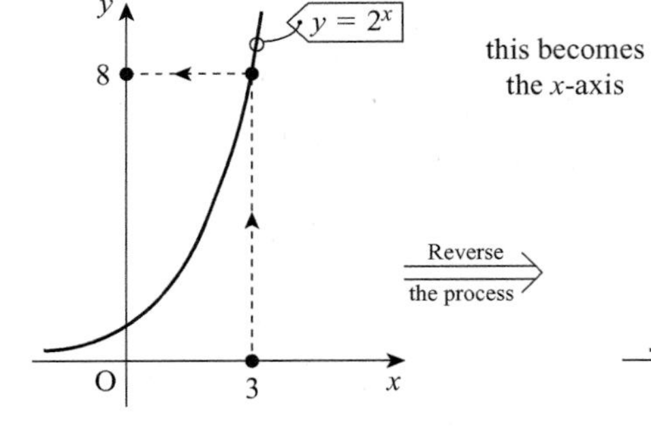

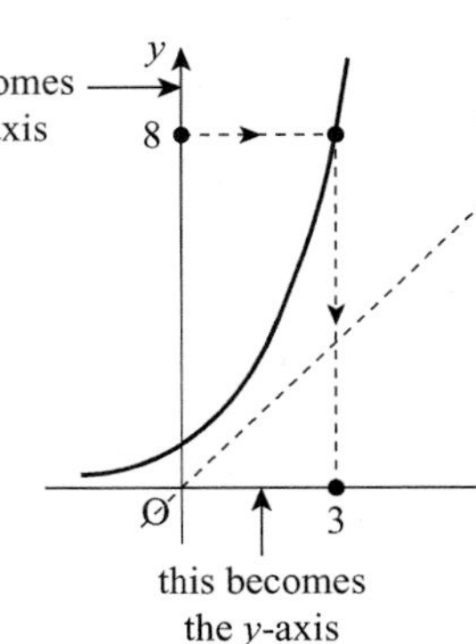

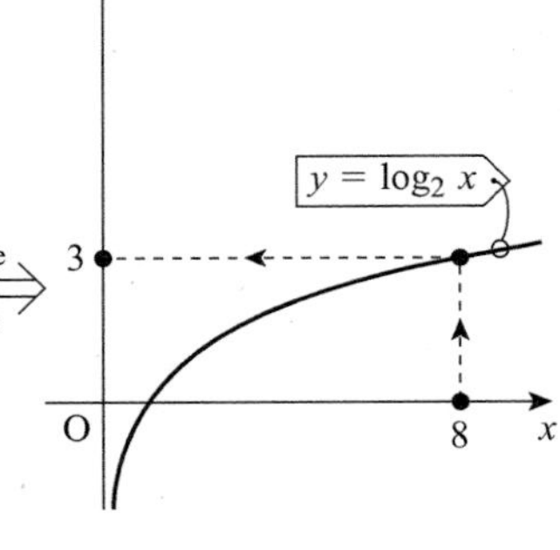

Input 3 ⟶ Output 8
3 ⟶ $2^3 = 8$
'2 to the power 3 gives 8'

Input 8 ⟶ Output 3
8 ⟶ $\log_2 8 = 3$
'log to the base 2 of 8 gives 3'

This is a power or exponential statement

$2^3 = 8 \iff \log_2 8 = 3$

This is a log or logarithmic statement

The logarithmic graph

Here is a typical log graph:

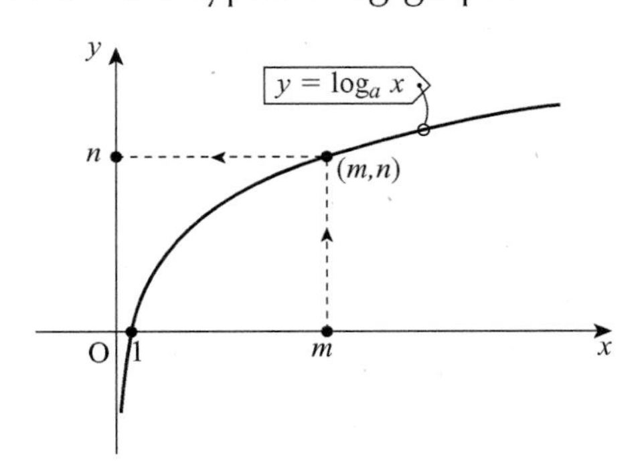

Notes:

1. (m, n) lies on the graph so
   $$n = \log_a m$$
   this 'log statement' can change to
   $$a^n = m$$
   which is a 'power statement'
2. For the graph shown, the base $a > 1$
3. Simple log graphs like the one shown pass through the point (1, 0). This is because: (1, 0) on the graph gives $0 = \log_a 1 \Leftrightarrow a^0 = 1$
4. The graph lies entirely to the right of the y-axis. $\log_a x$ is not defined for $x \leq 0$. This is because the value of any power of a is always positive.

Related log graphs I

Example

This is a sketch of part of the graph of $y = \log_3 x$

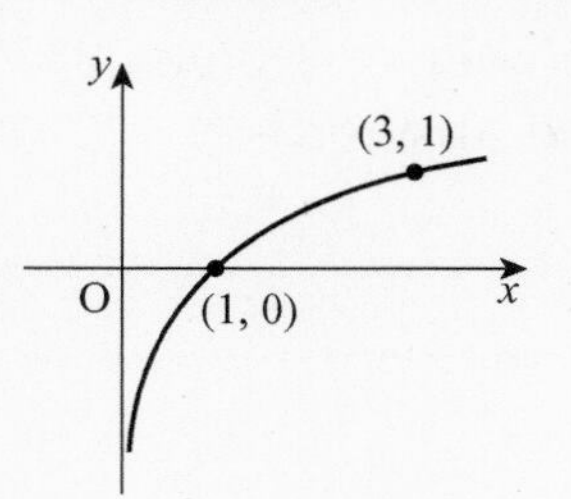

Sketch the graph $y = \log_3(x + 3) + 1$

Solution

$y = \log_3(x + 3) + 1$

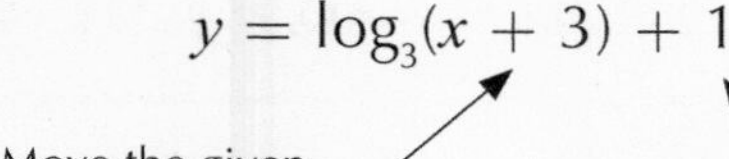

Here is the resulting graph...

(3, 1) ⟶ (0, 2)

(1, 0) ⟶ (–2, 1)

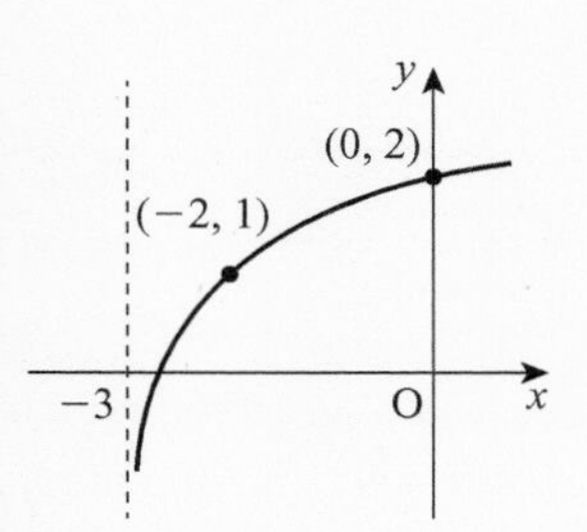

The special base *e*

There are two common bases used for exponential and logarithmic functions: 10 and e

The special number $e = 2{\cdot}71828182...$ used as a base gives:

the **natural exponential function**

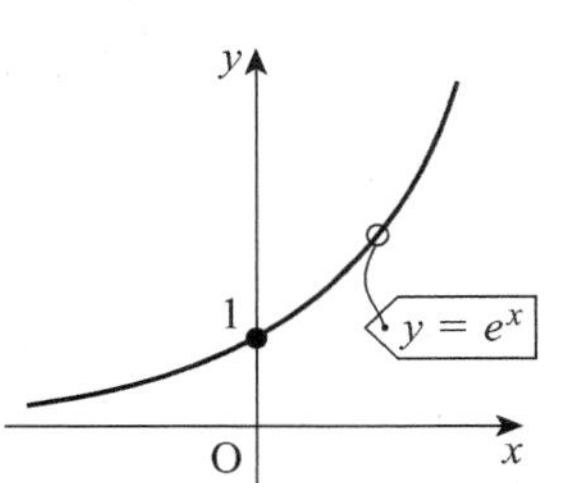

and its inverse function, the **natural logarithmic function**

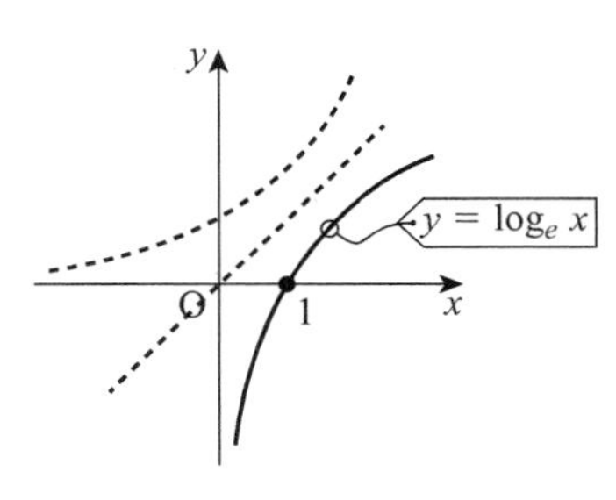

On your calculator, the keys 10^x and log give base 10 calculations.

The keys e^x and ln give base e calculations.

Example

Calculate $\log_e 23$ and hence write 23 as a power of e to 2 decimal places.

Solution

$\log_e 23 = 3{\cdot}135...$

so $23 = e^{3\cdot135...} \doteqdot e^{3\cdot14}$

TOP TIP

On your calculator, to calculate $\log_e 23$ press

Quick Test 6

1\.

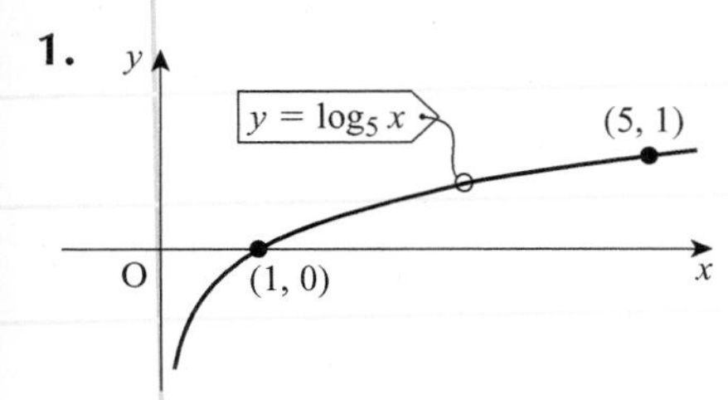

Sketch the graphs:

a) $y = \log_5 x + 1$

b) $y = \log_5(x + 1)$

c) $y = -\log_5 x$

d) $y = \log_5(x - 2) - 1$

2\. Change 'power statements' to 'log statements' and 'log statements' to 'power statements':

a) $92 = 10^x$ b) $4 = e^y$

c) $4 = \log_{10} t$ d) $\log_e A = B$

Working with logs

Understanding 'log statements'

Since $2^3 = 8$ changes to $\log_2 8 = 3$ you can read $\log_2 8 = 3$ as:
'What power of 2 gives 8? Answer:3'

For example:

$\log_3 81$: 'What power of 3 gives 81?'
Since $3^4 = 81$ the answer is 4
So $\log_3 81 = 4$

The rules for logs

Rules for indices		Corresponding rules for logs
$a^m \times a^n = a^{m+n}$	⟷	$\log_a(xy) = \log_a x + \log_a y$
$\dfrac{a^m}{a^n} = a^{m-n}$	⟷	$\log_a\left(\dfrac{x}{y}\right) = \log_a x - \log_a y$
$(a^m)^n = a^{mn}$	⟷	$\log_a(x^n) = n\log_a x$

and some particular examples are...

$a^0 = 1$	⟷	$\log_a 1 = 0$
$a^1 = a$	⟷	$\log_a a = 1$

TOP TIP

Learn how to swap power and log statements using different letters: $p = q^r \Leftrightarrow \log_q p = r$

Example

Simplify: $5\log_8 2 + \log_8 4 - \log_8 16$

Solution 1

$\log_8 2^5 + \log_8 4 - \log_8 16$

$= \log_8 \dfrac{2^5 \times 4}{16}$ — To add the logs, multiply the numbers. To subtract the logs, divide the numbers.

$= \log_8 \dfrac{32 \times 4}{16} = \log_8 8 = \mathbf{1}$

Solution 2

$5\log_8 2 + \log_8 2^2 - \log_8 2^4$

$= 5\log_8 2 + 2\log_8 2 - 4\log_8 2$

$= 3\log_8 2 = \log_8 2^3 = \log_8 8 = \mathbf{1}$

Simple log equations

Some simple equations can be solved using: $b^c = a \Leftrightarrow \log_b a = c$

Example

a) $\log_e x = 5$

b) $\log_{10} x = 2{\cdot}9$

c) $e^x = 4{\cdot}5$

d) $10^x = 2$

Solution

a) $\log_e x = 5 \Rightarrow x = e^5 \doteqdot 148{\cdot}4$ (using e^x)

b) $\log_{10} x = 2{\cdot}9 \Rightarrow x = 10^{2\cdot9} \doteqdot 794{\cdot}3$ (using 10^x)

c) $e^x = 4{\cdot}5 \Rightarrow x = \ln 4{\cdot}5 \doteqdot 1{\cdot}50$ (using ln)

d) $10^x = 2 \Rightarrow x = \log_{10}2 \doteqdot 0{\cdot}301$ (using log)

Taking the logs of both sides

For some problems a useful technique is to 'take the logs' of both sides of an equation.

Example 1

Evaluate $\log_3 2$

Solution

Let $x = \log_3 2$

Then $3^x = 2$

Now take logs of both sides (base e or 10)

So $\log_e 3^x = \log_e 2 \Rightarrow x \ln 3 = \ln 2$

$\Rightarrow x = \frac{\ln 2}{\ln 3} \doteqdot 0{\cdot}631$ $\left(\text{check } \frac{\log_{10} 2}{\log_{10} 3} \text{ gives the same}\right)$

Example 2

Solve $5^x = 4$

Solution

Take logs of both sides (either base e **or** base 10)

so $\log_{10}(5^x) = \log_{10} 4$

$\Rightarrow x \log_{10} 5 = \log_{10} 4$

$\Rightarrow x = \frac{\log_{10} 4}{\log_{10} 5} \doteqdot 0{\cdot}861$

Related log graphs II

Remind yourself how the graphs $y = af(x)$ and $y = f(x + b)$ are obtained from the graph $y = f(x)$

Example

The diagram shows the graph $y = a\log_2(x + b)$. Find the values of a and b.

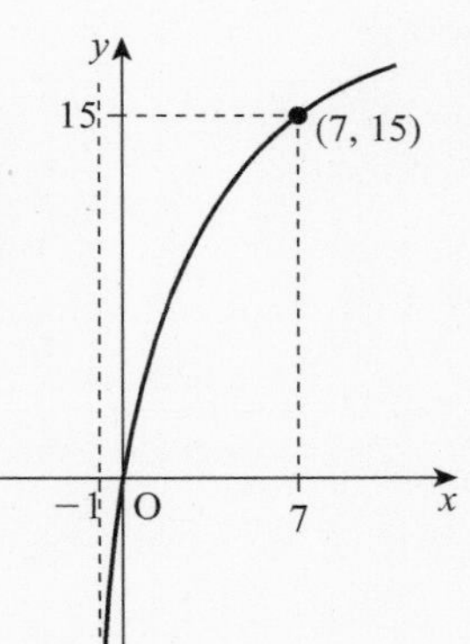

Solution

The graph $y = \log_a x$ passes through (1, 0)

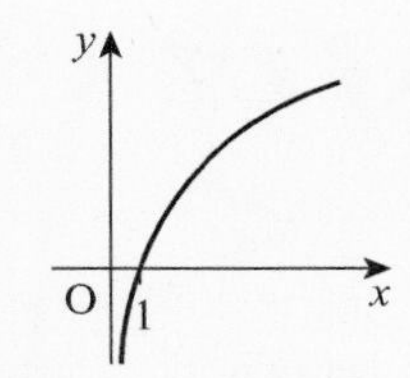

The given graph passes through (0, 0) and so $b = 1$ (the graph above moves 1 unit left) giving $y = a\log_2(x + 1)$

a is a y-axis scaling. Use (7, 15) to find a

$x = 7, y = 15$ satisfy the equation

so $15 = a\log_2(7 + 1)$

$\Rightarrow 15 = a\log_2 8 = a \times 3$

$\Rightarrow 15 = 3a \Rightarrow a = 5$

Quick Test 7

1. Evaluate $\log_3 27 + \log_2 \frac{1}{2}$
2. Simplify $\log_a x^2 - \log_a x + \log_a 1$
3. Solve a) $e^x = 3{\cdot}2$ b) $7^x = 5$ to 2 decimal places.
4. The diagram shows the graph of $y = f(x)$ where $f(x) = a\log_2(x - b)$

   Find the values of a and b.

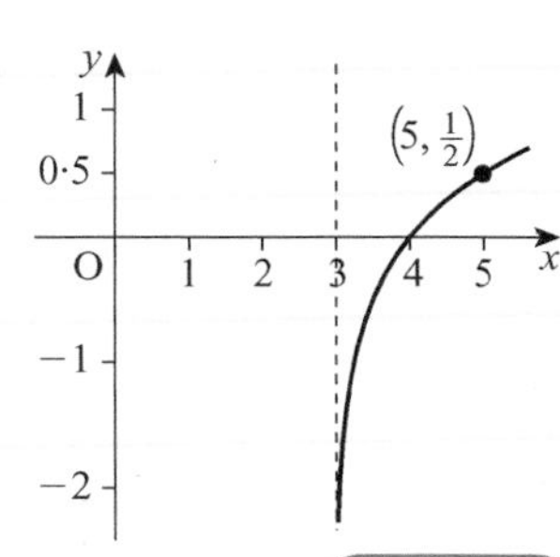

Problem solving with logs

Interpreting and solving problems

Example The amount A grams of a radioactive substance after t units of time is given by $A = A_0e^{-kt}$ where A_0 is the initial amount of the substance and k is a constant.

Calculate the half-life of neptunium if it takes 3 hours for 50 grams to reduce to 45·5 grams.

Solution What variable values do you know?

You know $t = 3$, $A_0 = 50$ and $A = 45{\cdot}5$ from which you can find the constant k:

$$45{\cdot}5 = 50e^{-3k} \implies \frac{45{\cdot}5}{50} = e^{-3k} \implies \log_e \frac{45{\cdot}5}{50} = -3k$$

$$\text{rearranging gives } k = \frac{\log_e \frac{45{\cdot}5}{50}}{-3} = 0{\cdot}0314\ldots$$

The formula is now: $A = A_0e^{-0.0314\ldots \times t}$

What does half-life mean?

This is the time it takes for the initial amount of the substance to reduce to half that amount.

You have to find t for A_0 to reduce to $0{\cdot}5A_0$:

$$0{\cdot}5A_0 = A_0e^{-0.0314\ldots \times t} \implies 0{\cdot}5 = e^{-0.0314\ldots \times t}$$

Change this to a log statement:

$$\log_e 0{\cdot}5 = -0{\cdot}0314\ldots \times t \implies t = \frac{\log_e 0{\cdot}5}{-0{\cdot}0314\ldots} = 22{\cdot}048\ldots$$

So the required half-life of this sample of neptunium is approximately 22 hours.

TOP TIP

Never use rounded values in your calculations. Only round at the end of the calculation.

Determining data set relationships

Often scientists look at experimental data to find the relationship between two variables. But the rule to calculate the values of y from the values of x may not be obvious even after the values are graphed.

Logs can be used to reveal certain types of relationships:

$y = ax^b$

Taking the log of both sides gives:

$$\log y = \log(ax^b) = \log a + \log x^b$$

so $\log y = b \log x + \log a$

compare $Y = mX + c$

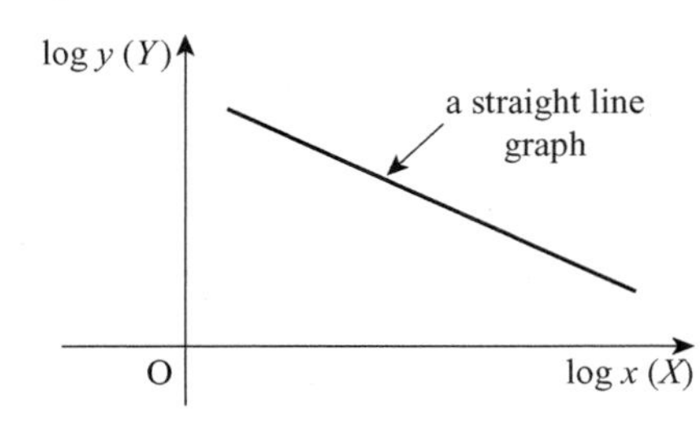

If plotting the values of $\log y$ against the values of $\log x$ results in a straight line graph then this confirms a relationship of the form

$$y = ax^b$$

for suitable values of constants a and b.

$y = ab^x$

Taking the log of both sides gives:

$$\log y = \log(ab^x) = \log a + \log b^x$$

so $\log y = (\log b)\, x + \log a$

compare $Y = m\, x + c$

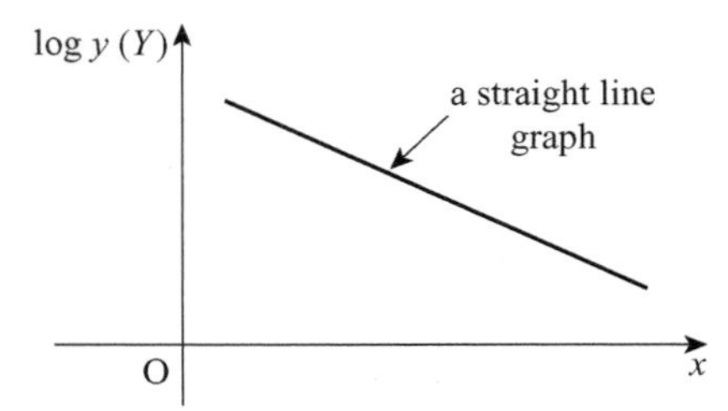

If plotting the values of $\log y$ against the values of x results in a straight line graph then this confirms a relationship of the form

$$y = ab^x$$

for suitable values of constants a and b.

Example

It is suspected that the relationship between x and y is of the form

$$y = ax^b$$

and the graph confirms this. Find the values of constants a and b.

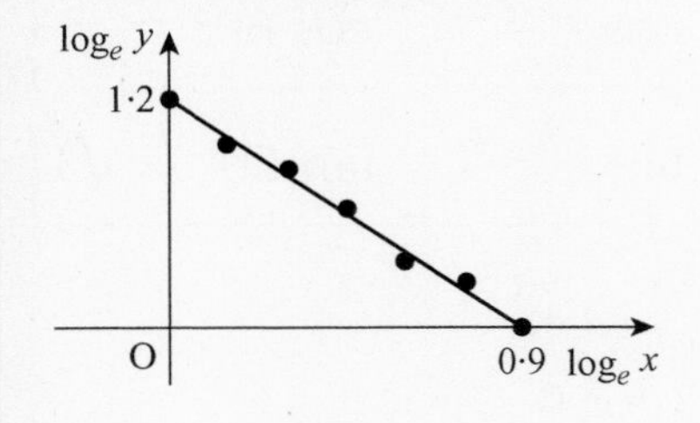

Solution

$y = ax^b$ gives $\log_e y = \log_e(ax^b) = \log_e a + \log_e x^b$

so $\log_e y = b \log_e x + \log_e a$

compare $Y = mX + c$

For the given graph, the gradient (m) gives b.
Use $A(0, 1{\cdot}2)$ and $B(0{\cdot}9, 0)$

So $m_{AB} = \frac{1{\cdot}2 - 0}{0 - 0{\cdot}9} = \frac{1{\cdot}2}{-0{\cdot}9} = -\frac{12}{9} = -\frac{4}{3}$ So $b = -\frac{4}{3}$

The y-intercept $(0, c)$ gives $\log_e a$ ($c = \log_e a$).

The intercept is $(0, 1{\cdot}2)$

so $\log_e a = 1{\cdot}2 \Rightarrow a = e^{1{\cdot}2} \doteqdot 3{\cdot}32$ (using e^x button)

This gives the relationship $\mathbf{y = 3{\cdot}32x^{-\frac{4}{3}}}$

Chapter 1

Using radians

What is a radian?

Lay a radius-length along the circumference of any circle. Then the angle formed at the centre is 1 radian.

Since circumference $= 2\pi r$ then semicircle $= \pi r$ So π radius-lengths make up a semicircle.

Looking at the angles formed at the centre

π radians = 180°

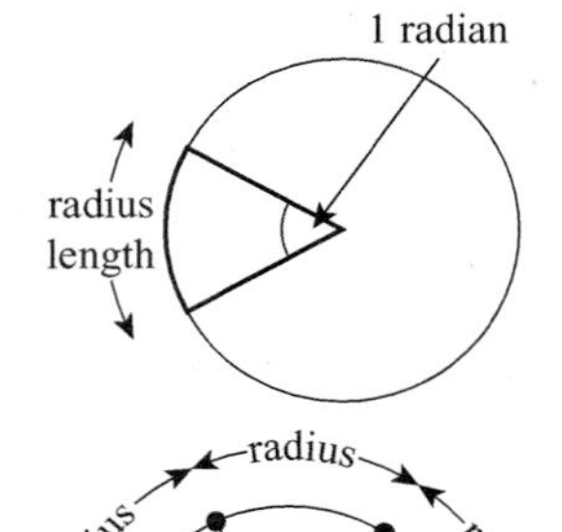

Some useful conversions:

radians		degrees
π	⟷	180°
$\frac{\pi}{6}$	⟷	30°
$\frac{\pi}{4}$	⟷	45°
$\frac{\pi}{3}$	⟷	60°
$\frac{\pi}{2}$	⟷	90°
$\frac{3\pi}{2}$	⟷	270°
2π	⟷	360°

Some exact values

For $\frac{\pi}{4}$ or 45° use half a square of side 1:

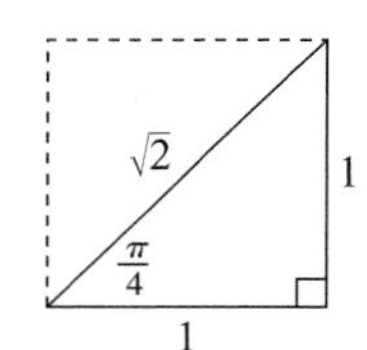

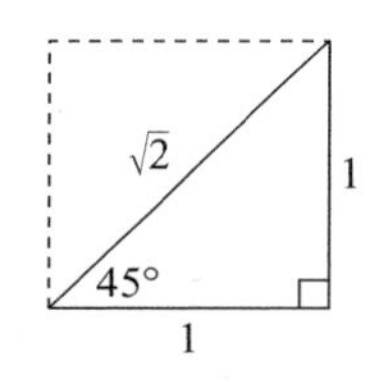

For $\frac{\pi}{6}$, $\frac{\pi}{3}$ or 30°, 60° use half an equilateral triangle of side 2:

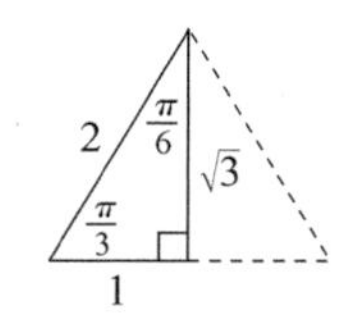

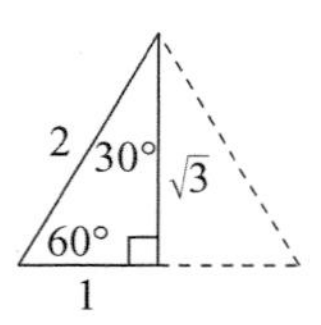

Now use 'SOHCAHTOA' to find **exact** values:

$\sin\frac{\pi}{4} = \frac{1}{\sqrt{2}}$	$\sin 45° = \frac{1}{\sqrt{2}}$	$\sin\frac{\pi}{6} = \frac{1}{2}$	$\sin 30° = \frac{1}{2}$	$\sin\frac{\pi}{3} = \frac{\sqrt{3}}{2}$	$\sin 60° = \frac{\sqrt{3}}{2}$
$\cos\frac{\pi}{4} = \frac{1}{\sqrt{2}}$	$\cos 45° = \frac{1}{\sqrt{2}}$	$\cos\frac{\pi}{6} = \frac{\sqrt{3}}{2}$	$\cos 30° = \frac{\sqrt{3}}{2}$	$\cos\frac{\pi}{3} = \frac{1}{2}$	$\cos 60° = \frac{1}{2}$
$\tan\frac{\pi}{4} = 1$	$\tan 45° = 1$	$\tan\frac{\pi}{6} = \frac{1}{\sqrt{3}}$	$\tan 30° = \frac{1}{\sqrt{3}}$	$\tan\frac{\pi}{3} = \sqrt{3}$	$\tan 60° = \sqrt{3}$

Example 1

Find the exact value of $1 - \sin^2\frac{\pi}{4}$

Solution

$$1 - \sin^2\frac{\pi}{4} = 1 - \left(\frac{1}{\sqrt{2}}\right)^2$$

$$= 1 - \frac{1}{2} = \mathbf{\frac{1}{2}}$$

Example 2

Solve $\sin x = 0{\cdot}3$ for $0 \le x \le \frac{\pi}{2}$

Solution

Put calculator into 'Radian Mode' and enter 0·3. Now use $\boxed{\sin^{-1}}$ giving $\boldsymbol{x = 0{\cdot}305}$ (to 3 sig. figs) radians.

TOP TIP

Know your own calculator! How do you change from Degree mode (D or DEG) to Radian mode (R or RAD)?

The trig graphs using radian measure

Reading values from the three trig graphs with the angles measured in radians gives:

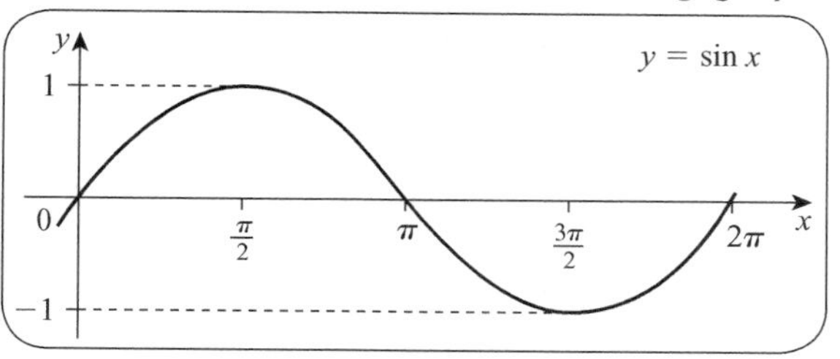

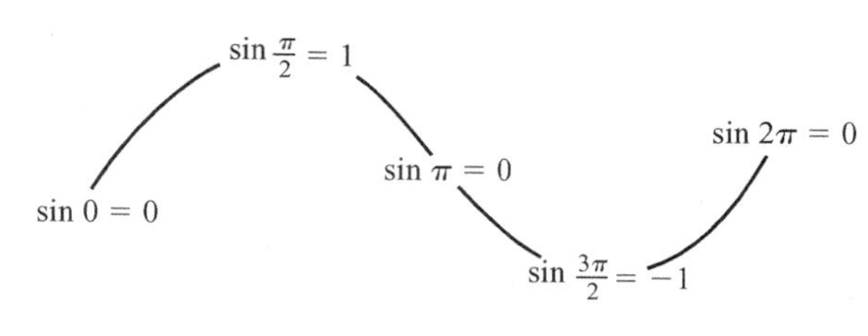

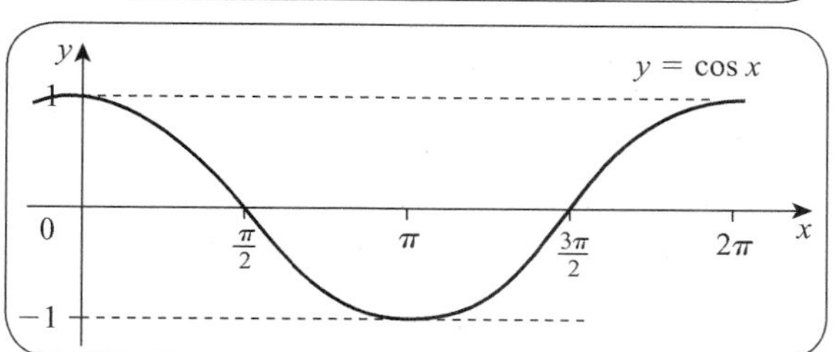

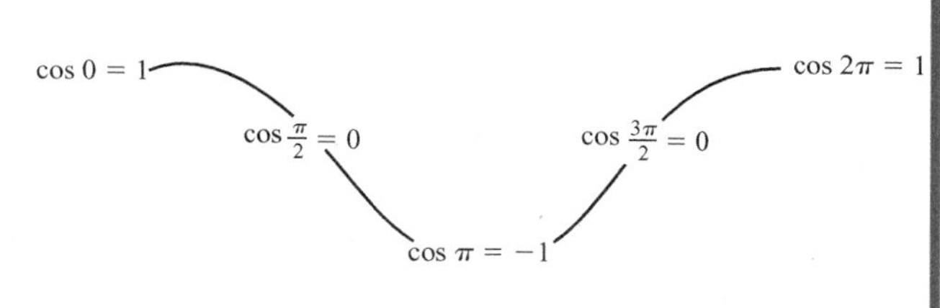

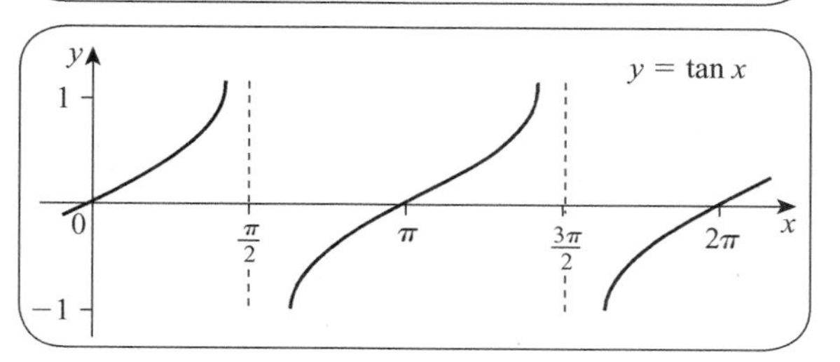

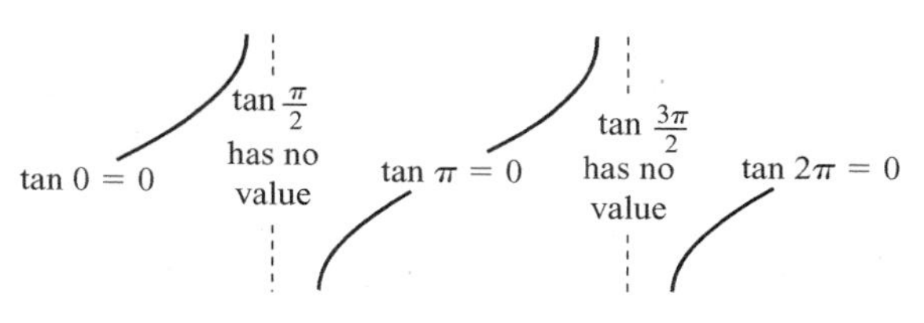

Example Solve $\sin x = 0$ for $0 \le x \le 2\pi$

Solution Using the $y = \sin x$ graph above
$x = 0, \pi, 2\pi$

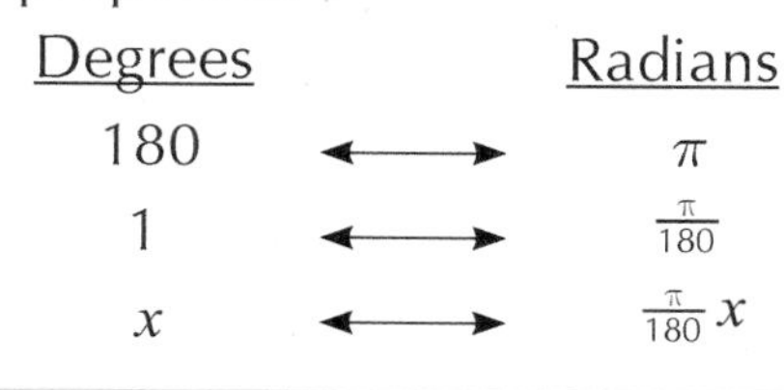

Conversions

Other conversations can be done using 'proportion':

Radians		Degrees	Degrees		Radians
π	⟷	180	180	⟷	π
1	⟷	$\frac{180}{\pi}$	1	⟷	$\frac{\pi}{180}$
x	⟷	$\frac{180}{\pi}x$	x	⟷	$\frac{\pi}{180}x$

Quick Test 8

1. Find the exact value of $3 - 2\cos^2\left(\frac{\pi}{6}\right)$
2. Solve $\cos x = 0{\cdot}8$ for $0 \le x \le \frac{\pi}{2}$ giving your answer to 2 decimal places.
3. Solve $\cos x = 1$ for $0 \le x \le 2\pi$
4. Change
   a) $\frac{7\pi}{4}$ into degrees
   b) 40° into radians.

Chapter 1

Related trig graphs

Graphs related to $y = \sin x$

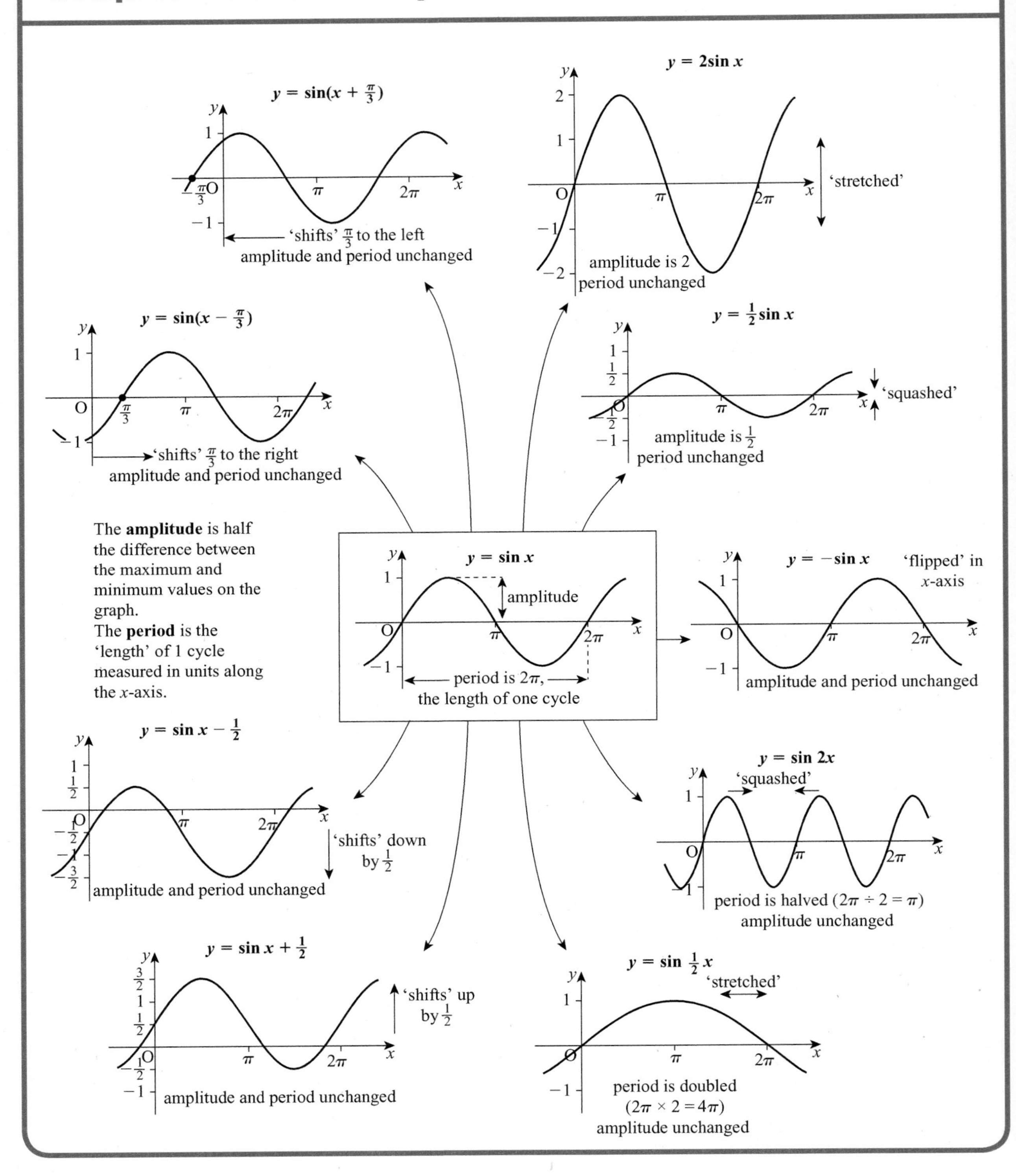

Summary of effects

Here is a description of the effect to the graphs $y = \cos x$ or $y = \sin x$

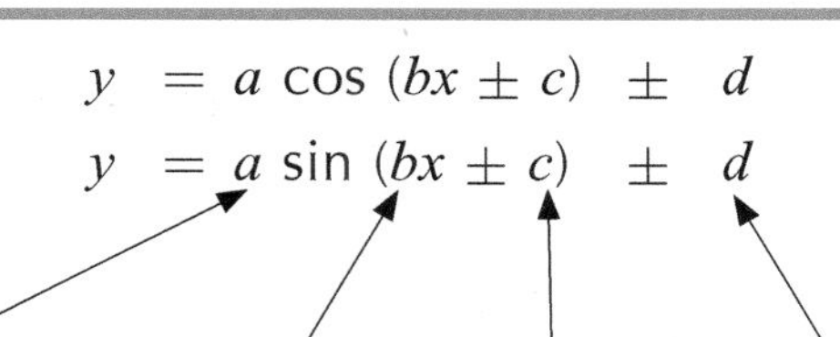

TOP TIP

For the left/right up/down 'shifts' c and d are positive.

If $a > 0$ the amplitude is a. If $a < 0$ the graph 'flips' in the x-axis. The amplitude is the magnitude of a (i.e. ignore the negative sign).

Alters the period to $\frac{2\pi}{b}$ $(b > 0)$

'Shifts' the graph by c units left for $+ c$ or right for $- c$

'Shifts' the graph by d units up for $+ d$ or down for $- d$

Example

The diagram shows part of the graph of $y = a\sin(x - b) + c$. Find the values of a, b and c.

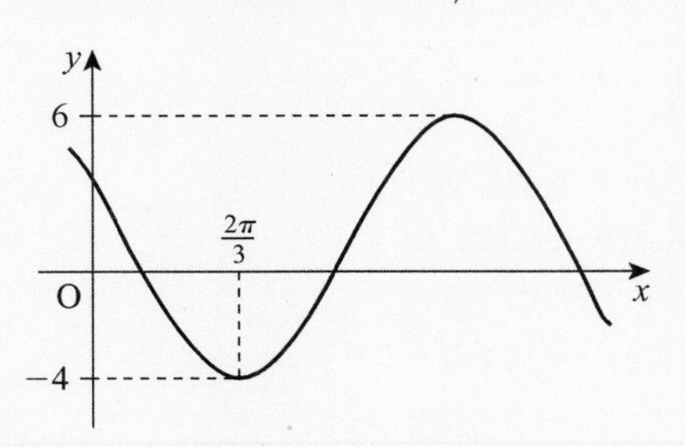

Solution

Compared with $y = \sin x$... The amplitude is 5 (half the difference between −4 and 6). However, the graph is 'flipped' in the x-axis so $\boldsymbol{a = -5}$. Also, the graph has moved up 1 (max/min 6 and −4 instead of 5 and −5) so $\boldsymbol{c = 1}$.

It has also moved to the right: $\frac{2\pi}{3}$ instead of $\frac{\pi}{2}$

...now $\frac{2\pi}{3} - \frac{\pi}{2} = \frac{4\pi}{6} - \frac{3\pi}{6} = \frac{\pi}{6}$

so $\boldsymbol{b} = \frac{\pi}{6}$. The graph is $y = -5\sin\left(x - \frac{\pi}{6}\right) + 1$

Sketching trig graphs

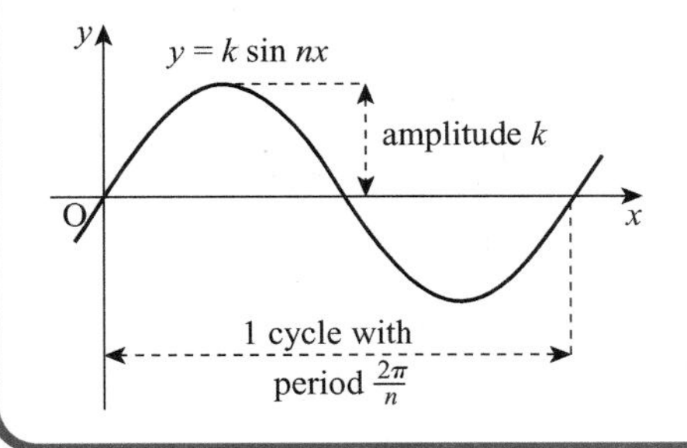

Example 1 Sketch $y = \frac{1}{2}\sin 2x$

Solution

Amplitude $= \frac{1}{2}$

2 cycles; period

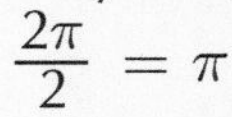

$\frac{2\pi}{2} = \pi$

Quick Test 9

1. Find the values of a and b for this graph:

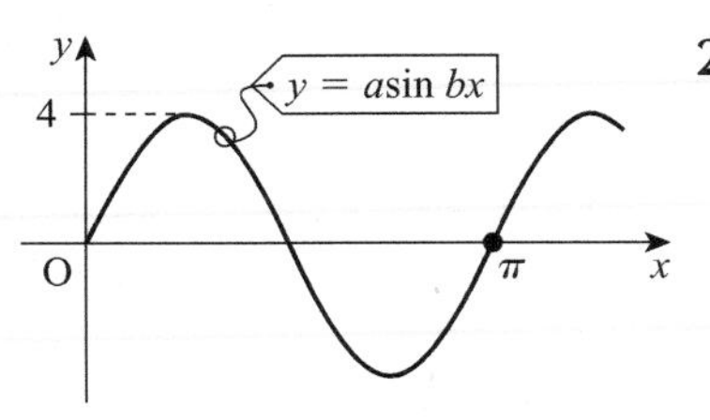

2. a) Sketch the graph $y = 5\cos\left(x - \frac{\pi}{4}\right)$ for $0 \leq x \leq 2\pi$

   b) Find the coordinates of the maximum and minimum points on your sketched graph.

Problem solving using trig

Working in 3D

The angle between a line and a plane:

If QR is the **projection** or 'vertical shadow' of the line PQ on the plane then

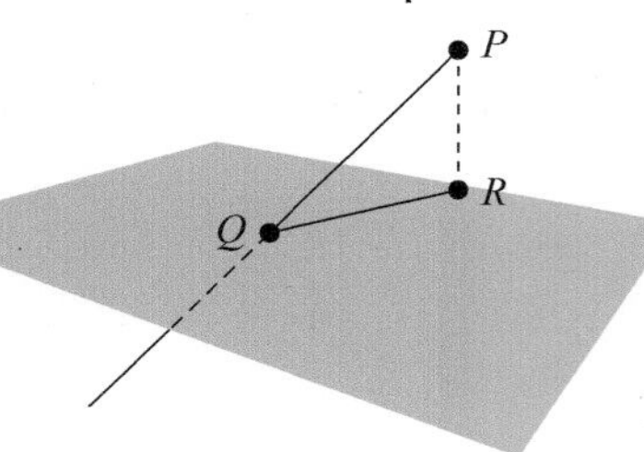

$\angle PQR$ is the angle between the line and plane.

The angle between two planes:

Choose a point B on the line l where the two planes meet. Draw AB on one plane and BC on the other, both perpendicular to line l.

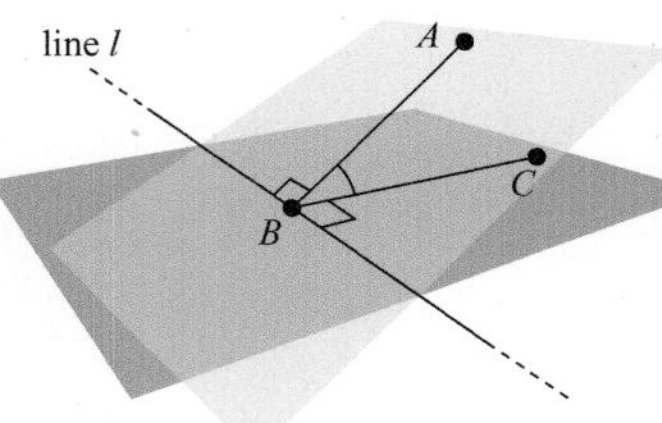

$\angle ABC$ is the angle between the two planes.

Example

A square-based pyramid has vertex P, 3 meters above Q, the centre of the base. The square base has side length 6 meters. Find the angle between the base and a sloping face in radians.

Solution

Choose M, the midpoint of one of the sides of the square base. Required angle is $\angle QMP = \frac{\pi}{4}$ (isosceles right-angled triangle).

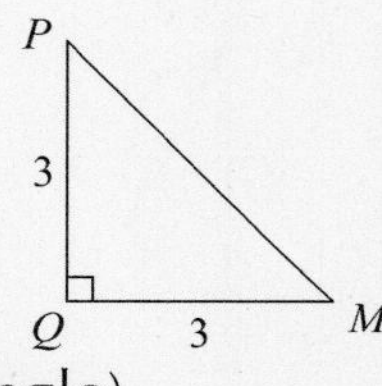

Basic trig identities

$$\frac{\sin\theta}{\cos\theta} = \tan\theta \qquad \sin^2\theta + \cos^2\theta = 1$$

rearranging gives

$$\sin\theta = \tan\theta\cos\theta \qquad \sin^2\theta = 1 - \cos^2\theta$$

$$\cos^2\theta = 1 - \sin^2\theta$$

$$\cos\theta = \frac{\sin\theta}{\tan\theta}$$

Example Prove $\tan^2 A = \frac{1-\cos^2 A}{1-\sin^2 A}$

Solution $\tan^2 A = \tan A \times \tan A = \frac{\sin A}{\cos A} \times \frac{\sin A}{\cos A}$

$$= \frac{\sin^2 A}{\cos^2 A} = \frac{1-\cos^2 A}{1-\sin^2 A}$$

Related angles useful for triangle problems

$\sin(-A)^\circ = -\sin A^\circ$ $\sin(180 - A)^\circ = \sin A^\circ$

$\cos(-A)^\circ = \cos A^\circ$ $\cos(180 - A)^\circ = -\cos A^\circ$

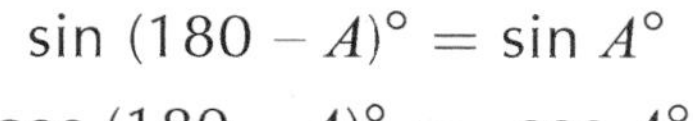

These and other similar results can be deduced from the quadrant diagram

S | A
180 − |
180 + | 360 −
T | C

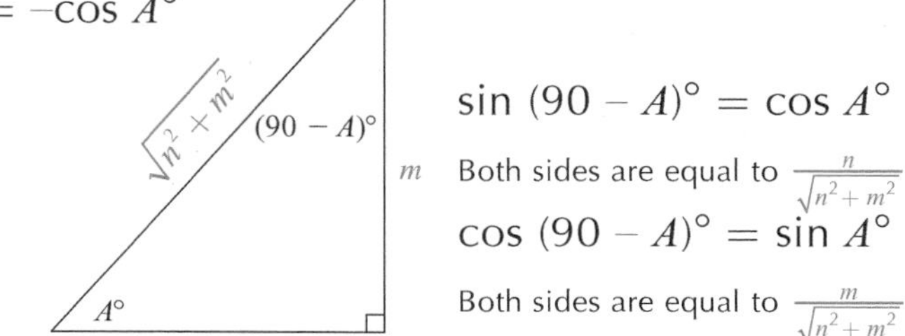

$\sin(90 - A)^\circ = \cos A^\circ$

Both sides are equal to $\frac{n}{\sqrt{n^2 + m^2}}$

$\cos(90 - A)^\circ = \sin A^\circ$

Both sides are equal to $\frac{m}{\sqrt{n^2 + m^2}}$

Note: Results are true for radian measure: $\sin(\pi - A) = \sin\pi$, $\cos\left(\frac{\pi}{2} - A\right) = \sin A$ etc.

Triangle problems

TOP TIP

Be careful not to mix degree and radian measure.

Useful results are:

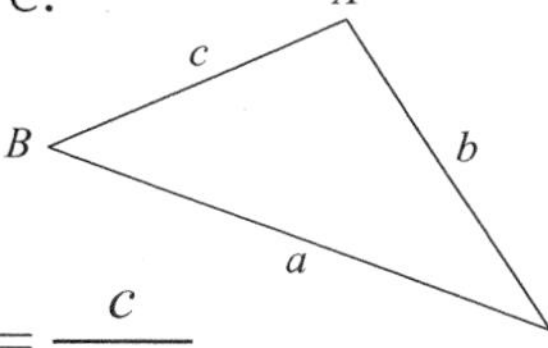

Sine Rule:

$$\frac{a}{\sin A} = \frac{b}{\sin B} = \frac{c}{\sin C}$$

Cosine Rule:

$$a^2 = b^2 + c^2 - 2bc\cos A$$

or $$b^2 = a^2 + c^2 - 2ac\cos B$$

or $$c^2 = a^2 + b^2 - 2ab\cos C$$

Area of triangle $= \frac{1}{2}ab\sin C$

or $\frac{1}{2}bc\sin A$

or $\frac{1}{2}ac\sin B$

Note that $\angle A = 180^\circ - (\angle B + \angle C)$...

Example

In triangle ABC prove that

$$\frac{a}{\sin 2x^\circ} = \frac{b}{\sin x^\circ}$$

B, A, C, a, b, x°

Solution

$\angle B = x^\circ$ (triangle ABC is isosceles)

so $\angle A = 180^\circ - (x^\circ + x^\circ) = 180^\circ - 2x^\circ$

The Sine Rule gives $\frac{a}{\sin A} = \frac{b}{\sin B}$

so $\frac{a}{\sin(180 - 2x)^\circ} = \frac{b}{\sin x^\circ}$

But $\sin(180 - 2x)^\circ = \sin 2x^\circ$

hence $\mathbf{\frac{a}{\sin 2x^\circ} = \frac{b}{\sin x^\circ}}$

Quick Test 10

1. A billboard is supported by a wooden support wedge as shown in the diagram. The wedge has dimensions as shown in the diagram on the right.

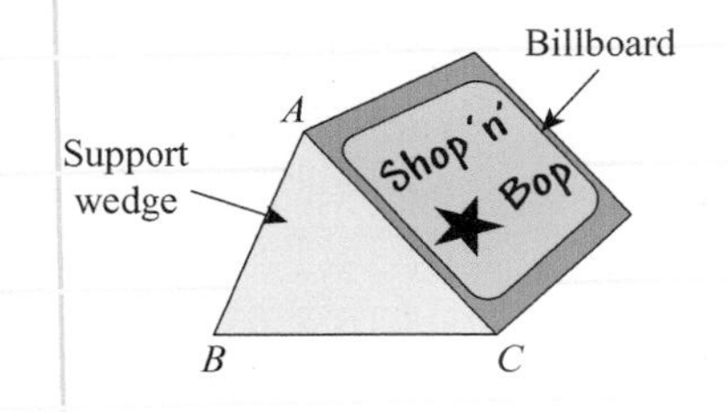

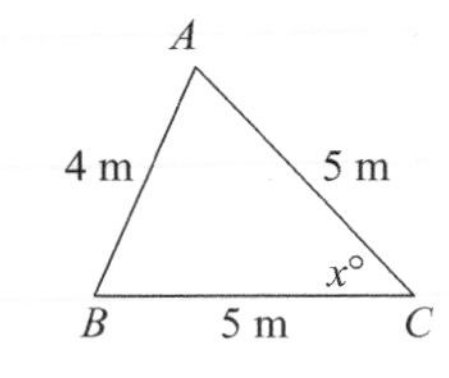

Show that $\cos\frac{1}{2}x = \frac{5}{4}\sin x$ where x° is the angle at which the billboard is inclined to the horizontal.

Hint: use the sine rule.

Trig formulae

The addition formulae

The following are the 'addition formulae' and are true for all values of A and B:

$\sin(A + B) = \sin A \cos B + \cos A \sin B$

$\sin(A - B) = \sin A \cos B - \cos A \sin B$

$\cos(A + B) = \cos A \cos B - \sin A \sin B$

$\cos(A - B) = \cos A \cos B + \sin A \sin B$

Example 1

Show that $\cos\left(\frac{\pi}{2} + x\right) = -\sin x$

Solution

$$\cos\left(\tfrac{\pi}{2} + x\right) = \cos\tfrac{\pi}{2}\cos x - \sin\tfrac{\pi}{2}\sin x$$
$$= 0 \times \cos x - 1 \times \sin x = -\sin x$$

Example 2

If A and B are acute angles with $\sin A = \frac{3}{5}$ and $\cos B = \frac{12}{13}$ find the **exact** value of $\cos(A - B)$.

Solution

Draw right-angled triangles showing $\angle A$ and $\angle B$ (use Pythagoras' Theorem to find third side).

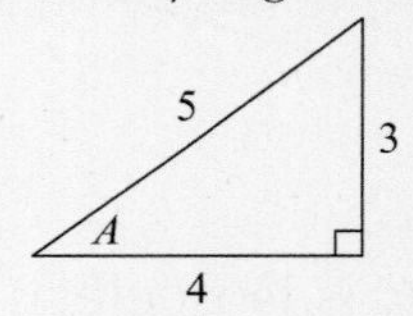

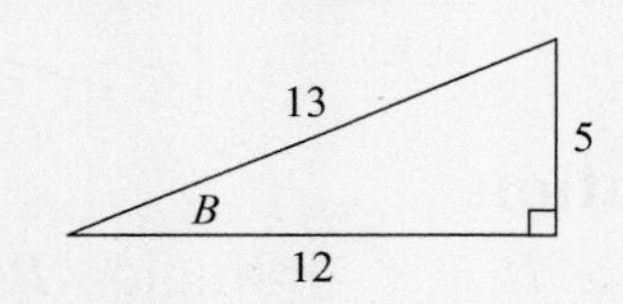

Then $\cos(A - B) = \cos A \cos B + \sin A \sin B$

$$= \frac{4}{5} \times \frac{12}{13} + \frac{3}{5} \times \frac{5}{13}$$
$$= \frac{48}{65} + \frac{15}{65} = \mathbf{\frac{63}{65}}$$

Example 3

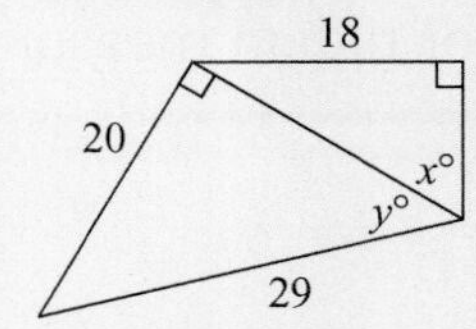

Show that the exact value of $\sin(x + y)°$ is $\frac{378 + 60\sqrt{13}}{609}$

Solution

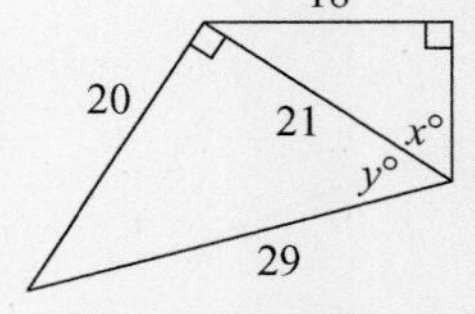

Pythagoras' Theorem is used to find the two missing lengths

$$\sqrt{21^2 - 18^2} = \sqrt{117}$$
$$= \sqrt{9 \times 13} = 3\sqrt{13}$$

$$\sin(x + y)° = \sin x° \cos y° + \cos x° \sin y°$$
$$= \frac{18}{21} \times \frac{21}{29} + \frac{3\sqrt{13}}{21} \times \frac{20}{29}$$
$$= \frac{18 \times 21}{21 \times 29} + \frac{3\sqrt{13} \times 20}{21 \times 29}$$
$$= \frac{378 + 60\sqrt{13}}{609}$$

The double angle sine formula

The double angle formula is:

$\sin 2A = 2\sin A \cos A$

Examples using this as a template are:

$$\sin 4\theta = 2\sin 2\theta \cos 2\theta$$
$$\sin P = 2\sin\frac{P}{2}\cos\frac{P}{2}$$

Example

Show that $(\cos x + \sin x)^2 = 1 + \sin 2x$

Solution

$$(\cos x + \sin x)^2 = (\cos x + \sin x)(\cos x + \sin x)$$
$$= \cos^2 x + \cos x \sin x + \sin x \cos x + \sin^2 x$$
$$= \underbrace{\cos^2 x + \sin^2 x}_{1} + \underbrace{2\sin x \cos x}_{\sin 2x}$$
$$= 1 + \sin 2x$$

The double angle cosine formulae

There are three versions of this double angle formula:

$\cos 2A = 2\cos^2 A - 1$
or **$\cos^2 A - \sin^2 A$**
or **$1 - 2\sin^2 A$**

Examples using these as a template are:

$$\cos\theta = 2\cos^2\frac{\theta}{2} - 1 = 1 - 2\sin^2\frac{\theta}{2}$$

Here are two useful rearrangements:

$$\cos^2 A = \frac{1}{2}(1 + \cos 2A)$$

$$\sin^2 A = \frac{1}{2}(1 - \cos 2A)$$

$$\cos^2\frac{P}{2} = \frac{1}{2}(1 + \cos P)$$

TOP TIP

The addition formulae and the double angle formulae will be given to you during your exam. But you should memorise them as well.

Example 1

Show that $\sin\frac{\pi}{12} = \sqrt{\frac{2-\sqrt{3}}{4}}$ Hint: use $\sin^2 A = \frac{1}{2}(1 - \cos 2A)$ with $A = \frac{\pi}{12}$

Solution

$$\sin^2\frac{\pi}{12} = \frac{1}{2}\left(1 - \cos\left(2\times\frac{\pi}{12}\right)\right) = \frac{1}{2}\left(1 - \cos\frac{\pi}{6}\right) = \frac{1}{2}\left(1 - \frac{\sqrt{3}}{2}\right)$$

$$= \frac{1}{2}\left(\frac{2}{2} - \frac{\sqrt{3}}{2}\right) = \frac{1}{2}\left(\frac{2-\sqrt{3}}{2}\right) = \frac{2-\sqrt{3}}{4}$$

So $\sin\frac{\pi}{12} = \sqrt{\frac{2-\sqrt{3}}{4}}$

Example 2

If $\tan\alpha = \frac{\sqrt{7}}{2}$ find the **exact** value of $\cos 2\alpha$ where $0 \le \alpha \le \frac{\pi}{2}$

Solution

Here is a right-angled triangle for which $\tan\alpha = \frac{\sqrt{7}}{2}$:

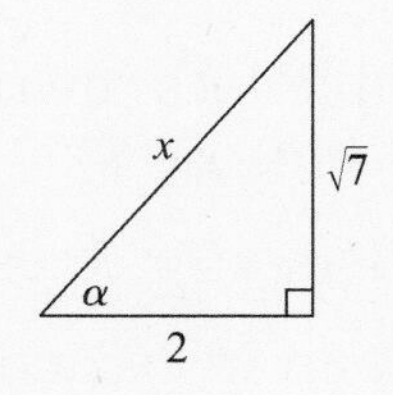

$$x^2 = 2^2 + (\sqrt{7})^2 = 4 + 7 = 11$$

So $x = \sqrt{11}$

$$\cos 2\alpha = 2\cos^2\alpha - 1 = 2\times\frac{2}{\sqrt{11}}\times\frac{2}{\sqrt{11}} - 1 = \frac{8}{11} - 1 = \frac{8}{11} - \frac{11}{11} = -\frac{3}{11}$$

Note: Other versions of $\cos 2\alpha$ e.g. $1 - 2\sin^2\alpha$ could have been used.

Quick Test 11

1. Show that $(\cos\theta + \sin\theta)(\cos\theta - \sin\theta) = \cos 2\theta$
2. If $0 < \alpha < \frac{\pi}{2}$ find the exact value of $\sin 2\alpha$ given that $\tan\alpha = \frac{1}{3}$
3. 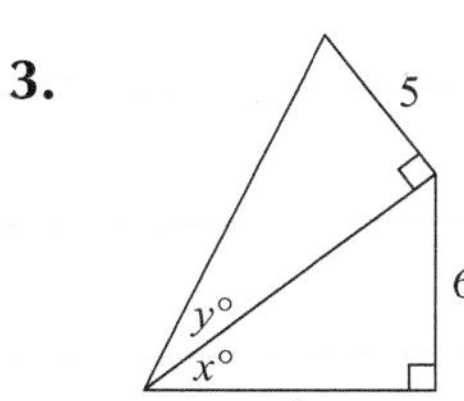

Show that $\cos\,(x + y)^\circ = \frac{1}{\sqrt{5}}$

The wave function

Linear combinations of sin x and cos x

Look at these graphs...

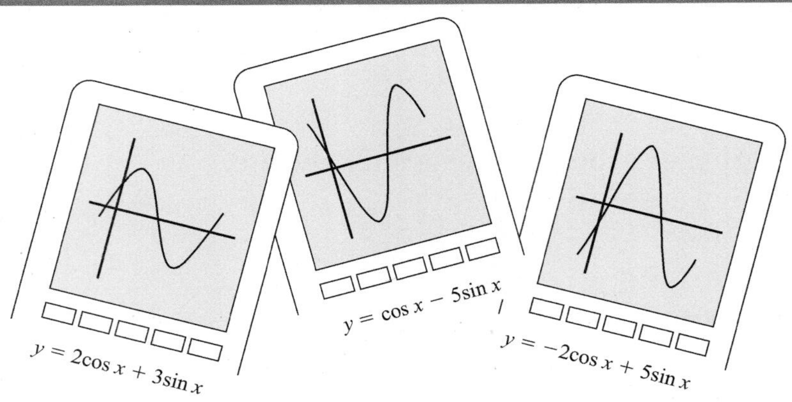

Graphs with equations of the form $y = a\cos x + b\sin x$ where a and b are constants are always sine or cosine graphs with differing amplitudes and shifted left or right.

They can therefore be expressed in any of the following forms:

$k\cos(x + \alpha)$

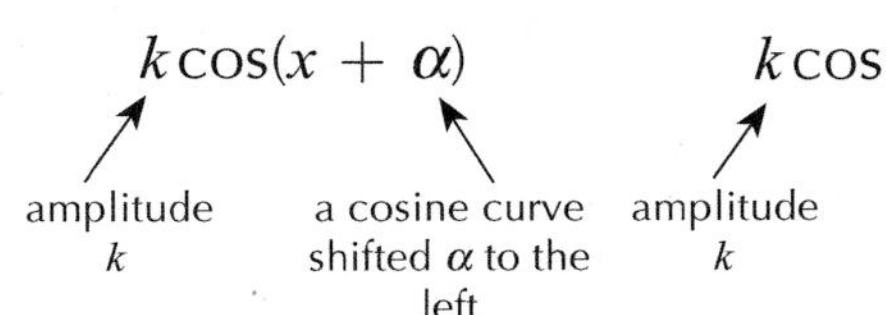

$k\cos(x - \alpha)$

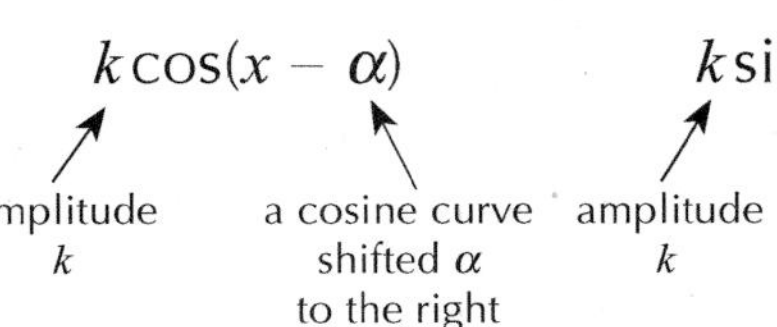

$k\sin(x + \alpha)$

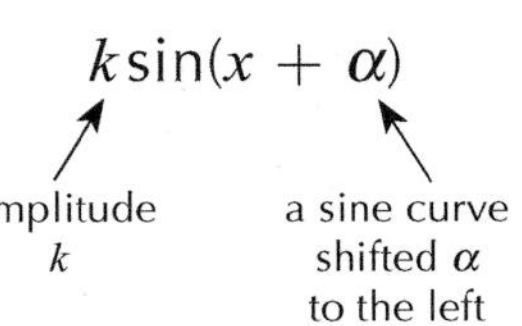

$k\sin(x - \alpha)$

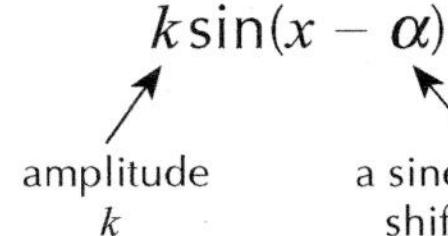

Note: the amplitude k will always be positive: $k > 0$

Comparing coefficients

As an example let's rewrite $2\cos x° + 3\sin x°$ in the form $k\cos(x - a)°$

Here you use the addition formula

$2\cos x° + 3\sin x° = k\cos(x - \alpha)°$

$2\cos x° + 3\sin x° = k[\cos x° \cos \alpha° + \sin x° \sin \alpha°]$

$2\cos x° + 3\sin x° = k\cos x° \cos \alpha° + k\sin x° \sin \alpha°$

Now find the coefficients of $\cos x°$ and the coefficients of $\sin x°$:

$2\cos x° + 3\sin x° = k\cos x° \cos \alpha° + k\sin x° \sin \alpha°$

$k\cos \alpha° = 2$ — Red coefficients are equal

$k\sin \alpha° = 3$ — Green coefficients are equal

You now have a pair of simultaneous equations in k and α

TOP TIP

k is always positive so these equations can tell you which quadrant the angle α is in.

Determining k and α

$$\left.\begin{aligned} k\cos\alpha^\circ &= 2 \\ k\sin\alpha^\circ &= 3 \end{aligned}\right\}$$

k is always positive so these equations tell you that $\sin\alpha^\circ$ and $\cos\alpha^\circ$ are both positive. This means α° is an angle in the 1st quadrant so $0^\circ < \alpha^\circ < 90^\circ$ (see page 62).

Divide both sides and use: $\frac{\sin\alpha}{\cos\alpha} = \tan\alpha$

$$\frac{\cancel{k}\sin\alpha^\circ}{\cancel{k}\cos\alpha^\circ} = \frac{3}{2} \Rightarrow \tan\alpha^\circ = 1\cdot5 \Rightarrow \alpha^\circ \doteqdot 56\cdot3^\circ \text{ (1st quadrant only)}$$

Square both sides and add then use: $\sin^2\alpha + \cos^2\alpha = 1$

$$(k\sin\alpha^\circ)^2 + (k\cos\alpha^\circ)^2 = 3^2 + 2^2$$
$$\Rightarrow k^2\sin^2\alpha^\circ + k^2\cos^2\alpha^\circ = 9 + 4$$
$$\Rightarrow k^2(\sin^2\alpha^\circ + \cos^2\alpha^\circ) = 13$$
$$\Rightarrow k^2 \times 1 = 13 \Rightarrow k^2 = 13 \Rightarrow k = \sqrt{13} \text{ (positive)}$$

Here is a radian example:

$$\left.\begin{aligned} k\sin\alpha &= 2 \\ k\cos\alpha &= -5 \end{aligned}\right\}$$

k is always positive so these equations tell you that $\sin\alpha$ is positive and $\cos\alpha$ is negative. This means α is an angle in the 2nd quadrant so $\frac{\pi}{2} < \alpha < \pi$

Divide both sides and use: $\frac{\sin\alpha}{\cos\alpha} = \tan\alpha$

$$\frac{\cancel{k}\sin\alpha}{\cancel{k}\cos\alpha} = \frac{2}{-5} \Rightarrow \tan\alpha = -\frac{2}{5} \Rightarrow \tan\alpha = -0\cdot4$$

$$\alpha = \pi - 0\cdot380\ldots$$
$$= 2\cdot761\ldots \doteqdot 2\cdot76 \text{ (2nd quadrant)}$$

Using $\tan\alpha = 0\cdot4$ the 1st quadrant angle is $0\cdot380\ldots$ (radians)

Square both sides and add then use: $\sin^2\alpha + \cos^2\alpha = 1$

$$(k\sin\alpha)^2 + (k\cos\alpha)^2 = 2^2 + (-5)^2$$
$$\Rightarrow k^2\sin^2\alpha + k^2\cos^2\alpha = 4 + 25$$
$$\Rightarrow k^2(\sin^2\alpha + \cos^2\alpha) = 29$$
$$\Rightarrow k^2 \times 1 = 29 \Rightarrow k^2 = 29 \Rightarrow k = \sqrt{29} \text{ (positive)}$$

The final form

Example Express $2\cos x^\circ + 3\sin x^\circ$ in the form $k\cos(x - \alpha)^\circ$ where $k > 0$

Solution From above $k = \sqrt{13}$ and $\alpha \doteqdot 56\cdot3$

so $2\cos x^\circ + 3\sin x^\circ \doteqdot \sqrt{13}\cos(x - 56\cdot3)^\circ$

Quick Test 12

1. Express $\cos x^\circ - 3\sin x^\circ$ in the form $k\cos(x + \alpha)^\circ$ where $k > 0$ and $0 \le \alpha < 360$
2. Express $-\cos x - \sqrt{3}\sin x$ in the form $k\sin(x + \alpha)$ where $k > 0$ and $0 \le \alpha < 2\pi$

Chapter 1

Vectors: basic properties

What is a vector?

A vector is a quantity with both **magnitude** and **direction**. It can be represented by a **directed line segment**. The arrow shows the sense of direction.

Directed line segment $\overrightarrow{AB}$ represents the vector $\boldsymbol{v}$.

Examples of vector quantities are: velocity, magnetic field strength, push/pull forces etc.

Vectors are described using components parallel to the x-, y- and z-axes. In the diagram, vector $\boldsymbol{v}$, represented by $\overrightarrow{AB}$, has components $\begin{pmatrix} 1 \\ 2 \\ 3 \end{pmatrix}$.

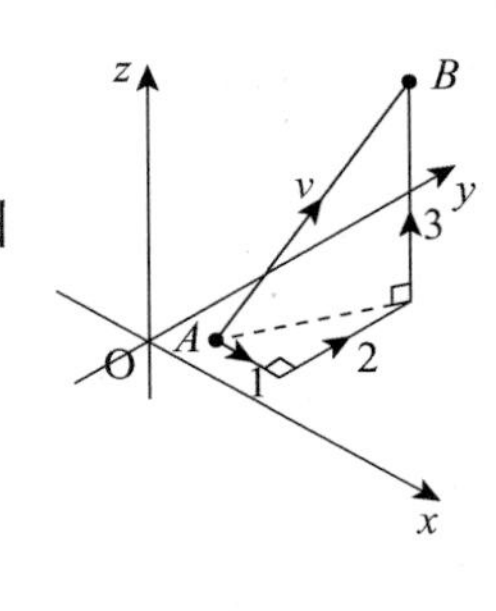

$$\boldsymbol{v} = \begin{pmatrix} 1 \\ 2 \\ 3 \end{pmatrix} \begin{matrix} \longleftarrow x\text{-component} \\ \longleftarrow y\text{-component} \\ \longleftarrow z\text{-component} \end{matrix}$$

Think of $\begin{pmatrix} 1 \\ 2 \\ 3 \end{pmatrix}$ as the 'instructions for a journey'.

It describes any journey that goes the same distance and direction as the journey from A to B.

Magnitude

The **magnitude** (length) of a vector $\boldsymbol{v} = \begin{pmatrix} a \\ b \\ c \end{pmatrix}$ is given by

$$|\boldsymbol{v}| = \sqrt{a^2 + b^2 + c^2}$$

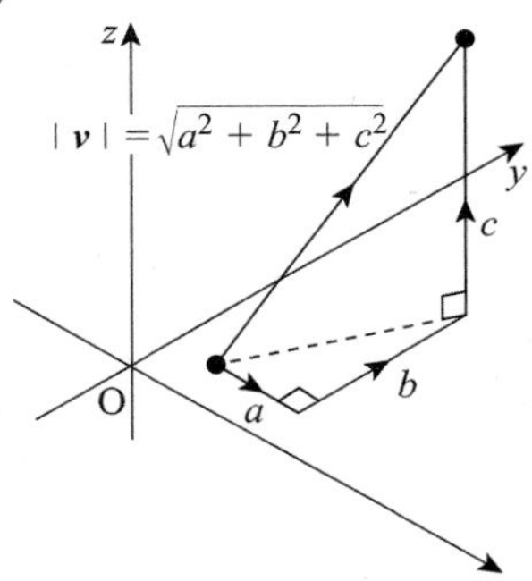

Example

Find the magnitude of $\boldsymbol{a} = \begin{pmatrix} -2 \\ 3 \\ 6 \end{pmatrix}$

Solution

$$|\boldsymbol{a}| = \sqrt{(-2)^2 + 3^2 + 6^2} = \sqrt{4 + 9 + 36}$$

$$= \sqrt{49}$$

$$= \mathbf{7}$$

Equal vectors

Each of these 'journeys' has the same distance and same direction and so represents the same vector $\boldsymbol{a}$.

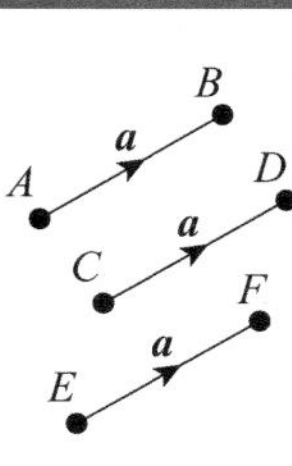

if $\overrightarrow{AB} = \overrightarrow{CD} = \overrightarrow{EF}$ (equal directed line segments)
then $AB = CD = EF$ (lines have equal lengths)
and $AB \parallel CD \parallel EF$ (lines are parallel)

If $\begin{pmatrix} x_1 \\ y_1 \\ z_1 \end{pmatrix} = \begin{pmatrix} x_2 \\ y_2 \\ z_2 \end{pmatrix}$ then $x_1 = x_2, y_1 = y_2, z_1 = z_2$

(equal vectors) (equal components)

The zero vector

$\mathbf{0} = \begin{pmatrix} 0 \\ 0 \\ 0 \end{pmatrix}$ The zero vector has magnitude $|\mathbf{0}| = 0$ but has **no** direction defined.

Addition

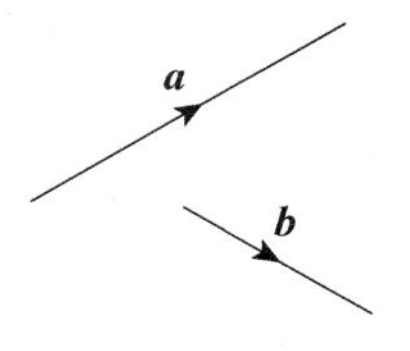

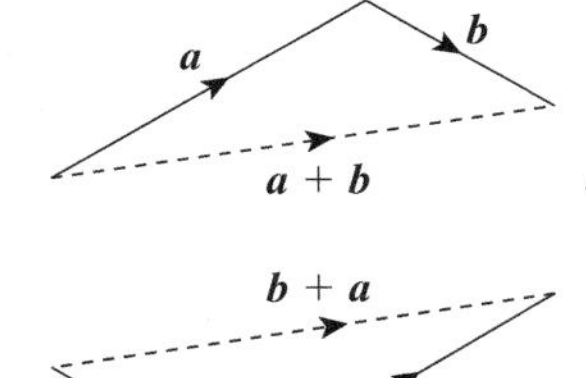

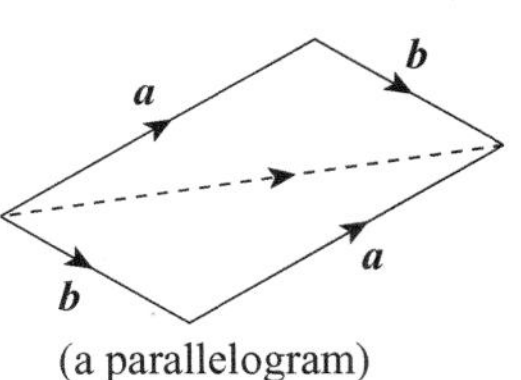

(a parallelogram)

Notice that $\boldsymbol{a} + \boldsymbol{b} = \boldsymbol{b} + \boldsymbol{a}$

To add these vectors place them 'nose-to-tail'.

Using components...

if $\boldsymbol{a} = \begin{pmatrix} x_1 \\ y_1 \\ z_1 \end{pmatrix}$ and $\boldsymbol{b} = \begin{pmatrix} x_2 \\ y_2 \\ z_2 \end{pmatrix}$ then $\boldsymbol{a} + \boldsymbol{b} = \begin{pmatrix} x_1 \\ y_1 \\ z_1 \end{pmatrix} + \begin{pmatrix} x_2 \\ y_2 \\ z_2 \end{pmatrix} = \begin{pmatrix} x_1 + x_2 \\ y_1 + y_2 \\ z_1 + z_2 \end{pmatrix}$ Add the corresponding components.

TOP TIP

All these properties hold for 2D vectors with just two components.

The negative of a vector

$-\boldsymbol{a}$ has the same magnitude as $\boldsymbol{a}$ but the opposite direction

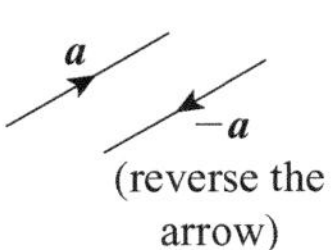

(reverse the arrow)

and using components...

if $\boldsymbol{a} = \begin{pmatrix} x_1 \\ y_1 \\ z_1 \end{pmatrix}$ then $-\boldsymbol{a} = \begin{pmatrix} -x_1 \\ -y_1 \\ -z_1 \end{pmatrix}$

Notice that $\boldsymbol{a} + (-\boldsymbol{a}) = \mathbf{0}$ the zero vector.

Subtraction

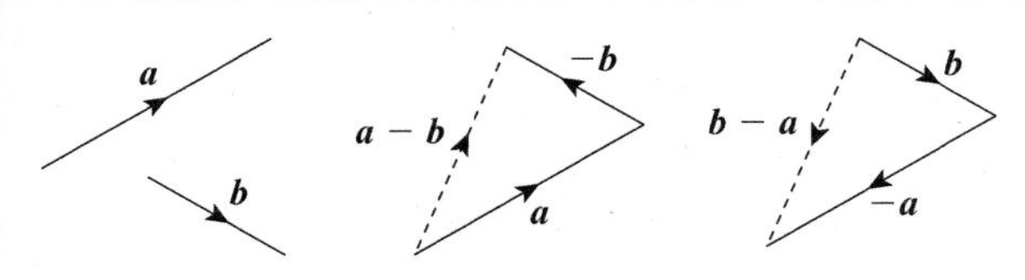

$\boldsymbol{a} - \boldsymbol{b}$ is the vector $\boldsymbol{a} + (-\boldsymbol{b})$
$\boldsymbol{b} - \boldsymbol{a}$ is the vector $\boldsymbol{b} + (-\boldsymbol{a})$
Notice that $\boldsymbol{b} - \boldsymbol{a} = -(\boldsymbol{a} - \boldsymbol{b})$

Using components...

if $\boldsymbol{a} = \begin{pmatrix} x_1 \\ y_1 \\ z_1 \end{pmatrix}$ and $\boldsymbol{b} = \begin{pmatrix} x_2 \\ y_2 \\ z_2 \end{pmatrix}$

Then $\boldsymbol{a} - \boldsymbol{b} = \begin{pmatrix} x_1 \\ y_1 \\ z_1 \end{pmatrix} - \begin{pmatrix} x_2 \\ y_2 \\ z_2 \end{pmatrix} = \begin{pmatrix} x_1 - x_2 \\ y_1 - y_2 \\ z_1 - z_2 \end{pmatrix}$

(Subtract the corresponding components.)

Example

Calculate the magnitudes of $\boldsymbol{a} + \boldsymbol{b}$ and $\boldsymbol{a} - \boldsymbol{b}$ where

$\boldsymbol{a} = \begin{pmatrix} -1 \\ 2 \\ 5 \end{pmatrix}$ and $\boldsymbol{b}$ $\begin{pmatrix} 4 \\ -2 \\ -1 \end{pmatrix}$

Solution

$$\boldsymbol{a} + \boldsymbol{b} = \begin{pmatrix} -1 \\ 2 \\ 5 \end{pmatrix} + \begin{pmatrix} 4 \\ -2 \\ -1 \end{pmatrix} = \begin{pmatrix} 3 \\ 0 \\ 4 \end{pmatrix}$$

so $|\boldsymbol{a} + \boldsymbol{b}| = \sqrt{3^2 + 0^2 + 4^2} = \sqrt{25} = 5$

$$\boldsymbol{a} - \boldsymbol{b} = \begin{pmatrix} -1 \\ 2 \\ 5 \end{pmatrix} - \begin{pmatrix} 4 \\ -2 \\ -1 \end{pmatrix} = \begin{pmatrix} -5 \\ 4 \\ 6 \end{pmatrix}$$

so $|\boldsymbol{a} - \boldsymbol{b}| = \sqrt{(-5)^2 + 4^2 + 6^2}$
$= \sqrt{25 + 16 + 36} = \sqrt{77}$

Quick Test 13

1. $\boldsymbol{v} = \begin{pmatrix} -3 \\ 4 \\ 2 \end{pmatrix}$ and $\boldsymbol{w} = \begin{pmatrix} 1 \\ 0 \\ -5 \end{pmatrix}$

a) Write down the components of $-\boldsymbol{v} - \boldsymbol{w}$

b) Find the exact value of $|\boldsymbol{w} - \boldsymbol{v}|$

Position vectors and applications

Multiplication by a scalar

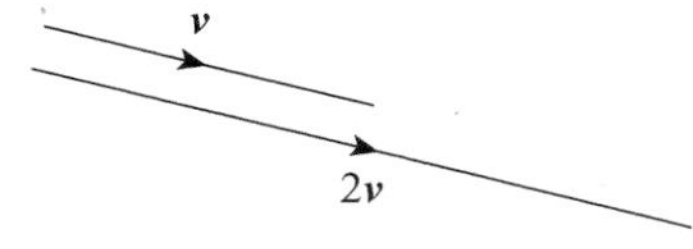

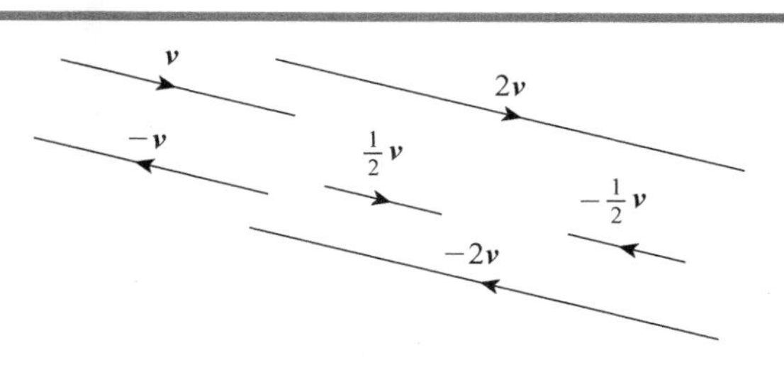

If $v = \begin{pmatrix} x_1 \\ y_1 \\ z_1 \end{pmatrix}$ then $2v = v + v = \begin{pmatrix} x_1 \\ y_1 \\ z_1 \end{pmatrix} + \begin{pmatrix} x_1 \\ y_1 \\ z_1 \end{pmatrix} = \begin{pmatrix} 2x_1 \\ 2y_1 \\ 2z_1 \end{pmatrix}$

In general $kv = k\begin{pmatrix} x_1 \\ y_1 \\ z_1 \end{pmatrix} = \begin{pmatrix} kx_1 \\ ky_1 \\ kz_1 \end{pmatrix}$

Then vector $\boldsymbol{v}$ has been multiplied by scalar k ('scalar' being number as opposed to vector).

$\boldsymbol{a} = k\boldsymbol{b}$ $(k \neq 0) \Longleftrightarrow$ $\boldsymbol{a}$ and $\boldsymbol{b}$ are parallel

If $k > 0$ they have the same direction.

If $k < 0$ they have opposite directions.

Example $\boldsymbol{a} = \begin{pmatrix} -1 \\ 1 \\ 2 \end{pmatrix}$, $\boldsymbol{b} = \begin{pmatrix} -2 \\ 3 \\ 0 \end{pmatrix}$ and $\boldsymbol{c} = \begin{pmatrix} 1 \\ -2 \\ -1 \end{pmatrix}$ Find the components of $2\boldsymbol{a} - 3\boldsymbol{b} - \boldsymbol{c}$

Solution $2\boldsymbol{a} - 3\boldsymbol{b} - \boldsymbol{c} = 2\begin{pmatrix} -1 \\ 1 \\ 2 \end{pmatrix} - 3\begin{pmatrix} -2 \\ 3 \\ 0 \end{pmatrix} - \begin{pmatrix} 1 \\ -2 \\ -1 \end{pmatrix} = \begin{pmatrix} 2\times(-1)-3\times(-2)-1 \\ 2\times1-3\times3-(-2) \\ 2\times2-3\times0-(-1) \end{pmatrix} = \begin{pmatrix} -2+6-1 \\ 2-9+2 \\ 4-0+1 \end{pmatrix} = \begin{pmatrix} 3 \\ -5 \\ 5 \end{pmatrix}$

What is a position vector?

$\overrightarrow{OP}$, written $\boldsymbol{p}$, is the **position vector** of the point P.

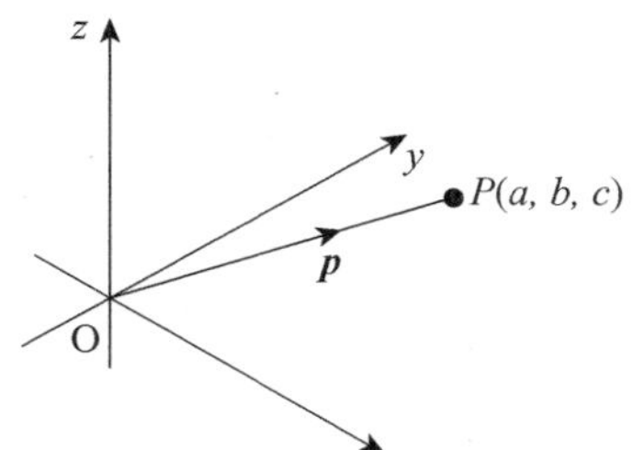

$P(a, b, c) \longleftrightarrow \boldsymbol{p} = \begin{pmatrix} a \\ b \\ c \end{pmatrix}$

a, b and c are the coordinates of the point P.

a, b and c are the components of the position vector $\boldsymbol{p}$.

Think of $P(a, b, c)$ as the address of point P and $\boldsymbol{p} = \begin{pmatrix} a \\ b \\ c \end{pmatrix}$ as the instructions to get from the origin O to point P.

Using position vectors to find components

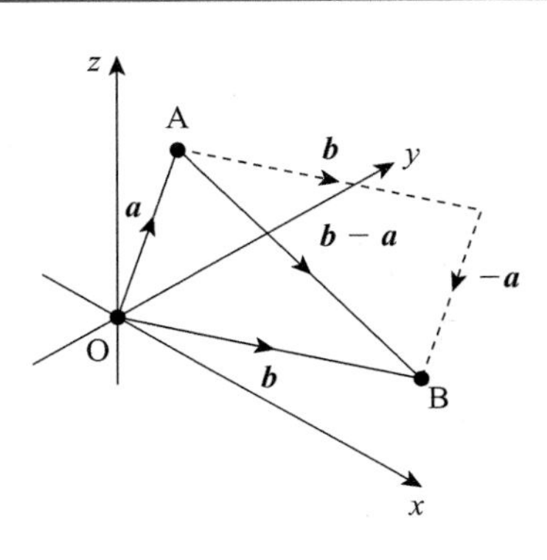

$\overrightarrow{AB} = \boldsymbol{b} - \boldsymbol{a}$

This is $\overrightarrow{AB}$ in terms of position vectors $\boldsymbol{a}$ and $\boldsymbol{b}$.

For other points:

$\overrightarrow{PQ} = \boldsymbol{q} - \boldsymbol{p}$

$\overrightarrow{RS} = \boldsymbol{s} - \boldsymbol{r}$ etc

Example

P has coordinates (1, –1, 3) and Q has coordinates (–2, 5, 6). Find the components of $\overrightarrow{PQ}$.

Solution

$\overrightarrow{PQ} = \boldsymbol{q} - \boldsymbol{p}$ (In terms of position vectors)

$$= \begin{pmatrix} -2 \\ 5 \\ 6 \end{pmatrix} - \begin{pmatrix} 1 \\ -1 \\ 3 \end{pmatrix} = \begin{pmatrix} \mathbf{-3} \\ \mathbf{6} \\ \mathbf{3} \end{pmatrix}$$

Collinear points

Collinear points lie on the same straight line.

To show A, B and C are collinear:

Step 1 Find the components of $\overrightarrow{AB}$ and $\overrightarrow{BC}$

Step 2 Write $\overrightarrow{AB} = k\overrightarrow{BC}$ (or $\overrightarrow{BC} = k\overrightarrow{AB}$) for some constant k

Step 3 State: 'Since AB and BC are parallel and have point B in common then A, B and C are collinear'.

Example

Show that $A(2, -1, -1)$, $B(4, 3, -5)$ and $C(5, 5, -7)$ are collinear

Solution

$$\overrightarrow{AB} = \boldsymbol{b} - \boldsymbol{a} = \begin{pmatrix} 4 \\ 3 \\ -5 \end{pmatrix} - \begin{pmatrix} 2 \\ -1 \\ -1 \end{pmatrix} = \begin{pmatrix} 2 \\ 4 \\ -4 \end{pmatrix}$$

$$\overrightarrow{BC} = \boldsymbol{c} - \boldsymbol{b} = \begin{pmatrix} 5 \\ 5 \\ -7 \end{pmatrix} - \begin{pmatrix} 4 \\ 3 \\ -5 \end{pmatrix} = \begin{pmatrix} 1 \\ 2 \\ -2 \end{pmatrix}$$

$$\overrightarrow{AB} = 2\overrightarrow{BC}$$

Since AB and BC are parallel and have point B in common then A, B and C are collinear.

For example $\overrightarrow{AB} = 2\overrightarrow{BC}$

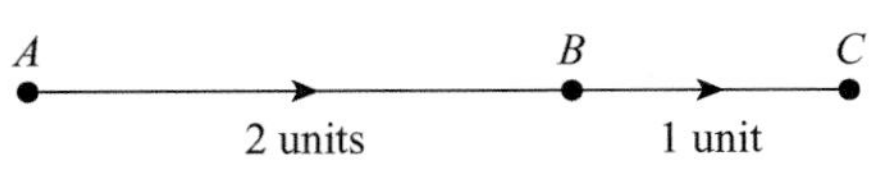

Note: In this case B divides AC in the ratio 2 : 1

For $\overrightarrow{AB} = k\overrightarrow{BC}$ the ratio is $k : 1$

Finding a point of internal division on a line

If P divides AB in the ratio $m : n$

then $\overrightarrow{AP} = \frac{m}{m+n}\overrightarrow{AB}$

So $(m + n)\overrightarrow{AP} = m\overrightarrow{AB}$

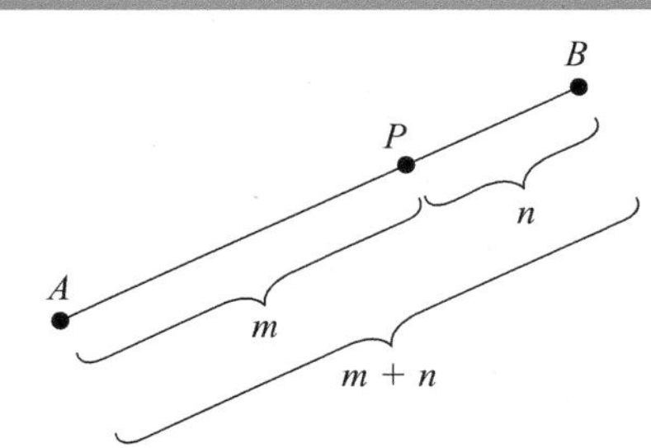

For example if Q divides PR in the ratio 3 : 2

then $\overrightarrow{PQ} = \frac{3}{5}\overrightarrow{PR}$

so $5\overrightarrow{PQ} = 3\overrightarrow{PR}$

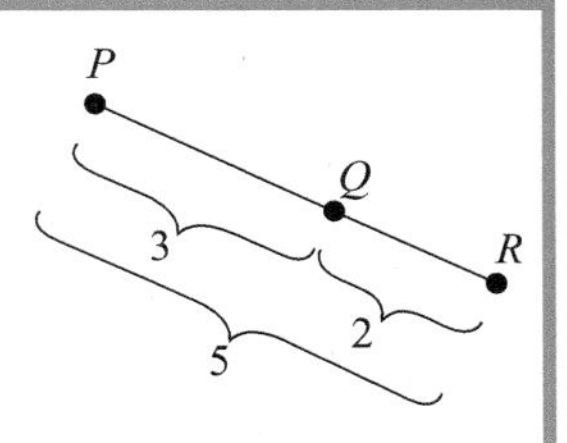

TOP TIP

The methods shown here in 3D can be used to solve similar examples in 2D.

Example

Find the coordinates of T the point which divides SU in the ratio 4 : 1, where S is the point (–1, 2, 6) and U is the point (4, –3, 1).

Solution

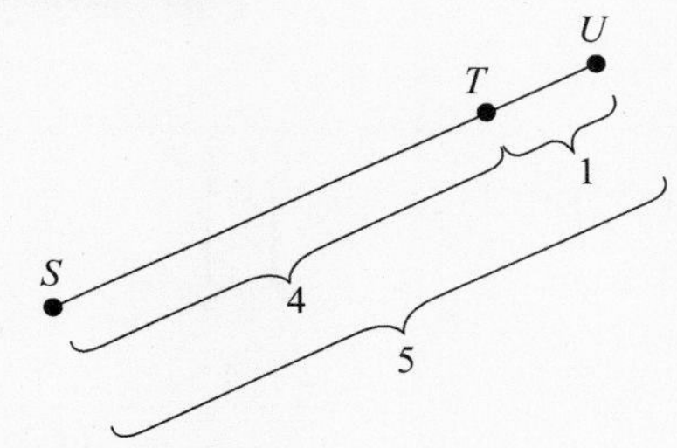

Think of 'journeys'.

The journey $\overrightarrow{ST}$ is $\frac{4}{5}$ of the whole journey $\overrightarrow{SU}$.

So $\overrightarrow{ST} = \frac{4}{5}\overrightarrow{SU}$ giving $5\overrightarrow{ST} = 4\overrightarrow{SU}$

Now use position vectors…

$5(\boldsymbol{t} - \boldsymbol{s}) = 4(\boldsymbol{u} - \boldsymbol{s})$

$5\boldsymbol{t} - 5\boldsymbol{s} = 4\boldsymbol{u} - 4\boldsymbol{s}$

$5\boldsymbol{t} = 4\boldsymbol{u} + \boldsymbol{s}$

The normal rules for algebra apply, so solve to get $\boldsymbol{t}$.

So $5\boldsymbol{t} = 4\begin{pmatrix} 4 \\ -3 \\ 1 \end{pmatrix} + \begin{pmatrix} -1 \\ 2 \\ 6 \end{pmatrix} = \begin{pmatrix} 15 \\ -10 \\ 10 \end{pmatrix}$

so $\boldsymbol{t} = \frac{1}{5}\begin{pmatrix} 15 \\ -10 \\ 10 \end{pmatrix} = \begin{pmatrix} 3 \\ -2 \\ 2 \end{pmatrix}$

Hence $T(\mathbf{3}, \mathbf{-2}, \mathbf{2})$ (coordinates asked for, not components).

Quick Test 14

1. The points $P(-1, 0, 2)$, $Q(3, 5, -2)$, $R(-1, 6, -7)$ and $S(-5, 1, -3)$ form a quadrilateral.
   a) Find the components of: (i) $\overrightarrow{PQ}$ (ii) $\overrightarrow{SR}$
   b) What sort of quadrilateral is $PQRS$?
2. Show that $A(2, -1, 4)$, $B(3, 2, 5)$ and $C(6, 11, 8)$ are collinear and find the ratio in which B divides AC.

Chapter 1

The scalar product and applications

The angle between two vectors

Place the two vectors tail-to-tail:

θ is the angle between $\boldsymbol{a}$ and $\boldsymbol{b}$

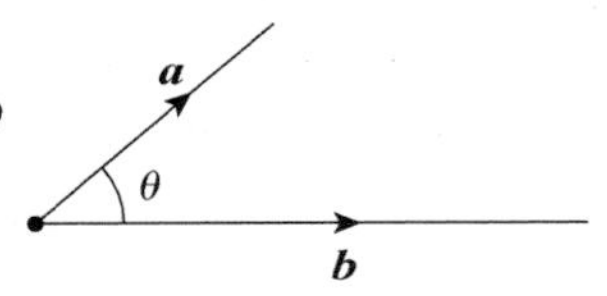

θ always lies in the range $0 \leq \theta \leq \pi$

Maximum angle is π

Minimum angle is 0

The scalar (dot) product

Using magnitudes and the angle...

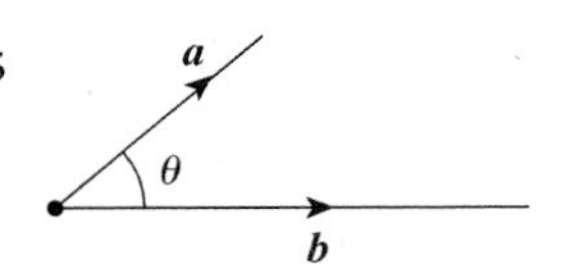

$$\boldsymbol{a}.\boldsymbol{b} = |\boldsymbol{a}|\,|\boldsymbol{b}|\cos\theta$$

Using components...

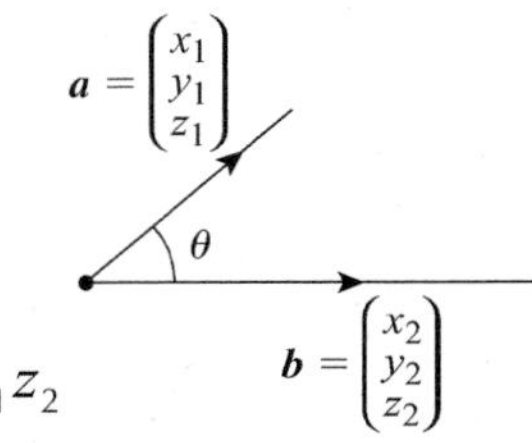

$$\boldsymbol{a} = \begin{pmatrix} x_1 \\ y_1 \\ z_1 \end{pmatrix} \qquad \boldsymbol{b} = \begin{pmatrix} x_2 \\ y_2 \\ z_2 \end{pmatrix}$$

$$\boldsymbol{a}.\boldsymbol{b} = x_1x_2 + y_1y_2 + z_1z_2$$

These two calculations yield the same **number** which is written $\boldsymbol{a}.\boldsymbol{b}$ and is called the **Scalar Product** or **Dot Product** of vectors $\boldsymbol{a}$ and $\boldsymbol{b}$.

Calculating the angle

Rearrange $\boldsymbol{a}.\boldsymbol{b} = |\boldsymbol{a}||\boldsymbol{b}|\cos\theta$ to get $\cos\theta = \dfrac{\boldsymbol{a}.\boldsymbol{b}}{|\boldsymbol{a}|\,|\boldsymbol{b}|}$

If you know the components $\boldsymbol{a} = \begin{pmatrix} x_1 \\ y_1 \\ z_1 \end{pmatrix}$ $\boldsymbol{b} = \begin{pmatrix} x_2 \\ y_2 \\ z_2 \end{pmatrix}$

then $\cos\theta = \dfrac{\boldsymbol{a}.\boldsymbol{b}}{|\boldsymbol{a}|\,|\boldsymbol{b}|} = \dfrac{x_1x_2 + y_1y_2 + z_1z_2}{\sqrt{x_1^2 + y_1^2 + z_1^2}\sqrt{x_2^2 + y_2^2 + z_2^2}}$

Remember that θ will be a 1st or 2nd quadrant angle since $0 \leq \theta \leq \pi$

Example

Calculate the angle θ between $\boldsymbol{a} = \begin{pmatrix} 2 \\ 3 \\ -1 \end{pmatrix}$ and $\boldsymbol{b} = \begin{pmatrix} 1 \\ -2 \\ 3 \end{pmatrix}$

Solution

$$|\boldsymbol{a}| = \sqrt{2^2 + 3^2 + (-1)^2} = \sqrt{14}$$

$$|\boldsymbol{b}| = \sqrt{1^2 + (-2)^2 + 3^2} = \sqrt{14}$$

$$\boldsymbol{a}.\boldsymbol{b} = \begin{pmatrix} 2 \\ 3 \\ -1 \end{pmatrix} \cdot \begin{pmatrix} 1 \\ -2 \\ 3 \end{pmatrix} = 2 \times 1 + 3 \times (-2) + (-1) \times 3 = -7$$

$$\text{so } \cos\theta = \frac{\boldsymbol{a}.\boldsymbol{b}}{|\boldsymbol{a}|\,|\boldsymbol{b}|} = \frac{-7}{\sqrt{14}\sqrt{14}} = -\frac{7}{14} = -\frac{1}{2}$$

$$\text{so} \quad \theta = \pi - \frac{\pi}{3} = \frac{2\pi}{3} \quad \text{(second quadrant only)}$$

Perpendicular vectors

If $\boldsymbol{a}$ and $\boldsymbol{b}$ are perpendicular then $\boldsymbol{a}.\boldsymbol{b} = 0$

(since $\cos \frac{\pi}{2} = 0$)

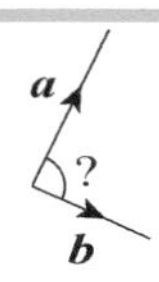

If $\boldsymbol{a}.\boldsymbol{b} = 0$ and both $\boldsymbol{a}$ and $\boldsymbol{b}$ are non-zero then $\boldsymbol{a}$ is perpendicular to $\boldsymbol{b}$

TOP TIP

To prove two vectors are perpendicular, you show their dot product is zero.

Solving perpendicularity problems

Example 1 A triangle ABC has vertices $A(-1, 3, 2)$, $B(1, -3, 3)$ and $C(0, 2, -6)$. Show that it is right-angled.

Solution

$$\overrightarrow{AB} = \boldsymbol{b} - \boldsymbol{a} = \begin{pmatrix} 1 \\ -3 \\ 3 \end{pmatrix} - \begin{pmatrix} -1 \\ 3 \\ 2 \end{pmatrix} = \begin{pmatrix} 2 \\ -6 \\ 1 \end{pmatrix}$$

$$\overrightarrow{AC} = \boldsymbol{c} - \boldsymbol{a} = \begin{pmatrix} 0 \\ 2 \\ -6 \end{pmatrix} - \begin{pmatrix} -1 \\ 3 \\ 2 \end{pmatrix} = \begin{pmatrix} 1 \\ -1 \\ -8 \end{pmatrix}$$

$$\overrightarrow{AB}.\overrightarrow{AC} = \begin{pmatrix} 2 \\ -6 \\ 1 \end{pmatrix} \cdot \begin{pmatrix} 1 \\ -1 \\ -8 \end{pmatrix} = 2 \times 1 + (-6) \times (-1) + 1 \times (-8) = 0$$

hence $\overrightarrow{AB}$ is perpendicular to $\overrightarrow{AC}$

$\angle BAC = 90°$

So ΔABC is right-angled at A.

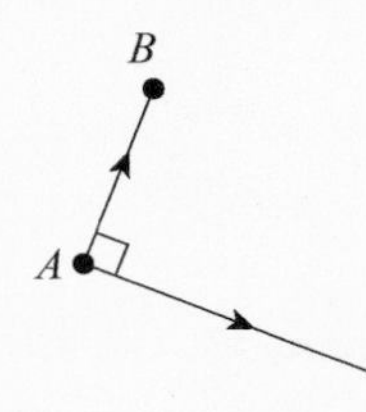

Example 2

If $\boldsymbol{v} = \begin{pmatrix} x \\ -2 \\ 9 \end{pmatrix}$ and $\boldsymbol{w} = \begin{pmatrix} x \\ 3x \\ 1 \end{pmatrix}$ are perpendicular, calculate the value of x

Solution

$$\boldsymbol{v}.\boldsymbol{w} = 0 \Rightarrow \begin{pmatrix} x \\ -2 \\ 9 \end{pmatrix} \cdot \begin{pmatrix} x \\ 3x \\ 1 \end{pmatrix} = 0 \Rightarrow x \times x + (-2) \times 3x + 9 \times 1 = 0$$

$$\Rightarrow x^2 - 6x + 9 = 0 \Rightarrow (x-3)^2 = 0 \Rightarrow x = 3$$

TOP TIP

In the exam, if you are given a vector written with unit vectors $\boldsymbol{i}, \boldsymbol{j}$ and $\boldsymbol{k}$ rewrite it in the usual component form.

Unit vectors

A **Unit Vector** has a magnitude of 1 unit. The three unit vectors parallel to the three axes are:

$$\boldsymbol{i} = \begin{pmatrix} 1 \\ 0 \\ 0 \end{pmatrix} \quad \boldsymbol{j} = \begin{pmatrix} 0 \\ 1 \\ 0 \end{pmatrix} \quad \boldsymbol{k} = \begin{pmatrix} 0 \\ 0 \\ 1 \end{pmatrix}$$

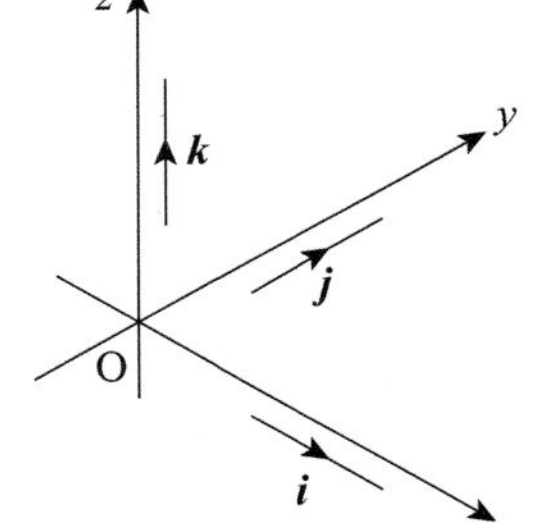

These form a set of **basis vectors** since any vector $\boldsymbol{v}$ can be written in terms of them.

If $\boldsymbol{v} = \begin{pmatrix} a \\ b \\ c \end{pmatrix} = a\begin{pmatrix} 1 \\ 0 \\ 0 \end{pmatrix} + b\begin{pmatrix} 0 \\ 1 \\ 0 \end{pmatrix} + c\begin{pmatrix} 0 \\ 0 \\ 1 \end{pmatrix}$

then $\boldsymbol{v} = a\boldsymbol{i} + b\boldsymbol{j} + c\boldsymbol{k}$

Example

$\boldsymbol{v} = \boldsymbol{i} - 3\boldsymbol{k}$ and $\boldsymbol{w} = 5\boldsymbol{i} - 2\boldsymbol{j} + \boldsymbol{k}$. Find a **unit** vector parallel to vector $\boldsymbol{v} - \boldsymbol{w}$.

Solution

$$\boldsymbol{v} - \boldsymbol{w} = \begin{pmatrix} 1 \\ 0 \\ -3 \end{pmatrix} - \begin{pmatrix} 5 \\ -2 \\ 1 \end{pmatrix} = \begin{pmatrix} -4 \\ 2 \\ -4 \end{pmatrix}$$

so $|\boldsymbol{v} - \boldsymbol{w}| = \sqrt{(-4)^2 + 2^2 + (-4)^2}$

$= \sqrt{36} = 6$

so $\frac{1}{6}(\boldsymbol{v} - \boldsymbol{w})$ has magnitude 1 unit.

$$\frac{1}{6}\begin{pmatrix} -4 \\ 2 \\ -4 \end{pmatrix} = \begin{pmatrix} -\frac{2}{3} \\ \frac{1}{3} \\ -\frac{2}{3} \end{pmatrix} = -\frac{2}{3}\boldsymbol{i} + \frac{1}{3}\boldsymbol{j} - \frac{2}{3}\boldsymbol{k}$$

is the required unit vector.

Quick Test 15

1. Find the size of angle KLM where K(–2, 5, 4), L(–1, 0, –3) and M(2, 2, 8).
2. The vectors $\begin{pmatrix} m \\ -2 \\ -1 \end{pmatrix}$ and $\begin{pmatrix} m-1 \\ 1 \\ 4 \end{pmatrix}$ are perpendicular. Find the two possible values for m.
3. Find a unit vector parallel to $\boldsymbol{u} + \boldsymbol{v}$ where: $\boldsymbol{u} = 3\boldsymbol{i} - \boldsymbol{j} + 2\boldsymbol{k}$ and $\boldsymbol{v} = -4\boldsymbol{i} - \boldsymbol{j}$

Working with vectors

Algebraic properties

Most of the 'normal' rules of algebra apply to vectors.

For instance:

$2(\boldsymbol{b} - \boldsymbol{a}) = 3(\boldsymbol{c} - \boldsymbol{a})$ rearranges to $\boldsymbol{a} = 3\boldsymbol{c} - 2\boldsymbol{b}$

$-5(\boldsymbol{v} - \boldsymbol{w}) = -5\boldsymbol{v} + 5\boldsymbol{w}$,
$-(\boldsymbol{a} + \boldsymbol{b}) = -\boldsymbol{a} - \boldsymbol{b}$ etc

For any three vectors $\boldsymbol{a}$, $\boldsymbol{b}$ and $\boldsymbol{c}$ that are non-zero

$$\boldsymbol{a}.(\boldsymbol{b} + \boldsymbol{c}) = \boldsymbol{a}.\boldsymbol{b} + \boldsymbol{a}.\boldsymbol{c}$$

Also

$$\boldsymbol{a}.\boldsymbol{a} = |\boldsymbol{a}||\boldsymbol{a}| \cos 0 = |\boldsymbol{a}||\boldsymbol{a}| \times 1 = |\boldsymbol{a}|^2$$

and

$$\boldsymbol{a}.\boldsymbol{b} = \boldsymbol{b}.\boldsymbol{a}$$

Warning

$\boldsymbol{a}^2$, $(\boldsymbol{a} + \boldsymbol{b})^2$, $\sqrt{\boldsymbol{a}}$ are all meaningless.

For the basis vectors $\boldsymbol{i}$, $\boldsymbol{j}$ and $\boldsymbol{k}$ you get:

$$\boldsymbol{i}.\boldsymbol{i} = \boldsymbol{j}.\boldsymbol{j} = \boldsymbol{k}.\boldsymbol{k} = 1\times1\times\cos 0 = 1\times1\times1 = 1$$

$$\boldsymbol{i}.\boldsymbol{j} = \boldsymbol{j}.\boldsymbol{k} = \boldsymbol{i}.\boldsymbol{k} = 1\times1\times\cos\frac{\pi}{2} = 1\times1\times0 = 0$$

Example

All the edges of this square-based pyramid have length 3 units. Calculate $\boldsymbol{a}.(\boldsymbol{b} + \boldsymbol{c})$

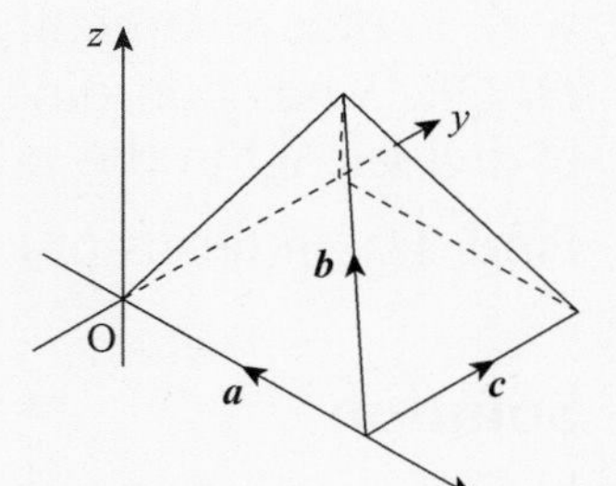

Solution

$$\boldsymbol{a}.(\boldsymbol{b} + \boldsymbol{c}) = \boldsymbol{a}.\boldsymbol{b} + \boldsymbol{a}.\boldsymbol{c}$$

Sloping faces are equilateral.

$$= |\boldsymbol{a}||\boldsymbol{b}|\cos 60° + |\boldsymbol{a}||\boldsymbol{c}|\cos 90°$$

The base is a square.

$$= 3\times3\times\frac{1}{2} + 3\times3\times0 = \frac{\mathbf{9}}{\mathbf{2}}$$

TOP TIP

Always remember that $\boldsymbol{a}.\boldsymbol{b}$ is a number not a vector

Using: $\boldsymbol{a}.\boldsymbol{b} = |\boldsymbol{a}||\boldsymbol{b}| \cos\theta$
and: $|\boldsymbol{i}| = |\boldsymbol{j}| = |\boldsymbol{k}| = 1$ (unit vectors)

Vector pathways

Example

The diagram shows a square-based pyramid *OPQRS* with vertex *S* vertically above the centre of the square base *OPQR*. *S* has coordinates (3, 3, 6). *M* is the midpoint of *PQ*. *N* divides *SQ* in the ratio 3 : 1.

Find $\overrightarrow{MN}$ in terms of $\mathbf{i}$, $\mathbf{j}$ and $\mathbf{k}$

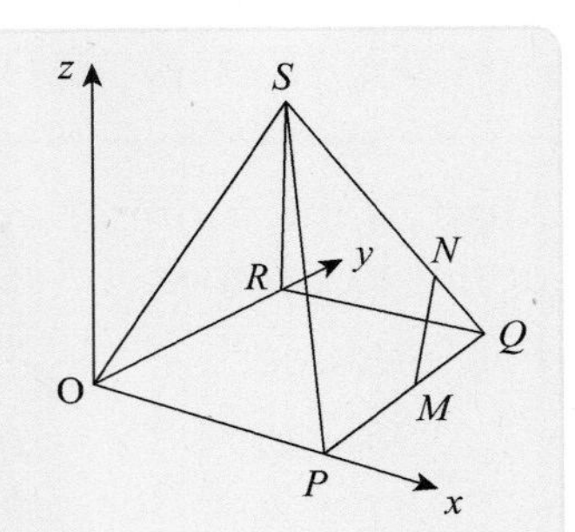

Solution

Here's a pathway from *O* to *S*:

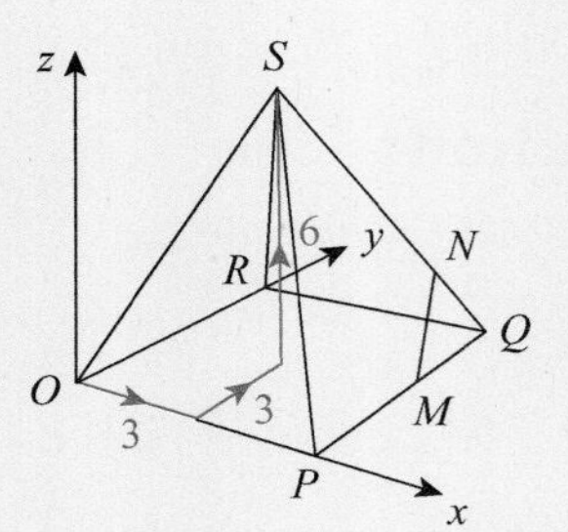

You can now see that each side of the square base is 6 units long:

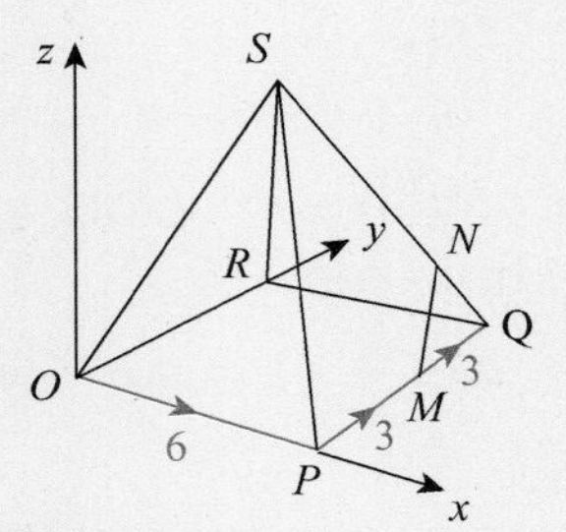

You can now fill in some missing coordinates:

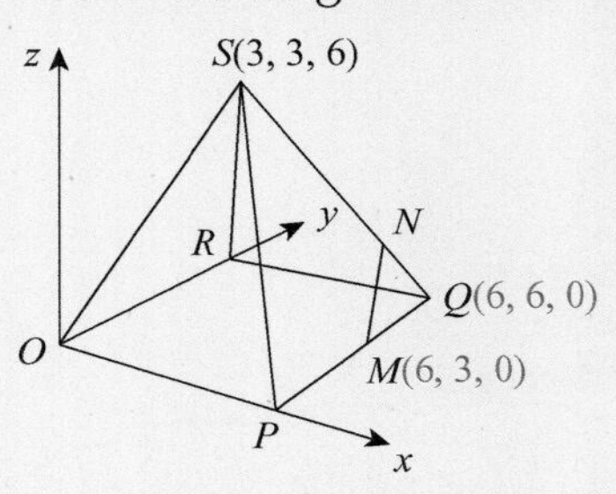

Since *N* divides *SQ* in the ratio 3 : 1

then $\overrightarrow{SN} = 3\overrightarrow{NQ}$

$\Rightarrow \mathbf{n} - \mathbf{s} = 3(\mathbf{q} - \mathbf{n})$

$\Rightarrow \mathbf{n} - \mathbf{s} = 3\mathbf{q} - 3\mathbf{n}$

$\Rightarrow \mathbf{n} + 3\mathbf{n} = 3\mathbf{q} + \mathbf{s}$

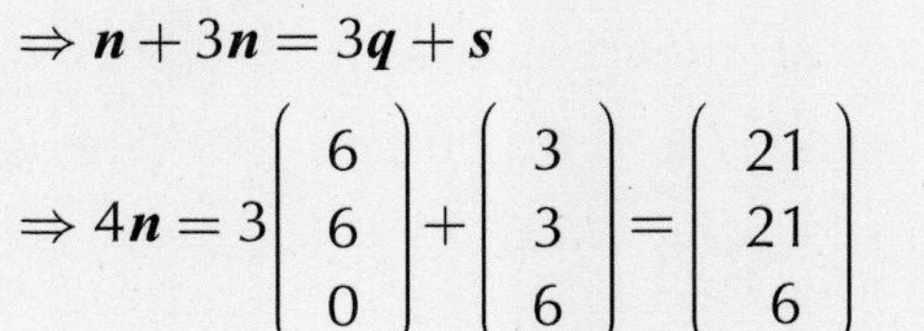

$$\Rightarrow 4\mathbf{n} = 3\begin{pmatrix} 6 \\ 6 \\ 0 \end{pmatrix} + \begin{pmatrix} 3 \\ 3 \\ 6 \end{pmatrix} = \begin{pmatrix} 21 \\ 21 \\ 6 \end{pmatrix}$$

$$\Rightarrow \mathbf{n} = \frac{1}{4}\begin{pmatrix} 21 \\ 21 \\ 6 \end{pmatrix} = \begin{pmatrix} \frac{21}{4} \\ \frac{21}{4} \\ \frac{3}{2} \end{pmatrix}$$

So $N(\frac{21}{4}, \frac{21}{4}, \frac{3}{2})$

$$\overrightarrow{MN} = n - m = \begin{pmatrix} \frac{21}{4} \\ \frac{21}{4} \\ \frac{3}{2} \end{pmatrix} - \begin{pmatrix} 6 \\ 3 \\ 0 \end{pmatrix} = \begin{pmatrix} -\frac{3}{4} \\ \frac{9}{4} \\ \frac{3}{2} \end{pmatrix} = -\frac{3}{4}\mathbf{i} + \frac{9}{4}\mathbf{j} + \frac{3}{2}\mathbf{k}$$

Forces in equilibrium

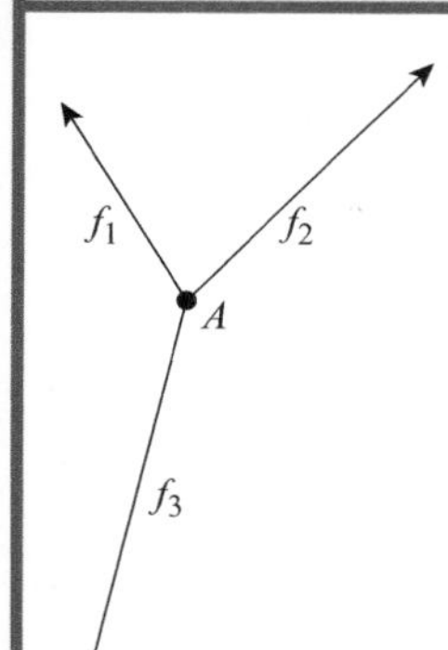

There are three forces acting on an object at A. They are represented by vectors $\boldsymbol{f}_1$, $\boldsymbol{f}_2$ and $\boldsymbol{f}_3$.

If the object is stationary then the system is said to be in equilibrium.

In this case the vector sum of the forces is zero:

$$\boldsymbol{f}_1 + \boldsymbol{f}_2 + \boldsymbol{f}_3 = \boldsymbol{0}$$

Example

Three forces $\boldsymbol{f}_p = \begin{pmatrix} 3 \\ y \\ -1 \end{pmatrix}$, $\boldsymbol{f}_q = \begin{pmatrix} 3 \\ -2 \\ z \end{pmatrix}$ and $\boldsymbol{f}_r = \begin{pmatrix} x \\ 4 \\ -1 \end{pmatrix}$ acting on an object are in equilibrium. Find the values of x, y and z.

Solution

Since the system is in equilibrium $\boldsymbol{f}_p + \boldsymbol{f}_q + \boldsymbol{f}_r = \boldsymbol{0}$

So $\begin{pmatrix} 3 \\ y \\ -1 \end{pmatrix} + \begin{pmatrix} 3 \\ -2 \\ z \end{pmatrix} + \begin{pmatrix} x \\ 4 \\ -1 \end{pmatrix} = \begin{pmatrix} 0 \\ 0 \\ 0 \end{pmatrix} \Rightarrow \begin{cases} 3+3+x=0 \Rightarrow 6+x=0 \Rightarrow x=-6 \\ y-2+4=0 \Rightarrow y+2=0 \Rightarrow y=-2 \\ -1+z-1=0 \Rightarrow z-2=0 \Rightarrow z=2 \end{cases}$

Quick Test 16

TOP TIP

It is difficult to write in bold font. When you write down a vector $\boldsymbol{v}$ you should underline it $\underline{v}$

1. Each edge of this cube has length 1 unit.

   a) Find the exact value of $|\boldsymbol{b}|$ and $|\boldsymbol{c}|$

   b) Find the exact value of $\boldsymbol{a}.(\boldsymbol{b} + \boldsymbol{c})$

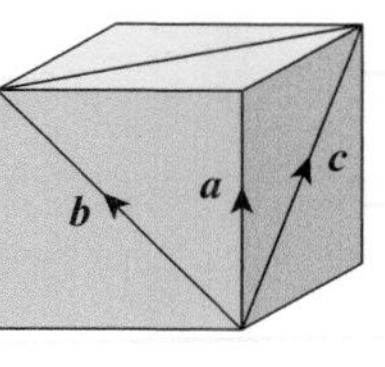

2. $OABC$, $DEFG$ is a cuboid.

   The vertex F is the point (5, 6, 2).

   M is the midpoint of DG.

   N divides AB in the ratio 1 : 2.

   a) Find the coordinates of M and N.

   b) Write $\overrightarrow{MN}$ in terms of unit vectors $\boldsymbol{i}$, $\boldsymbol{j}$ and $\boldsymbol{k}$.

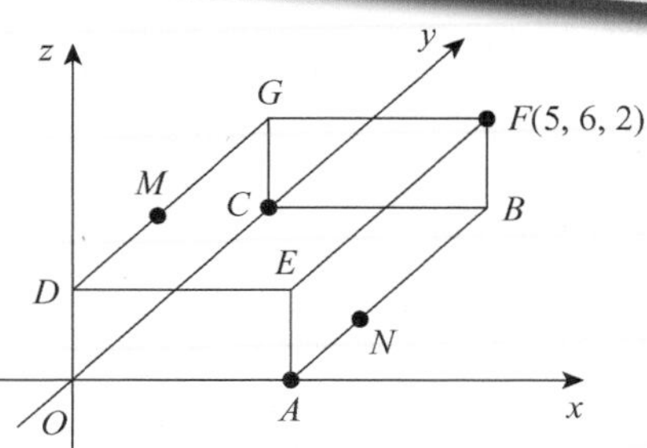

3. Four forces act on an object and are in equilibrium.

   Three of the forces are: $\boldsymbol{F}_1 = \begin{pmatrix} -2 \\ 5 \\ 0 \end{pmatrix}$, $\boldsymbol{F}_2 = \begin{pmatrix} 3 \\ -3 \\ -1 \end{pmatrix}$ and $\boldsymbol{F}_3 = \begin{pmatrix} 2 \\ 1 \\ -1 \end{pmatrix}$

   a) Find the components of the fourth force $\boldsymbol{F}_4$

   b) Find the magnitude of $\boldsymbol{F}_4$

Check-up questions (Chapter 1)

1.1 Applying algebraic skills to logarithms and exponentials

1. Simplify $\log_2 3ab - \log_2 3b$
2. Express $\log_b a^3 + \log_b a^2$ in the form $m\log_b a$
3. Solve $\log_2 (y - 3) = 1$

1.2 Applying trig skills to manipulating expressions

1. Express $\sin x° + 3\cos x°$ in the form $k\cos(x - a)°$ where $k > 0$ and $0 \leq a < 360$
2. Show that $(1 - \sin x)^2 + 2\sin x = 2 - \cos^2 x$
3. Use the information in the diagram to find the exact value of $\sin(a + b)$

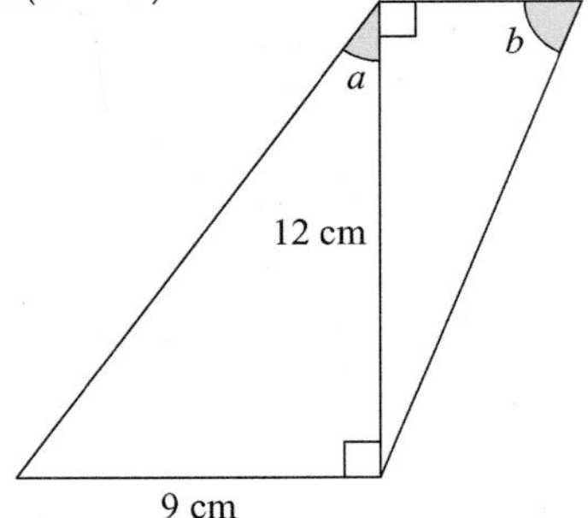

1.3 Applying algebra and trig skills to functions

1. Sketch the graph $y = k\cos\left(x + \frac{\pi}{4}\right)$ for $0 \leq x < 2\pi$ and $k > 0$.
   Show the minimum and maximum values and where the graph cuts the x-axis.
2. The graph of $y = f(x)$ is shown.
   It passes through the points $A(0, 2)$, $B(1, 0)$, $C(2, 1)$ and $D(3, 0)$.
   Sketch the graph $y = f(x + 1) - 2$

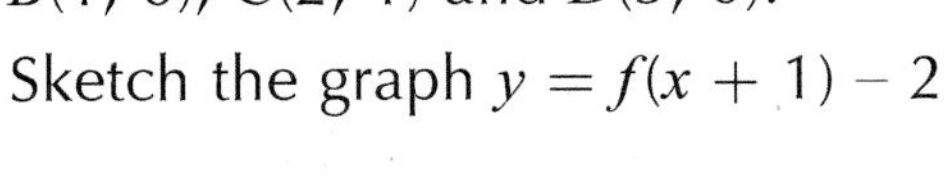

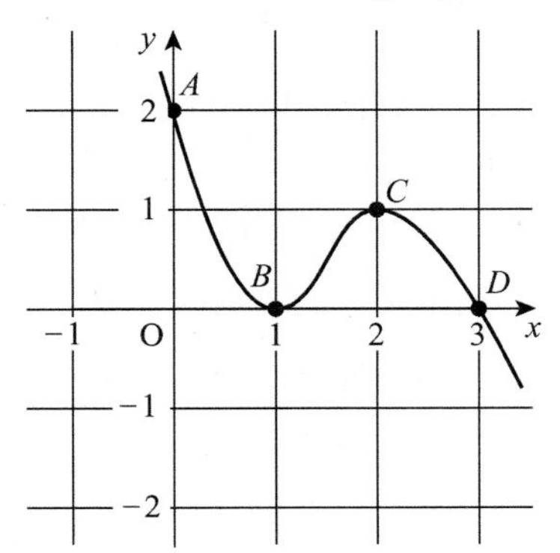

3. Find $f^{-1}(x)$ where $f(x) = \frac{1}{2}x + 3$

4. The diagram shows part of the graph of $y = k \cos ax + b$. Write down the values of k, a and b.

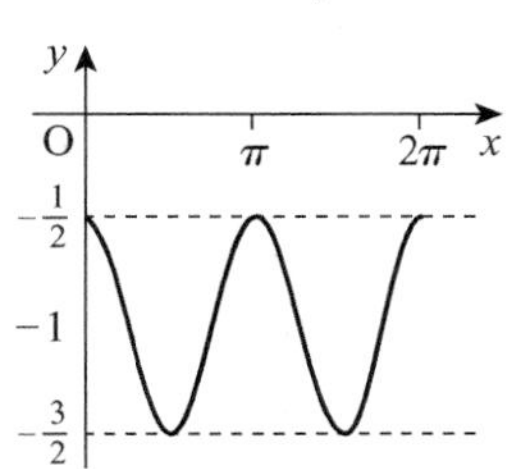

5. The graph $y = \log_m (x + n)$ is shown. Find the values of m and n.

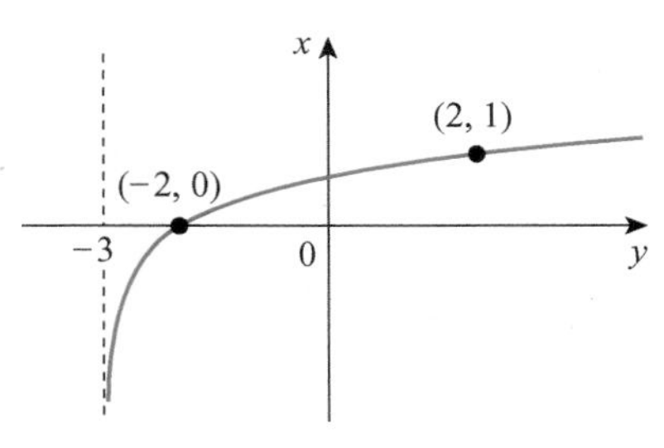

6. Functions f and g are given by $f(x) = 2 - x$ and $g(x) = -\sqrt{x}$

a) Find an expression for $f(g(x))$

b) State a suitable domain for $f(g(x))$

1.4 Applying geometric skills to vectors

1. A golf club is installing a new drain in one of its car parks. There are two new pipes in the system as shown. The pipes are joined at point $Q(-3, 4, 0)$ with entrance point $P(-5, 5, 1)$ and exit point $R(3, 1, -3)$. [All coordinates are relative to a suitable set of axes.]

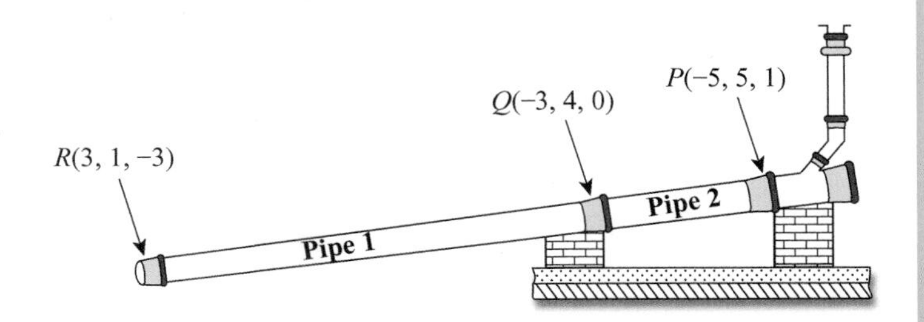

a) Are the pipes in a straight line?

b) How much longer is Pipe 1 than Pipe 2?

You must justify your answers.

2. Find the coordinates of point T which divides line AB in the ratio 2 : 1 as shown.

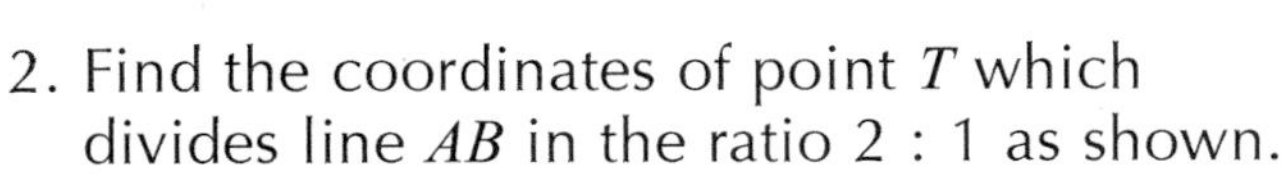

3. In the diagram $\overrightarrow{AB} = -8\boldsymbol{i}$

$\overrightarrow{AC} = 4\boldsymbol{j}$

$\overrightarrow{AD} = -4\boldsymbol{i} + 2\boldsymbol{j} + 3\boldsymbol{k}$

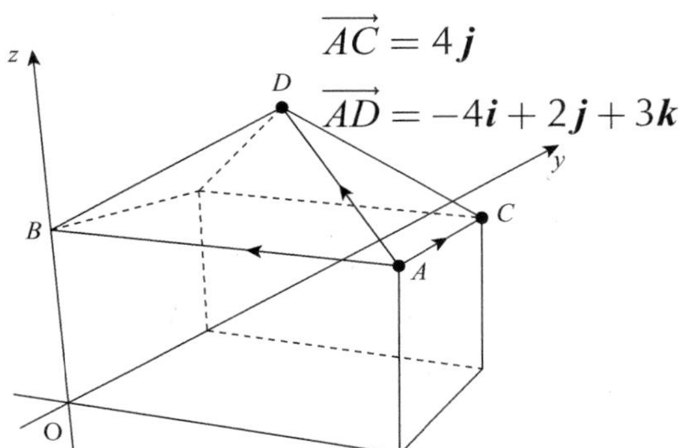

Find $\overrightarrow{DB}$ in component form.

4. The points shown in the diagram are: $D(-1, 5, -3)$, $E(2, 0, 3)$ and $F(0, -2, 1)$

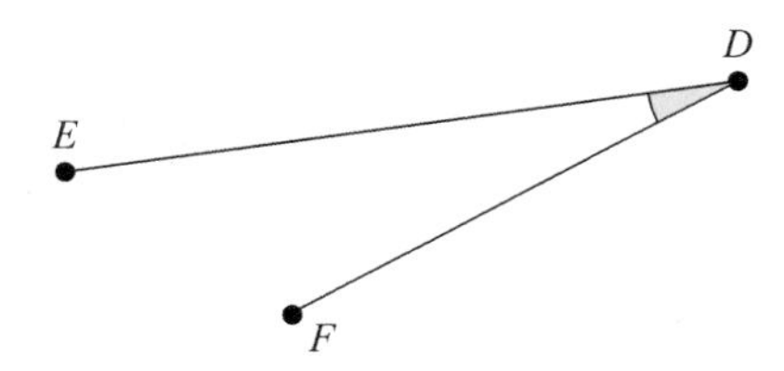

Calculate the size of angle EDF.

Sample end-of-course exam questions (Chapter 1)

Non-calculator

1. $g(x) = 3x$ and $h(x) = \sin 2x$ are two functions defined on suitable domains.
   What is the value of $h\left(g\left(\frac{\pi}{12}\right)\right)$?

2. If the exact value of $\tan x$ is $\frac{1}{\sqrt{3}}$, where $0 \leq x \leq \frac{\pi}{2}$, find the exact value of $\sin 2x$.

3. $OABC$ is a tetrahedron. A is the point $(4, -2, 0)$, B is $(3, 5, 0)$ and C is $(1, 4, 6)$. P divides AC in the ratio $2 : 1$.

   a) Find the coordinates of P.

   b) Express $\overrightarrow{BP}$ in terms of the unit vectors $\boldsymbol{i}$, $\boldsymbol{j}$ and $\boldsymbol{k}$.

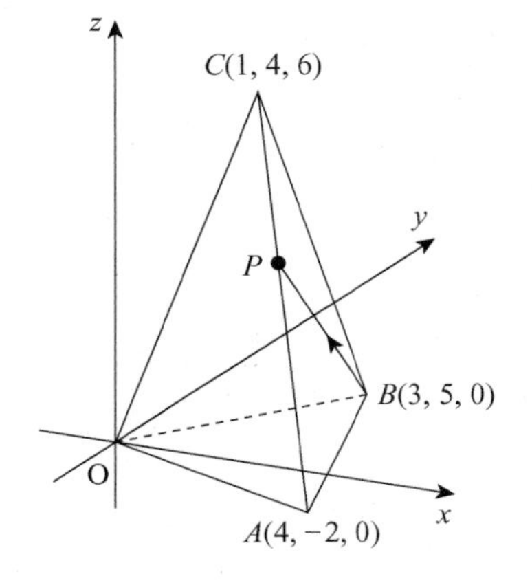

4. Express $2(\sqrt{3}\cos x - \sin x)$ in the form $k\cos(x + a)$ where $k > 0$ and $0 \leq a < 2\pi$.

5. ABC is a right-angled triangle as shown in the diagram, with vertices $A(1, -2, -k)$, $B(k, k, 0)$ and $C(4, -3, 3 - k)$.

   a) Find the value of k.

   D is the point $(13, -6, 11)$.

   b) Show that A, C and D are collinear and find the ratio in which C divides AD.

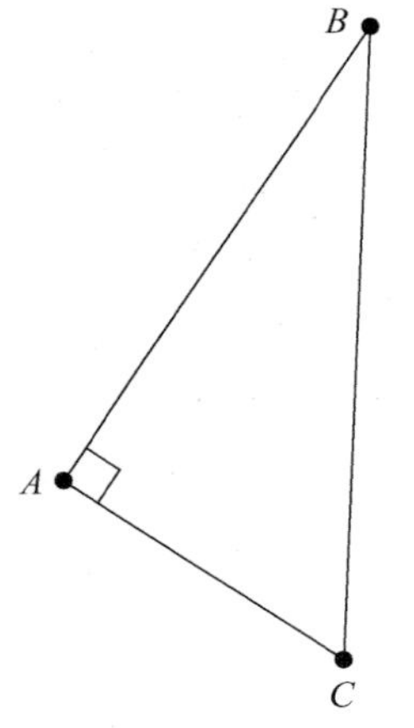

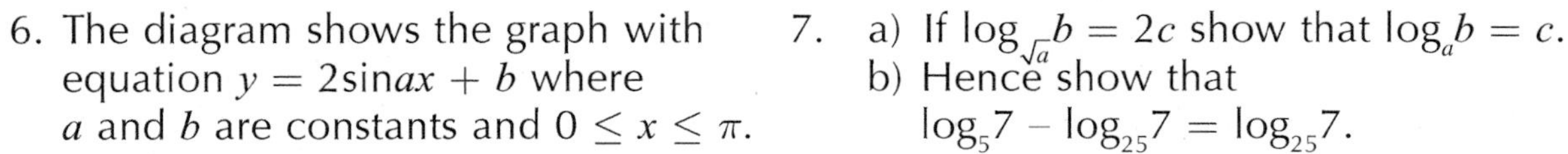

6. The diagram shows the graph with equation $y = 2\sin ax + b$ where a and b are constants and $0 \leq x \leq \pi$.

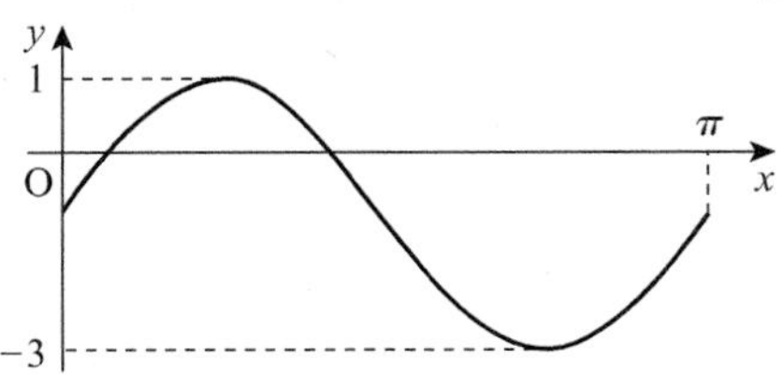

   What are the values of a and b?

7. a) If $\log_{\sqrt{a}} b = 2c$ show that $\log_a b = c$.
   b) Hence show that $\log_5 7 - \log_{25} 7 = \log_{25} 7$.

Calculator allowed

1. a) Simplify $\frac{2x^2-7x+6}{x^2-4}$.

   b) Solve $\log_3(2x^2-7x+6)-\log_3(x^2-4)=2$.

2. The diagram shows a cuboid surmounted by a pyramid. One of the triangular faces of the pyramid has vertices $A(3, 2, 5)$, $B(6, 0, 3)$ and $C(6, 4, 3)$.

   (a) Express $\overrightarrow{CA}$ and $\overrightarrow{CB}$ in component form.

   (b) Calculate the angle between the two edges CA and CB.

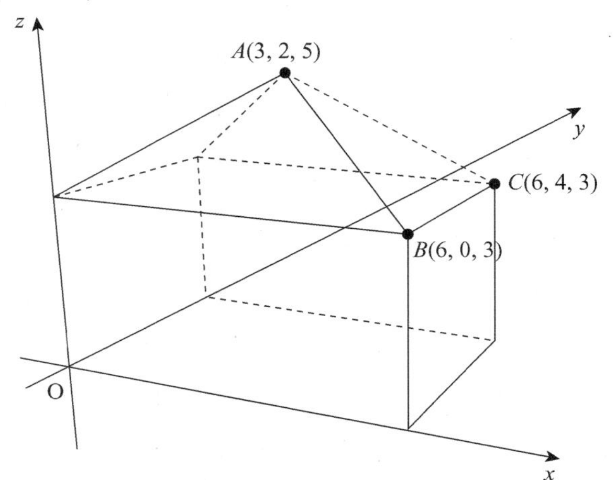

3. The formula $P=6e^{0.0138t}$ is used to predict the population P of the world, in billions, t years after January 1, 2000.

   (a) What was the population of the world on January 1, 2000?

   (b) At the start of which year will the world's population be more than double that of the population on January 1, 2000?

4. Two right-angled triangles PQR and RPS have lengths as shown in the diagram.

   Angle $PRQ = a°$ and angle $PRS = b°$.

a) Show that the exact value of $\cos(a+b)°$ is $-\frac{33}{65}$.

b) Calculate the exact value of $\sin(a+b)°$.

c) Hence, calculate the exact value of $\tan(a+b)°$.

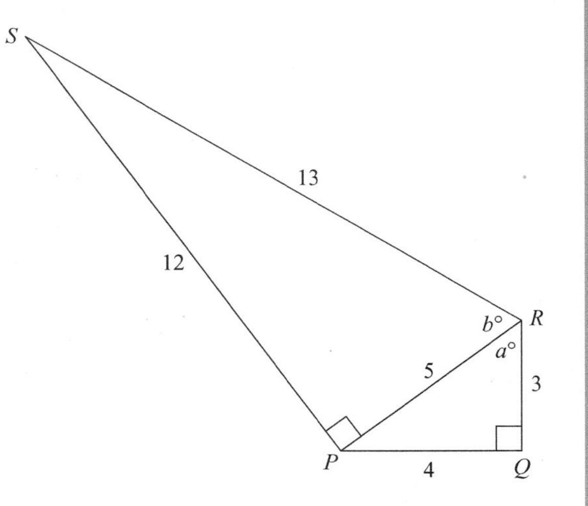

5. The functions f and g are defined on a suitable domain

   by $f(x)=\frac{2}{x+1}$ and $g(x)=2x+1$.

   Prove that $f(g(x))=\frac{1}{2}f(x)$.

Quadratic theory revisited I

Roots of quadratic equations

If a quadratic equation cannot be solved by factorising then you can use **The Quadratic Formula**.

There appear to be two real **roots** or **solutions** but this may not be true. It depends on the number $b^2 - 4ac$ that appears under the square root sign. This number is called **The Discriminant**.

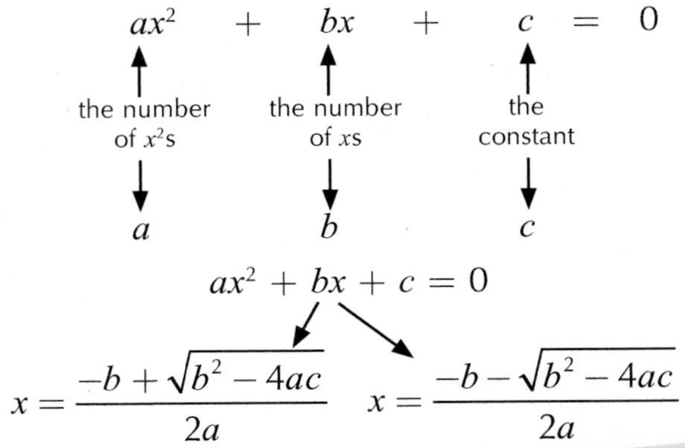

TOP TIPS

The square root of a negative number is not a real number.

ALSO

The words: **root** and **solution** mean exactly the same in the context of solving equations.

The discriminant

The graph of $y = ax^2 + bx + c$ meets the x-axis at 2 points.	The graph of $y = ax^2 + bx + c$ meets the x-axis at 1 point.	The graph of $y = ax^2 + bx + c$ does not meet the x-axis
	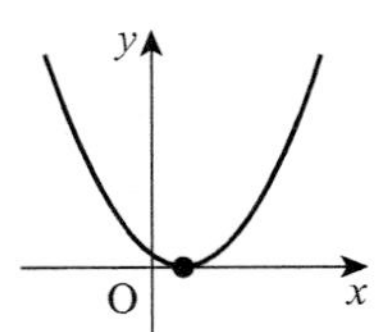	
$ax^2 + bx + c = 0$ has **2 distinct real roots**	$ax^2 + bx + c = 0$ has **only 1 real root** (equal roots)	$ax^2 + bx + c = 0$ has **no real roots**
$b^2 - 4ac > 0$ **discriminant is positive**	$b^2 - 4ac = 0$ **discriminant is zero**	$b^2 - 4ac < 0$ **discriminant is negative**

Calculating the discriminant allows you to determine the **nature** of the roots:

$b^2 - 4ac > 0$ $\xrightarrow[\text{positive}]{\text{discriminant}}$ two distinct real roots

$b^2 - 4ac = 0$ $\xrightarrow[\text{zero}]{\text{discriminant}}$ one real root (two equal roots)

$b^2 - 4ac < 0$ $\xrightarrow[\text{negative}]{\text{discriminant}}$ no real roots

Example

Determine the nature of the roots of these equations:

a) $x^2 - 4x + 3 = 0$ b) $x^2 - 4x + 4 = 0$ c) $x^2 - 4x + 5 = 0$

Solution

In each case compare the equation with $ax^2 + bx + c = 0$

a) $a = 1, b = -4, c = 3$

$b^2 - 4ac = (-4)^2 - 4 \times 1 \times 3 = 4$

Discriminant is positive so there are **two distinct real roots.**

b) $a = 1, b = -4, c = 4$

$b^2 - 4ac = (-4)^2 - 4 \times 1 \times 4 = 0$

Discriminant is zero so there is **one real root** (equal roots).

c) $a = 1, b = -4, c = 5$

$b^2 - 4ac = (-4)^2 - 4 \times 1 \times 5 = -4$

Discriminant is negative so there are **no real roots.**

Tangency

The discriminant can be used to show that lines are tangents to curves:

For example, to find where a line $y = mx + d$ meets a parabola $y = ax^2 + bx + c$ you proceed as follows:

$$\left.\begin{aligned} y &= ax^2 + bx + c \\ y &= mx + d \end{aligned}\right\}$$ For points of intersection replace y by $mx + d$ in the quadratic equation.

This will give a quadratic equation.

Calculate the discriminant of this equation. If it is **zero** then there is only **one solution.** This means **one point of intersection** and so the line is **a tangent** to the curve.

Example

Show that $y = 4x - 2$ is a tangent to the parabola $y = x^2 + 2$

Solution

To find the points of intersection solve

$$\left.\begin{aligned} y &= x^2 + 2 \\ y &= 4x - 2 \end{aligned}\right\} \quad \text{so} \quad \begin{aligned} x^2 + 2 &= 4x - 2 \\ x^2 - 4x + 4 &= 0 \end{aligned}$$

Discriminant $= (-4)^2 - 4 \times 1 \times 4 = 0$

so there is one solution and hence one point of intersection. **The line $y = 4x - 2$ is a tangent.**

Quadratic inequalities

If $f(x)$ is any quadratic expression like $ax^2 + bx + c$ then to solve quadratic inequalities like

$$f(x) > \text{ or } f(x) < 0$$

follow these steps:

Step 1 Solve the corresponding quadratic equation $f(x) = 0$ to find the x-axis intercepts for the graph $y = f(x)$.

Step 2 Sketch the graph $y = f(x)$ clearly showing the x-axis intercepts.

Step 3 Write down the solution using the sketch of the graph $y = f(x)$. Examples are given in this table:

Inequality:	What to look for:	For this graph the solution is:	For this graph the solution is:
$ax^2 + bx + c > 0$	Where is the graph **above** the x-axis?	$x < p$ or $x > q$	$p < x < q$
$ax^2 + bx + c < 0$	Where is the graph **below** the x-axis?	$p < x < q$	$x < p$ or $x > q$

Note: If the inequality uses the signs $\leq$ or $\geq$ then so will the solution.

Example

Find the real values of x satisfying $2x^2 + x - 1 > 0$

Solution

First solve $2x^2 + x - 1 = 0$

so $(2x - 1)(x + 1) = 0$

$2x - 1 = 0$ or $x + 1 = 0$

$x = \frac{1}{2}$ $\qquad$ $x = -1$

Sketch of $y = 2x^2 + x - 1$

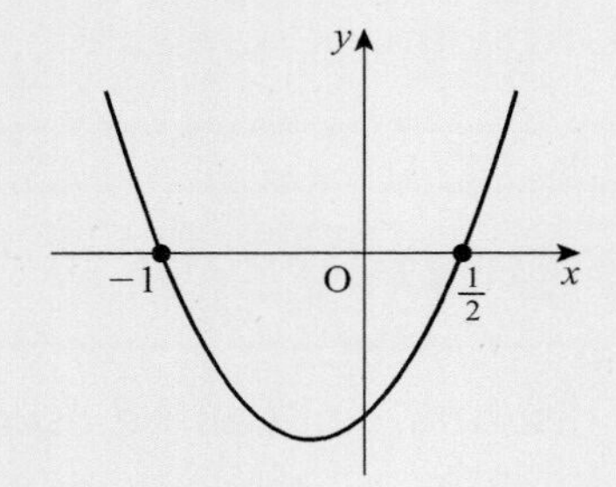

Required solution is $\boldsymbol{x < -1}$ **or** $\boldsymbol{x > \frac{1}{2}}$

(graph is **above** x-axis for these values).

Quadratic theory revisited II

Restrictions on the discriminant

It may be necessary to impose conditions on the nature of the roots of a quadratic equation. This can be done by applying restrictions to the discriminant. Here's how:

Condition on roots		Restriction on discriminant
Real ('one or two')	$\Longrightarrow$	Positive or zero ($b^2 - 4ac \geq 0$)
Equal ('one')	$\Longrightarrow$	Zero ($b^2 - 4ac = 0$)
Non-real ('none')	$\Longrightarrow$	Negative ($b^2 - 4ac < 0$)

Example

For what values of p does $x^2 - 2x + p = 0$ have real roots?

Solution

Compare $x^2 - 2x + p = 0$

with $ax^2 + bx + c = 0$

This gives $a = 1$, $b = -2$ and $c = p$

Discriminant $= b^2 - 4ac$

$= (-2)^2 - 4 \times 1 \times p$

$= 4 - 4p$

The condition 'real roots' requires the discriminant to be restricted to positive values or zero.

So $4 - 4p \geq 0$ giving $-4p \geq -4$

so $\boldsymbol{p \leq 1}$

Example

Find values of k so that $\frac{2(3x+1)}{3x^2+1} = k$ has two equal roots.

Solution

Rearrange the equation…

$6x + 2 = 3kx^2 + k$ giving $3kx^2 - 6x + k - 2 = 0$ and comparing this quadratic equation with $ax^2 + bx + c = 0$ gives $a = 3k$, $b = -6$, $c = k - 2$

Discriminant $= b^2 - 4ac = (-6)^2 - 4 \times 3k(k - 2)$

$= 36 - 12k^2 + 24k$

$= 36 + 24k - 12k^2$

The condition 'equal roots' gives the restriction 'discriminant = 0'.

So $36 + 24k - 12k^2 = 0$ giving $12(3 + 2k - k^2) = 0$ so $12(3 - k)(1 + k) = 0$

This gives two possible values for k namely $\boldsymbol{k = 3}$ **or** $\boldsymbol{k = -1}$

Building equations from the roots

It is possible to 'design' a quadratic equation that has two particular roots.
If the roots are $x = a$ and $x = b$ then $(x - a)(x - b) = 0$ is one such equation.

Example

Find a quadratic equation that has roots $\frac{1}{3}$ and -2

Solution

One such equation is

$$\left(x - \frac{1}{3}\right)(x + 2) = 0$$

This gives $x^2 + 2x - \frac{1}{3}x - \frac{2}{3} = 0$

Multiplying both sides of the equation by 3 gives:

$$3x^2 + 6x - x - 2 = 0$$

$$\mathbf{3x^2 + 5x - 2 = 0}$$

TOP TIP

This technique extends to cubic equations with three roots.

Where are the roots?

The graph is below the x-axis at $x = a$ and above the x-axis at $x = b$. If the graph is a continuous curve then it must cross the x-axis somewhere between a and b, at $x = \alpha$ say:

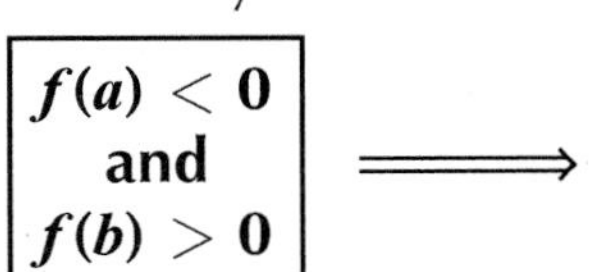

$f(\alpha) = 0$ **for some value** α **with** $a < \alpha < b$

i.e. $f(x) = 0$ has a root between a and b.

Example

Show that $x^3 - 3x + 1 = 0$ has a root between 1 and 2.

Solution

Let $f(x) = x^3 - 3x + 1$

then $f(1) = -1$ so $f(1) < 0$

and $f(2) = 3$ so $f(2) > 0$

Hence there is a root, α say, with $1 < \alpha < 2$.

TOP TIP

Zooming in on a graph using a graphing calculator is very useful for locating roots but you must always write down evidence in your written solution.

Intersection of two parabolas

To find where two curves $y = f(x)$ and $y = g(x)$ intersect, you solve the equation $f(x) = g(x)$

Example

Find the coordinates of the intersection points A and B as shown in the diagram.

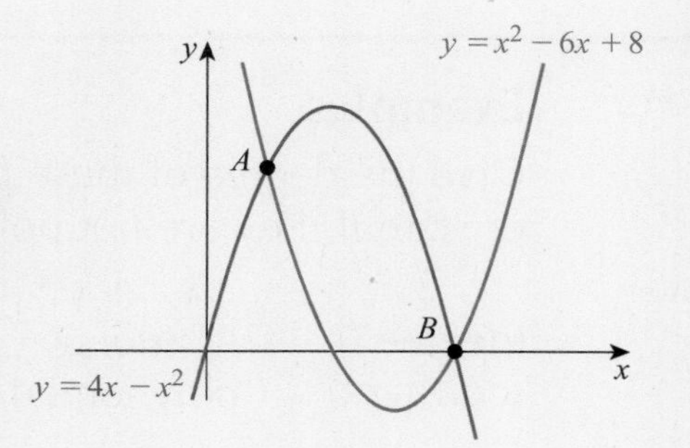

Solution

$x^2 - 6x + 8 = 4x - x^2$

$2x^2 - 10x + 8 = 0$

$x^2 - 5x + 4 = 0$

$(x - 1)(x - 4) = 0$

$x = 1$ or $x = 4$

Using $y = 4x - x^2$

when $x = 1$ $y = 4 \times 1 - 1^2 = 4 - 1 = 3$ so $A(1, 3)$

when $x = 4$ $y = 4 \times 4 - 4^2 = 16 - 16 = 0$ so $B(4, 0)$

Quick Test 17

1. Use the discriminant to show that $y = 3x + 2$ is a tangent to the curve $y = x^2 - 11x + 51$
2. For what values of k does $kx^2 - 3x + k = 0$ have real roots?
3. Solve $20 + x - x^2 > 0$
4. Find a quadratic equation with roots $x = -\frac{1}{3}$ and $x = \frac{1}{4}$
5. Two functions are defined by $f(x) = 3x^2 - x - 2$ and $g(x) = x^2 + 4x + 1$ with domain the set of real numbers. Find the intersection points of the graphs $y = f(x)$ and $y = g(x)$

Chapter 2

Polynomials and synthetic division

What is a polynomial?

A **polynomial** consists of sums and/or differences of terms like:

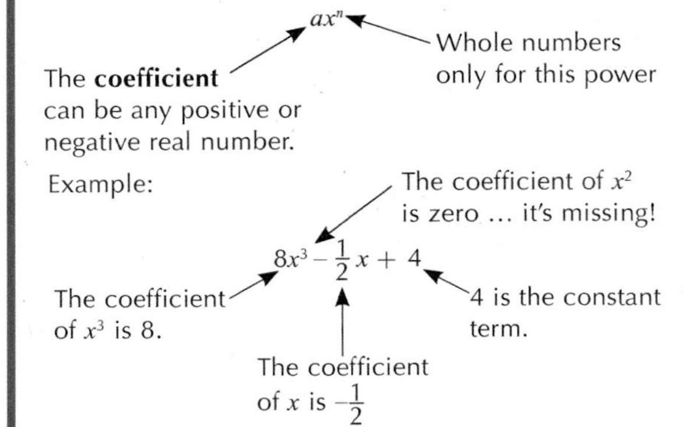

The highest power of x is the **degree** of the polynomial (3 in the example above).

Examples

Give the degree of these polynomials or state if they are not polynomials:

3	$x^{-1} + x^{-2}$	$3x^2$
(degree 0) (constant)	(not a polynomial)	(degree 2) (quadratic)
$15 - 2x$	$2x^2 - x + 3$	
(degree 1) (linear)	(degree 2) (quadratic)	
$\sqrt{x}$	$4x^3 + 2x$	$3x^{\frac{3}{2}}$
(not a polynomial)	(degree 3) (cubic)	(not a polynomial)

TOP TIP

It is important that missing terms (with coefficients zero) are each recorded with a 0 otherwise your results will be wrong!

Synthetic division

As an illustrative example, let's divide the degree 3 polynomial $f(x) = x^3 - 3x^2 + 6$ by $x - 2$. $x - 2$ is called the divisor.

Step 1 Write down the coefficients of $f(x)$:
$f(x) = 1x^3 - 3x^2 + 0x + 6$

$$\begin{array}{r|rrrr} & 1 & -3 & 0 & 6 \\ \hline \end{array}$$

Step 2 Find the value of x that makes the divisor zero: 2

$$\begin{array}{r|rrrr} 2 & 1 & -3 & 0 & 6 \\ \hline \end{array}$$

Step 3 Bring down the first coefficient:

$$\begin{array}{r|rrrr} 2 & 1 & -3 & 0 & 6 \\ \hline & 1 & & & \end{array}$$

Step 4 Multiply by the divisor value and add the result to the next coefficient and keep repeating this step:

$$\begin{array}{r|rrrr} 2 & 1 & -3 & 0 & 6 \\ & & 2 & -2 & -4 \\ \hline & 1 & -1 & -2 & 2 \end{array}$$

(×2, add)

What do all these numbers mean? The colour coding below explains how to interpret these numbers:

$$\begin{array}{r|rrrr} 2 & 1 & -3 & 0 & 6 \\ & & 2 & -2 & -4 \\ \hline & 1 & -1 & -2 & 2 \end{array}$$

$$\begin{aligned} f(x) &= 1x^3 - 3x^2 + 0x + 6 \\ &= (x - 2)(1x^2 - 1x - 2) + 2 \end{aligned}$$

Divisor — Quotient — Remainder

Compare this number example:
Divide 7 by 2: $2 \lfloor 7$ → 3 rl
Giving: $7 = 2 \times 3 + 1$

Using synthetic division

Here is a colour coded diagram showing $f(x)$ divided by $x - h$:

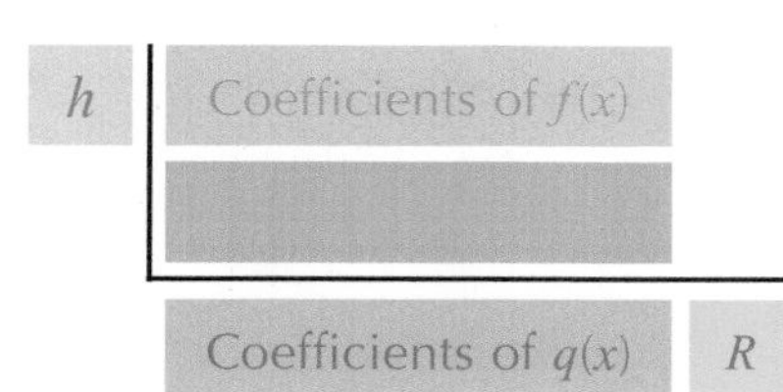

Giving: $f(x) = (x - h)q(x) + R$

There are different uses for synthetic division:

Use 1 To calculate the value of $f(h)$. Notice that $f(h) = (h - h)q(x) + R = 0 + R = R$

Use 2 To find the quotient $q(x)$ and remainder R when $f(x)$ is divided by $x - h$

Use 3 To find a factor $x - h$ of $f(x)$. This will happen if you find $R = 0$

Example 1 $f(x) = 2x^3 - 3x + 10$ Find $f(-3)$

Solution

–3	2	0	–3	10
		–6	18	–45
	2	–6	15	–35

So $f(-3) = -35$

Example 2 $f(x) = 2x^3 - 3x + 10$ Find the quotient and remainder when $f(x)$ is divided by $x - 2$

Solution

2	2	0	–3	10
		4	8	10
	2	4	5	20

So $f(x) = (x - 2)(2x^2 + 4x + 5) + 20$

Quotient: $2x^2 + 4x + 5$

Remainder: 20

Example 3 $f(x) = 2x^3 - 3x + 10$ Show that $x + 2$ is a factor of $f(x)$

Solution

–2	2	0	–3	10
		–4	8	–10
	2	–4	5	0

The remainder is 0 when $f(x)$ is divided by $x + 2$ so $x + 2$ is a factor of $f(x)$

TOP TIP

When dividing by $x + h$ you use $-h$ for the synthetic division.

Notes:

- The fact that $f(x)$ is divided by $x - h$ $\Rightarrow R = f(h)$ is known as **The Remainder Theorem**
- The fact that $R = 0 \Leftrightarrow f(h) = 0 \Rightarrow x - h$ is a factor is known as **The Factor Theorem**

Quick Test 18

1. Which of these are not polynomials? x^{-1}, $3x^2$, $\frac{1}{2}x$, $\frac{1}{2}\sqrt{x}$, $\frac{1}{2}$
2. Show that both $x - 3$ and $x + 2$ are factors of $x^4 - 7x^2 - 6x$
3. Find the quotient and remainder when $4x^3 - 12x + 7$ is divided by $x + 2$
4. Use synthetic division to calculate the exact value of $f\left(-\frac{1}{3}\right)$ where $f(x) = x^4 + 2x^3 + x^2 + 1$ (no calculators!)

Factors, roots and graphs

Factorising polynomials

If you divide 6 by 3 the quotient is 2 and the remainder is 0. $6 = 3 \times 2$ so 3 is a factor of 6.

Similarly if you divide $f(x)$ by $x - h$ and find the remainder is 0 then $f(x) = (x - h) \times q(x)$ so $x - h$ is a factor of $f(x)$.

Using synthetic division:

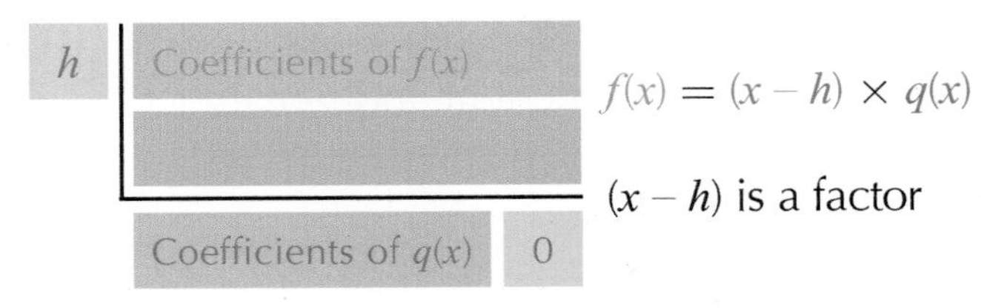

$f(x) = (x - h) \times q(x)$

$(x - h)$ is a factor

You are using the Factor Theorem:

Remainder is zero.	$\Longleftrightarrow$	$x - h$ is a factor.

Note: Be systematic when hunting for a value of h that gives a zero remainder. Try 1, then −1, then 2, then −2, and so on.

Example

Factorise $f(x) = x^3 + 2x^2 - 5x - 6$

Solution Divide $f(x)$ by $x - 1$:

1	1	2	−5	−6
		1	3	−2
	1	3	−2	−8

remainder is not zero so $x - 1$ is not a factor

Now let's try $x + 1$ as a candidate:

−1	1	2	−5	−6
		−1	−1	6
	1	1	−6	0

remainder is zero so $x + 1$ is a factor

So $f(x) = (x + 1)(x^2 + x - 6)$ ← quadratic factorising

$= (x + 1)(x - 2)(x + 3)$

TOP TIP

When hunting for factors look at the constant term of $f(x)$. You only need to try values of h that are factors of this constant.

Factors, roots and graphs

If $x - h$ is a factor of $f(x)$ then h is a root of $f(x) = 0$ and the graph $y = f(x)$ cuts the x-axis at $(h, 0)$.

Example: Let $f(x) = x^3 + 2x^2 - 5x - 6$

$f(x) = (x + 3)(x + 1)(x - 2)$

The roots of $f(x) = 0$ are: −3 −1 2

The graph $y = f(x)$

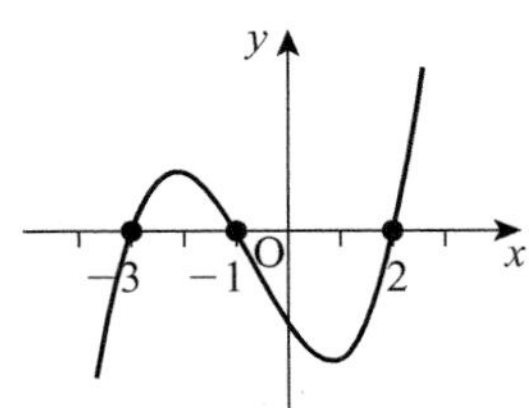

Finding an unknown coefficient

Example $x^3 + 2x^2 + kx - 6$ has a factor of $x + 1$. Calculate the value of k.

Solution

−1	1	2	k	−6
		−1	−1	$-k + 1$
	1	1	$k - 1$	$-k - 5$

Since $x + 1$ is a factor, then the remainder will be 0.

So: $-k - 5 = 0 \Rightarrow -k = 5 \Rightarrow k = -5$

Problem solving

Example $f(x) = 12x^3 - 28x^2 - 9x + 10$

a) Show that $\frac{1}{2}$ is a root of $f(x) = 0$ and find the other two roots.

b) Hence, find where the graph $y = f(x)$ cuts the x-axis.

a) Solution

Let $f(x) = 12x^3 - 28x^2 - 9x + 10$

Synthetic division gives:

$$\begin{array}{r|rrrr} \frac{1}{2} & 12 & -28 & -9 & 10 \\ & & 6 & -11 & -10 \\ \hline & 12 & -22 & -20 & 0 \end{array}$$

Since $f\left(\frac{1}{2}\right) = 0$, then $\frac{1}{2}$ is a root.

Using the table above...

$$f(x) = \left(x - \tfrac{1}{2}\right)(12x^2 - 22x - 20)$$
$$= \left(x - \tfrac{1}{2}\right) \times 2(6x^2 - 11x - 10)$$
$$= (2x - 1)(6x^2 - 11x - 10)$$

so

$$f(x) = (2x - 1)(3x + 2)(2x - 5)$$

and so $f(x) = 0$ gives

$2x - 1 = 0$ or $3x + 2 = 0$ or $2x - 5 = 0$

$\boldsymbol{x = \frac{1}{2}}$ $\boldsymbol{x = -\frac{2}{3}}$ $\boldsymbol{x = \frac{5}{2}}$

b) Solution

So $y = f(x)$ crosses the x-axis at the points:

$\left(\frac{1}{2}, 0\right)$, $\left(-\frac{2}{3}, 0\right)$ and $\left(\frac{5}{2}, 0\right)$

Quick Test 19

1. a) Show that $x - 1$ is a factor of $f(x) = 2x^4 - 3x^3 - x^2 + 3x - 1$.
   b) Hence, factorise $f(x)$ into two factors.
   c) Now show that $x + 1$ is a factor of one of these factors.
   d) Hence, express $f(x)$ in fully factorised form.
2. Find where the graph $y = f(x)$ crosses the x-axis if $f(x) = 6x^3 - 7x^2 - x + 2$

TOP TIP

To factorise **fully** you must break up any quadratic or cubic factors into further factor pairs if possible.

It's like 30 = 2 × 15 where 15 can be broken into 3 × 5 so 30 = 2 × 3 × 5 and is now **fully factorised.**

Solving trig equations

Solving trig equations using radians

Step 1 Rearrange the equation to one of the forms:
$\sin\theta = k$ or $\cos\theta = k$ or $\tan\theta = k$ where k is a constant (positive or negative).

Step 2 Which quadrants can θ be in ? Use the sign of k (positive or negative) and this diagram:

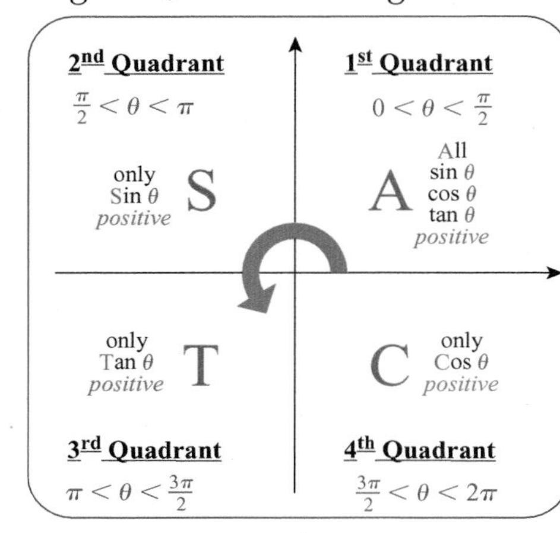

Step 3 Find the related 1st quadrant angle α. To find this angle you assume k is positive and solve the equation for the 1st quadrant.

Step 4 Now use the value of α to find the possible values for θ knowing the quadrants from Step 2 above:

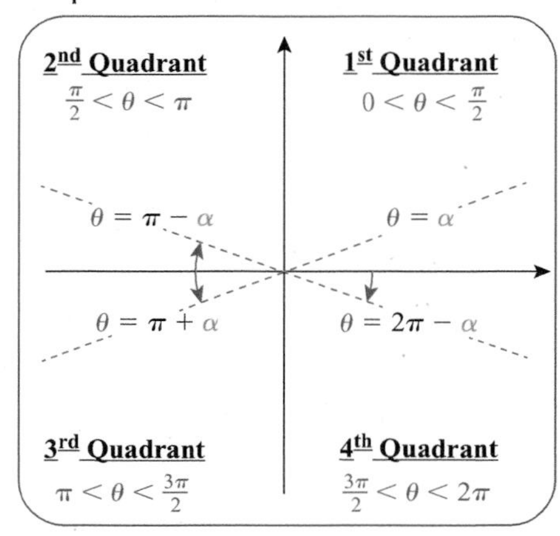

Example 1

Solve $2\sin x + 1 = 0$, $0 \leq x < 2\pi$

Solution

1. $2\sin x + 1 = 0$ rearranges to $\sin x = -\frac{1}{2}$
2. $\sin x$ is negative for angles in the 3rd or 4th quadrants
3. 1st quadrant angle is $\frac{\pi}{6}$ (since $\sin\frac{\pi}{6} = \frac{1}{2}$)
4. $x = \pi + \frac{\pi}{6}$ (3rd quadrant) or $x = 2\pi - \frac{\pi}{6}$ (4th quadrant)

$x = \frac{6\pi}{6} + \frac{\pi}{6}$ or $x = \frac{12\pi}{6} - \frac{\pi}{6}$

$x = \frac{7\pi}{6}$ or $x = \frac{11\pi}{6}$

Note:
A graphical check may be made using the 'sine graph':

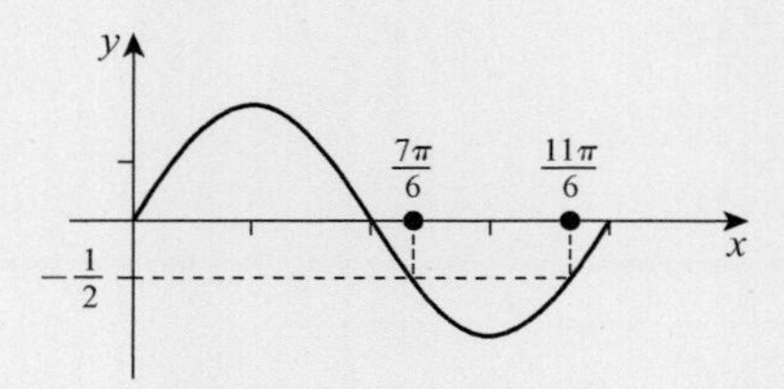

The diagram indicates the two solutions to $\sin x = -\frac{1}{2}$.

Example 2

Solve $3\cos\left(2x - \frac{\pi}{6}\right) = 2$ for $0 \leq x < \pi$

TOP TIP

Pay close attention to the allowed values for x.

Solution

Rearrange to $\cos\left(2x - \frac{\pi}{6}\right) = \frac{2}{3}$

So the angle $2x - \frac{\pi}{6}$ is in the 1st or 4th quadrants since cosine is positive in these quadrants.

Notes:

- If $k = 0, -1$ or 1 use the trig graphs to solve the equations $\sin\theta = k$ or $\cos\theta = k$.
- Remember the **exact values** (see page 26).
- Make sure your calculator is in radian mode for radian work.

Set calculator to Radian Mode. Enter $\frac{2}{3}$ and $\boxed{\cos^{-1}}$ giving 0·841 radians as the 1st quadrant angle.

so $2x - \frac{\pi}{6} = 0{\cdot}841$ (1st quad) or $2x - \frac{\pi}{6} = 2\pi - 0{\cdot}841$ (4th quad)

giving $2x = 0{\cdot}841 + \frac{\pi}{6} = 1{\cdot}364\ldots$ or $2x = 2\pi - 0{\cdot}841 + \frac{\pi}{6} = 5{\cdot}965\ldots$

so $\mathbf{x = 0{\cdot}682}$ or $\mathbf{x = 2{\cdot}983}$ (to 3 dec. places)

Quadratic trig equations

Factorise the quadratic expression and then set each factor to zero.

You are now able to solve these two simpler equations.

Example

$$3\cos^2 x + 7\cos x - 6 = 0$$
$$(3\cos x - 2)(\cos x + 3) = 0$$

Compare: $3c^2 + 7c - 6 = 0$, $(3c - 2)(c + 3) = 0$

$3\cos x - 2 = 0$ or $\cos x + 3 = 0$

$\cos x = \frac{2}{3}$ (x is in 1st or 4th quads) etc

$\cos x = -3$ (no solutions since $-1 \leq \cos x \leq 1$)

Double angle formulae and equations

Here is the general method to solve a trig equation in which a double angle appears:

trig expression = trig expression

↓ rearrange

trig expression = 0

↓ use a double angle formula

quadratic trig expression = 0

↓ factorise

(factor 1)(factor 2) = 0

factor 1 = 0 → **solve** or **factor 2 = 0** → **solve**

Example 1

Solve $\cos 2x^\circ + 3 = 5(1 + \cos x^\circ)$ for $0 \leq x \leq 360$

Solution

$$\cos 2x^\circ + 3 = 5 + 5\cos x^\circ$$
$$\cos 2x^\circ - 5\cos x^\circ - 2 = 0$$
$$2\cos^2 x^\circ - 1 - 5\cos x^\circ - 2 = 0$$
$$2\cos^2 x^\circ - 5\cos x^\circ - 3 = 0$$

(a quadratic in $\cos x^\circ$)

$$(2\cos x^\circ + 1)(\cos x^\circ - 3) = 0$$

$2\cos x^\circ + 1 = 0$ or $\cos x^\circ - 3 = 0$

$2\cos x^\circ = -1$, $\cos x^\circ = -\frac{1}{2}$ (x° is in 2nd or 3rd quads) (1st quad angle is 60°)

$\cos x^\circ = 3$, no solutions (since $-1 \leq \cos x^\circ \leq 1$)

so $x^\circ = 180^\circ - 60^\circ$ or $180^\circ + 60^\circ$

$\mathbf{x^\circ = 120^\circ \text{ or } 240^\circ}$

Chapter 2

Equations involving the wave function

Wave functions equations

To solve $a\cos x + b\sin x = c$

Step 1 Express the left side in one of the forms...

$k\cos(x \pm \alpha)$ or $k\sin(x \pm \alpha)$

Step 2 Divide through by k to give

$\cos(x \pm \alpha) = \dfrac{c}{k}$ or $\sin(x \pm \alpha) = \dfrac{c}{k}$

and solve in the 'usual' way.

Example Find algebraically the values of x between 0 and 180 for which $12\sin x° - 5\cos x° = 10$.

Solution

Expressing $12\sin x° - 5\cos x°$ in the form $k\sin(x - \alpha)°$ gives $13\sin(x - 22{\cdot}6)°$

The equation becomes

$$13\sin(x - 22{\cdot}6)° = 10$$

$$\text{so} \quad \sin(x - 22{\cdot}6)° = \frac{10}{13}$$

($(x - 22{\cdot}6)°$ is in 1st or 2nd quadrants)
(1st quadrant angle is $50{\cdot}3°$)

Check on graphic calculator.

The graphs $y = 13\sin(x - 22{\cdot}6)°$ **and** $y = 10$ are shown.
The answers appear reasonable.

so $x° - 22{\cdot}6° = 50{\cdot}3°$ or $180° - 50{\cdot}3° = 129{\cdot}7°$

giving $x° = 50{\cdot}3° + 22{\cdot}6°$ or $129{\cdot}7° + 22{\cdot}6°$

so $x° = \mathbf{72{\cdot}9°}$ or $\mathbf{152{\cdot}3°}$

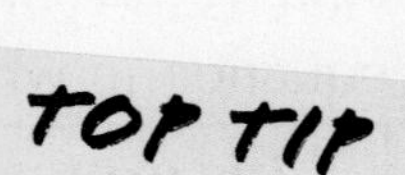

Be careful that your calculator is in the correct mode: RAD or DEG.

Maximum and minimum values

To determine the Maximum/Minimum value of $a\cos x + b\sin x$ and to find the corresponding values of x:

Write $a\cos x + b\sin x$ in one of the forms $k\cos(x \pm \alpha)$ or $k\sin(x \pm \alpha)$.

The maximum value is k.
The minimum value is $-k$.

Set the angles $x + \alpha$ or $x - \alpha$ equal to the 'normal' angle for which sin/cos is at a max/min (see the diagrams for examples) and then solve to find x.

A sine curve $y = k\sin(x - \alpha)$

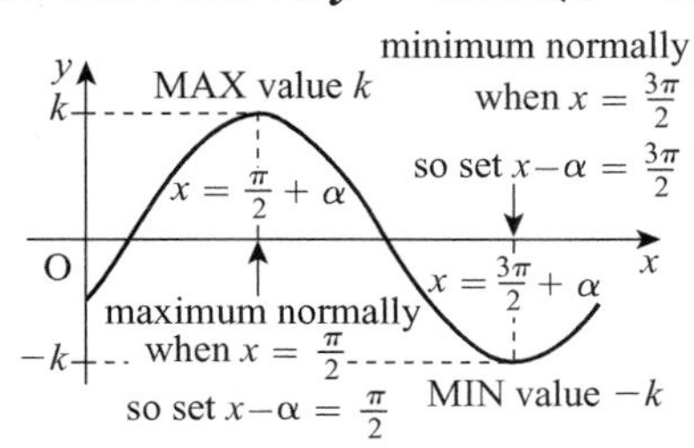

A cosine curve $y = k\cos(x - \alpha)$

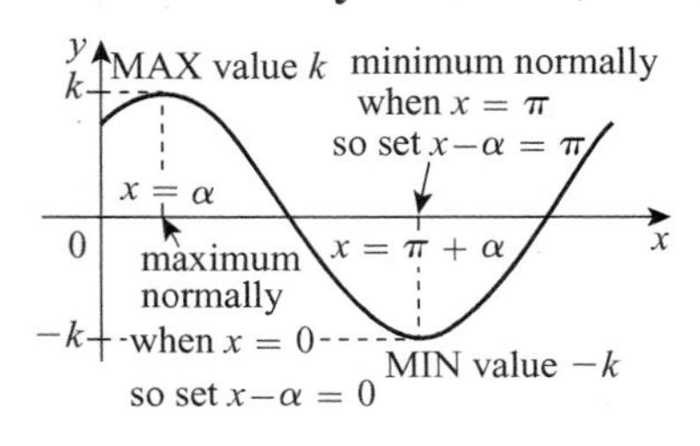

Problem solving with the wave function

Example 1

Find the maximum value of L where $L = \cos 2\theta + \sqrt{3}\sin 2\theta$ and the corresponding values for θ where $0 \le \theta \le 2\pi$.

Solution

$\cos 2\theta + \sqrt{3}\sin 2\theta$ can be expressed as $k\cos(2\theta - \alpha)$ giving $k = 2$ and $\alpha = \frac{\pi}{3}$ so $L = 2\cos\left(2\theta - \frac{\pi}{3}\right)$

The maximum value of L is **2** and this happens when $2\theta - \frac{\pi}{3} = 0, 2\pi, \cdots$

$\Rightarrow \quad 2\theta = \frac{\pi}{3}, 2\pi + \frac{\pi}{3}, \cdots$

$\Rightarrow \quad 2\theta = \frac{\pi}{3}, \frac{7\pi}{3}, \cdots$

$\Rightarrow \quad \theta = \frac{\pi}{6}, \frac{7\pi}{6}, \cdots$

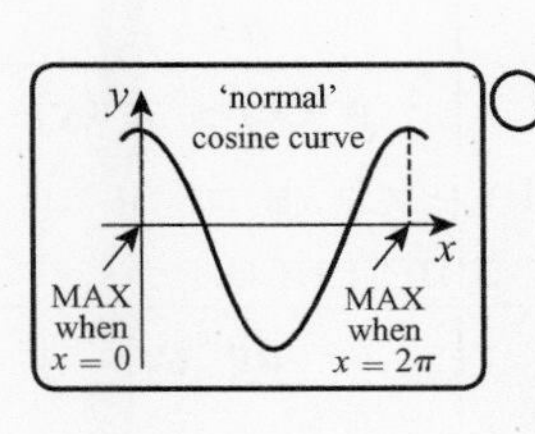

So $\theta = \frac{\pi}{6}$ **and** $\theta = \frac{7\pi}{6}$ are the required values (all other values are outside the range $0 \le \theta \le 2\pi$).

TOP TIP

For Paper 2 questions try to check solutions using a graphing calculator if possible. You may be able to pick up calculation errors in your solution by doing this.

Quick Test 20

1. Solve these equations
   a) $3\tan x° + 4 = 0$ for $0 \le x \le 360$
   b) $3\sin x + 1 = 0$ for $0 \le x \le 2\pi$
2. Find the exact solutions of these equations for $0 \le x \le 2\pi$:
   a) $\sqrt{3}\tan x = 1$ b) $2\cos x + 1 = 0$
3. Find algebraically the coordinates of P and Q, the points of intersection of the two graphs.

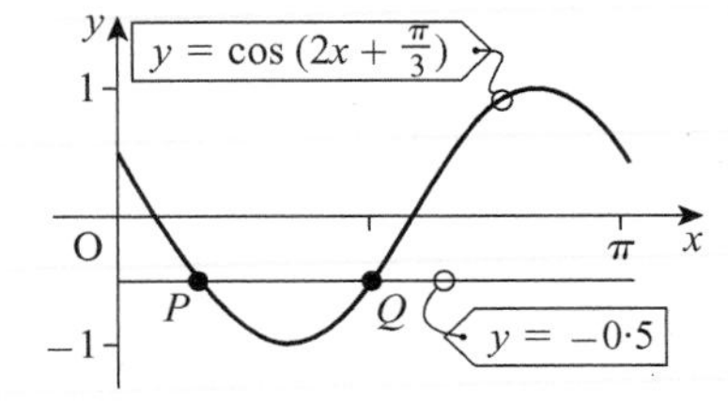

4. Solve: a) $4\cos^2 x° + 4\cos x° - 3 = 0$ for $0 \le x \le 360$ (to 1 decimal place)
   b) $3\sin x + 2 = \cos 2x$ for $0 \le x \le 2\pi$
5. $g(x) = \sqrt{3}\sin x° - \cos x°$
   a) Express $g(x)$ in the form $k\sin(x - \alpha)°$ where $k > 0$ and $0 \le \alpha < 360$.
   b) Hence, solve algebraically $g(x) = 0{\cdot}8$ for $0 \le x < 360$.

Chapter 2

Basic rules and techniques

What is differentiation?

The gradient at a point on a graph is given by the gradient of the tangent line at that point:

By calculating the gradient at every point on the graph $y = f(x)$ you have **differentiated** the function f and produced a new gradient function f' derived from f:

$$y = f(x) \xrightarrow{\text{differentiate}} \frac{dy}{dx} = f'(x)$$

formula for function f — formula for gradient function f'

The basic rules

TOP TIP

You should thoroughly revise the laws of indices as they are essential when differentiating expressions with powers.

$f(x)$	$f'(x)$
x^n	nx^{n-1}
$g(x) \pm h(x)$	$g'(x) \pm h'(x)$
(Differentiate each term of a sum or difference.)	
$ag(x)$	$ag'(x)$
(When a term is multiplied by a constant then differentiate as normal and multiply the result by the same constant.)	

Special cases:

1.

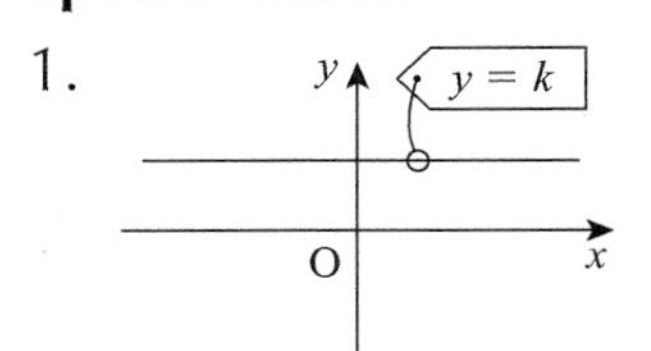

$y = k$ gives $\dfrac{dy}{dx} = 0$

2.

$y = mx$ gives $\dfrac{dy}{dx} = m$

Example

Differentiate:

a) $y = 5x^3 - 3x^2$

Solution

Differentiate each term.

$$\frac{dy}{dx} = 5 \times 3x^2 - 3 \times 2x^1$$

$$= \mathbf{15x^2 - 6x}$$

b) $f(x) = \dfrac{2}{\sqrt{x}} - \dfrac{3}{x}$

Solution

First, prepare the 'formula' for differentiating by writing it as a difference of powers of x:

$$f(x) = \frac{2}{x^{\frac{1}{2}}} - \frac{3}{x^1} = 2x^{-\frac{1}{2}} - 3x^{-1}$$

Now use the differentiation rules:

$$f'(x) = 2 \times \left(-\frac{1}{2}\right)x^{-\frac{1}{2}-1} - 3 \times (-1)x^{-1-1}$$

$$= -x^{-\frac{3}{2}} + 3x^{-2}$$

Differentiation techniques

You need to first 'prepare' an expression before attempting to differentiate it. Your aim is to write the expression as sums and/or differences of terms of the form ax^n

technique 1 Remove root signs

e.g. $\sqrt{x} = x^{\frac{1}{2}}$ $\frac{1}{\sqrt{x}} = x^{-\frac{1}{2}}$

technique 2 Remove brackets
e.g. $(2x - 1)(x + 2)$
$= 2x^2 + 3x - 2$

technique 3 Fractions with a single term on the denominator can be split:

e.g. $\frac{x^3 + x - 1}{x^2} = \frac{x^3}{x^2} + \frac{x}{x^2} - \frac{1}{x^2}$
$= x + x^{-1} - x^{-2}$
(using the Laws of Indices)

Example 1
Calculate the **exact** value of $f'(9)$ where

$$f(x) = \frac{x - 3x^2}{\sqrt{x}}$$

Solution

$$f(x) = \frac{x}{x^{\frac{1}{2}}} - \frac{3x^2}{x^{\frac{1}{2}}} = x^{\frac{1}{2}} - 3x^{\frac{3}{2}}$$

So $f'(x) = \frac{1}{2}x^{-\frac{1}{2}} - \frac{9}{2}x^{\frac{1}{2}} = \frac{1}{2\sqrt{x}} - \frac{9\sqrt{x}}{2}$

giving $f'(9) = \frac{1}{2\sqrt{9}} - \frac{9\sqrt{9}}{2} = \frac{1}{6} - \frac{27}{2}$

$= \frac{1}{6} - \frac{81}{6} = -\frac{80}{6} = -\mathbf{\frac{40}{3}}$

A decimal approximation is not acceptable for the **exact** value.

Example 2
Find the gradient of the curve $y = x^3$ at the point $P(1, 1)$.

Solution

$\frac{dy}{dx} = 3x^2$

When $x = 1$

$\frac{dy}{dx} = 3 \times 1^2 = 3$

The required **gradient is 3**.

Example 3
Find the points on the curve $y = x^3$ where the gradient is 12.

Solution

You require $\frac{dy}{dx} = 12$ so $3x^2 = 12$, $x^2 = 4$

giving $x = 2$ or -2

Now substitute into $y = x^3$ for y-coordinates.

The required points are **(2, 8) and (–2, –8).**

TOP TIP
The laws of indices tell you that $x^{\frac{3}{2}}$ means: find the square root of x then cube the result.

Quick Test 21

1. Find $\frac{dy}{dx}$ where: a) $y = -\frac{1}{x^3}$ b) $y = \frac{2}{\sqrt{x}}$ c) $y = -\frac{1}{\sqrt{x}}$
2. Differentiate: a) $y = \frac{3x}{\sqrt{x}} - \frac{2}{x}$ b) $f(x) = \frac{2x+1}{\sqrt{x}}$
3. Find: a) $f'(4)$ where $f(x) = 5\sqrt{x} - x$ b) $f'(9)$ where $f(x) = \frac{x^2 - x}{\sqrt{x}}$
4. Find the points on the graph of $y = \frac{1}{2}x^4$ where the gradient is -2.

Tangents and stationary points

Equations of tangents

The gradient of the tangent to the graph $y = f(x)$ at the point (x, y) is given by

$$\frac{dy}{dx} \text{ or } f'(x)$$

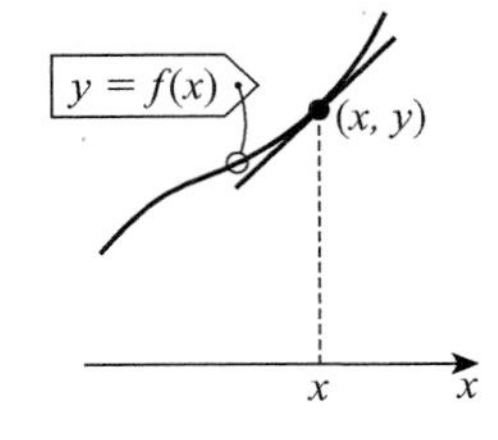

In particular, at the point (a, b) the gradient of the tangent is $f'(a)$.

To find the equation of a tangent line at the point (a, b) on the graph $y = f(x)$:

Step 1 Find $f'(x)$

Step 2 Calculate $m = f'(a)$

Step 3 Equation is $y - b = m(x - a)$

Example

Find the equation of the tangent to $y = x^2$ at the point (3, 9).

Solution

$\frac{dy}{dx} = 2x$. When $x = 3$ $\frac{dy}{dx} = 2 \times 3 = 6$

Gradient = 6. Point is (3,9).

Equation is $y - 9 = 6(x - 3)$

$\Rightarrow \quad y = 6x - 9$

TOP TIP

If you are asked for a **stationary point**, give the coordinates. If you are asked for a **stationary value**, give the y-coordinate only.

Finding stationary points

Points on a graph $y = f(x)$ where the gradient is zero are called **stationary points**.

$$f'(a) = 0$$

$(a, f(a))$	$f(a)$
is a	is a
stationary **point**.	stationary **value**.

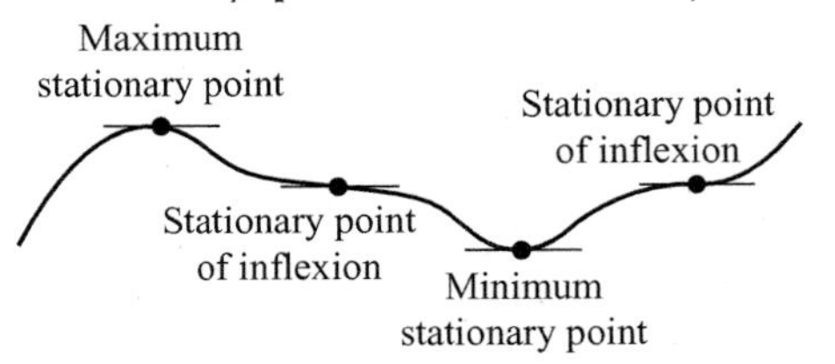

To find the stationary points on the graph $y = f(x)$:

Step 1 Find $f'(x)$

Step 2 Set $f'(x) = 0$

Step 3 Solve $f'(x) = 0$

Each solution $x = a$ gives a stationary point.

Step 4 Calculate $y = f(a)$ for each solution $x = a$. $(a, f(a))$ is a stationary point.

Example

Find the stationary points on the graph

$$y = x^4 - 4x^3 + 3$$

Solution

$y = x^4 - 4x^3 + 3$ gives $\frac{dy}{dx} = 4x^3 - 12x^2$

To find stationary points, set $\frac{dy}{dx} = 0$

so $4x^3 - 12x^2 = 0$

Now solve: $4x^2(x - 3) = 0$

$x^2 = 0$ or $x - 3 = 0$

$x = 0$ $\quad x = 3$

when $x = 0$ $\quad y = 0^4 - 4 \times 0^3 + 3 = 3$ giving (0, 3)

when $x = 3$ $\quad y = 3^4 - 4 \times 3^3 + 3 = -24$ giving (3, −24)

So there are two stationary points, namely (0, 3) and (3, −24), on this graph.

Identifying types of stationary points

There are three types of stationary points:

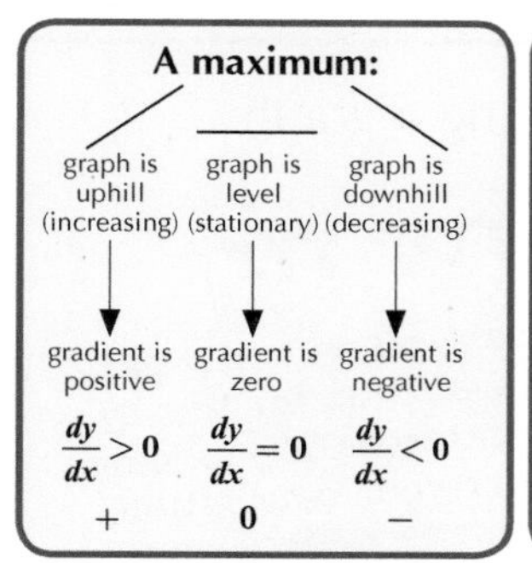

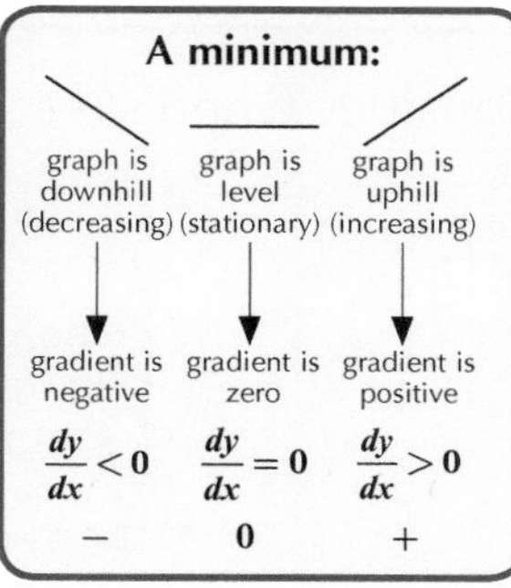

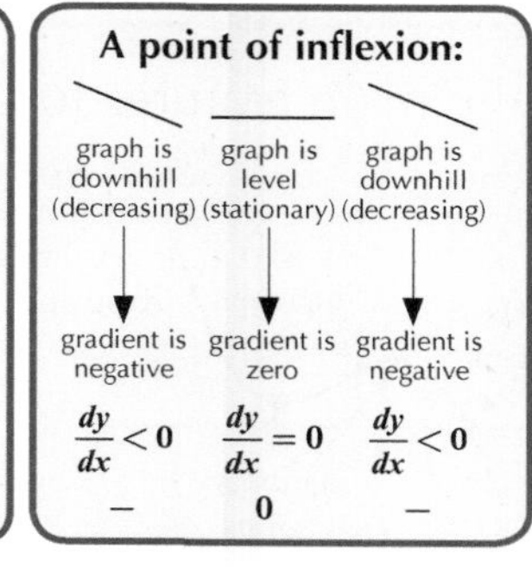

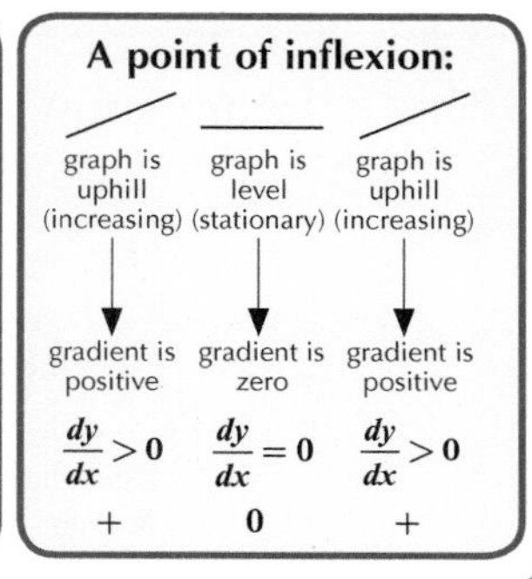

Table of signs

Having identified where the stationary points are, say, at $x = a$ and $x = b$, then:

Step 1 Draw a number line and place the x-values, $x = a$ and $x = b$, in order on the number line.

Step 2 Underneath draw a two-row table:

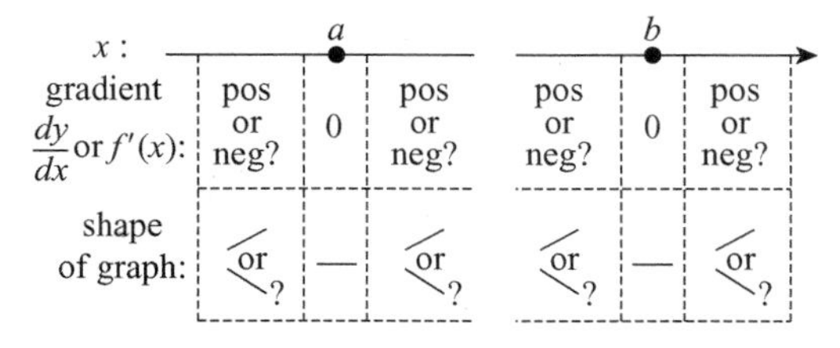

Example

Determine the **nature** of the stationary points from the example at the bottom of page 68.

Solution

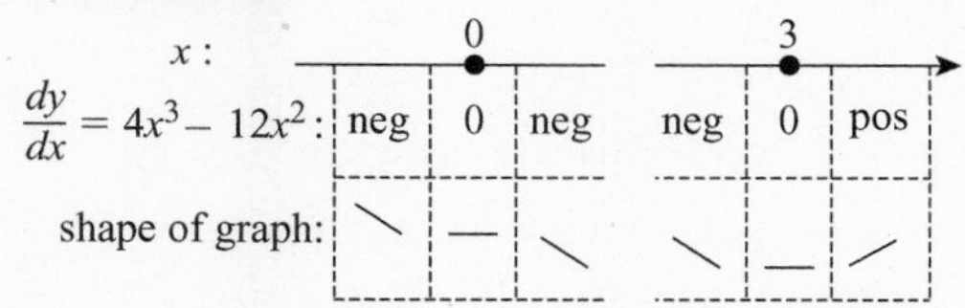

So (0, 3) is a stationary point of inflexion and (3, –24) is a minimum stationary point.

Quick Test 22

1. On the curve $y = \frac{1}{3}x^3 - 2x + 1$ tangents are drawn at the points $A\left(-1, \frac{8}{3}\right)$, $B(0, 1)$ and $C\left(1, -\frac{2}{3}\right)$.

   a) Which pair of tangents are parallel?
   Hint: Find their gradients.

   b) Find the equations of the three tangents.

2. Find algebraically the coordinates of the stationary points and determine their nature for the curve: $y = 4x^5 + 5x^4 - 2$

TOP TIP

When asked for the **nature** of stationary points you must give evidence. This should be a 'table of signs'.

Chapter 2

Graph sketching and gradients

Sketching a graph

This diagram shows the main features to consider when sketching a graph:

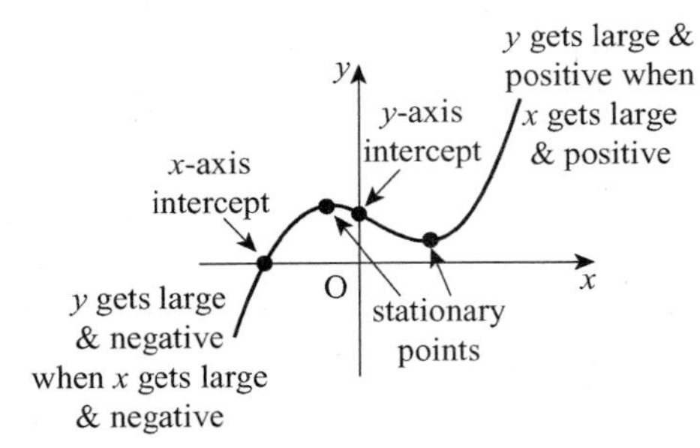

TOP TIP

Remember you will only gain marks in the exam for this type of graphical work if you show evidence of your calculations.

When you want to sketch the graph $y = f(x)$ follow these steps:

Step 1 Find where the graph cuts the x-axis: Set $y = 0$ and solve $f(x) = 0$.

Step 2 Find where the graph cuts the y-axis: Set $x = 0$ and find the value of $f(0)$.

Step 3 Find the stationary points and determine their nature:

Solve the equation $\frac{dy}{dx} = 0$ and then set up a table of signs.

Step 4 Check the behaviour of y for large positive/negative x values.

Example

Sketch the graph $y = x^3(4 - x)$

Step 1 For x-axis intercepts set $y = 0$:

$x^3(4 - x) = 0 \Longrightarrow x = 0$ or $x = 4$

Intercepts are (0, 0) and (4, 0)

Step 2 For y-axis intercepts set $x = 0$:

$y = 0^3 \times (4 - 0) = 0$

Intercept is (0, 0)

Step 3 $y = x^3(4 - x) = 4x^3 - x^4$

$\Longrightarrow \dfrac{dy}{dx} = 12x^2 - 4x^3 = 4x^2(3 - x)$

For stationary points set $\frac{dy}{dx} = 0$

So $4x^2(3 - x) = 0 \Longrightarrow x = 0$ or $x = 3$

When $x = 3$ $y = 3^3 \times (4 - 3) = 27$

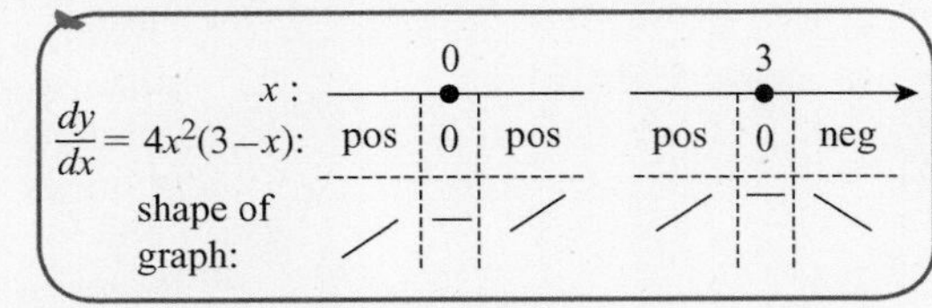

(0, 0) is a point of inflexion

(3, 27) is maximum stationary point

Here is a diagram showing the information from steps 1, 2 and 3:

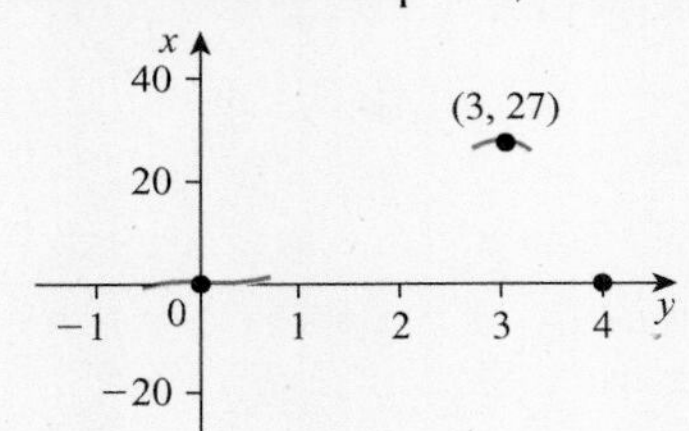

Complete the graph by joining up the known points with a smooth curve:

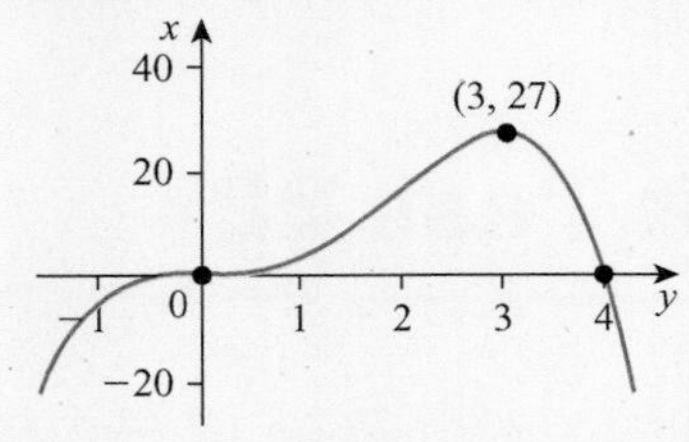

Step 4 Check that as x gets large and positive $y = x^3(4 - x)$ gets large and negative (try $x = 100$).
Check that as x gets large and negative $y = x^3(4 - x)$ gets large and negative (try $x = -100$).

Sketching a gradient graph

You should be able to sketch the graph $\frac{dy}{dx} = f'(x)$ if you are given the graph $y = f(x)$

Here is a typical cubic function graph lined up with its gradient graph:

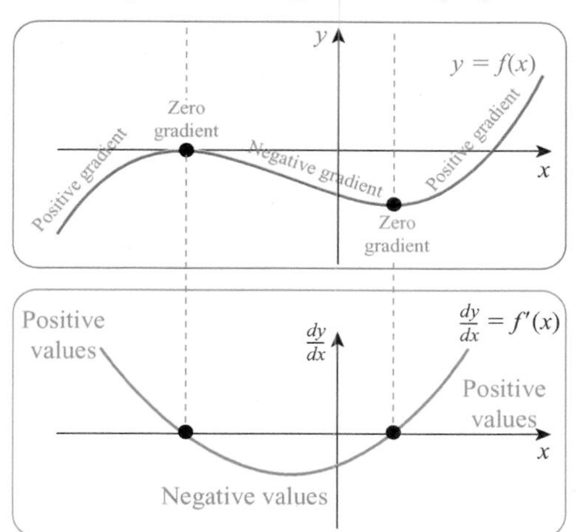

Notes:

- The graph $y = f(x)$ shows the values of the function f.
  The graph $\frac{dy}{dx} = f'(x)$ shows the values of the gradient of the original graph.
- Differentiating a cubic (degree 3) will give a quadratic (degree 2) which explains the parabola shape shown for the gradient graph.
- Notice the graph is divided into sections using the stationary points:

graph $y = f(x)$:	uphill	stationary	downhill
gradient graph $\frac{dy}{dx} = f'(x)$:	above x-axis	on the x-axis	below x-axis

Special cases:

TOP TIP
You may also be asked to sketch $y = f(x)$ having been given the gradient function graph. You should be able to reverse all the thinking!

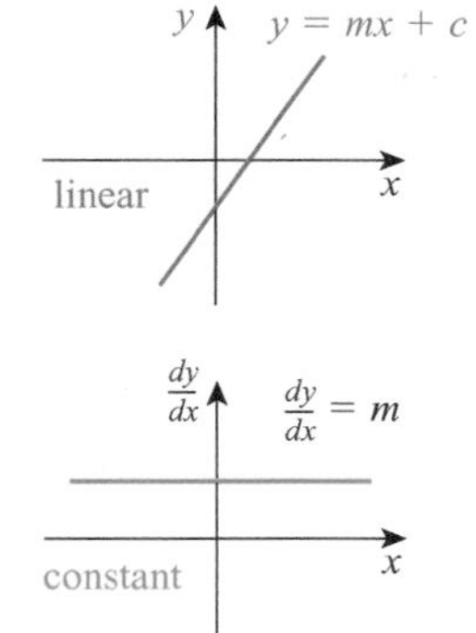

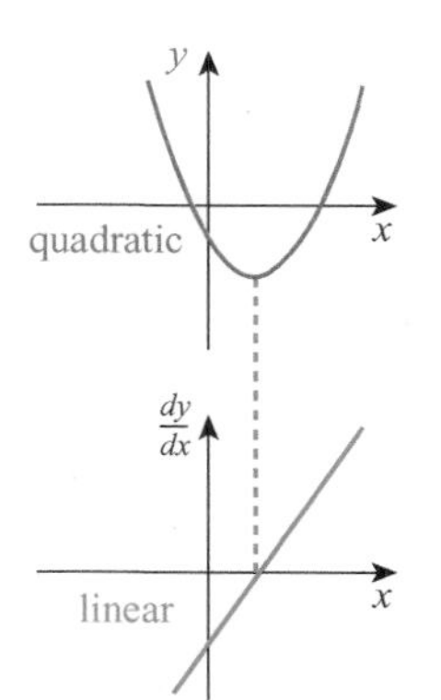

TOP TIP
If you are given the value of the gradient on $y = f(x)$ when $x = a$ then this tells you the height of your gradient graph at the point where $x = a$

Quick Test 23

1. Sketch the graph $y = x^4 - 2x^2 - 3$ given that the graph has two points of intersection with the x-axis, namely $(-\sqrt{3}, 0)$ and $(\sqrt{3}, 0)$.

2. For this graph of $y = f(x)$ sketch the gradient graph $\frac{dy}{dx} = f'(x)$.

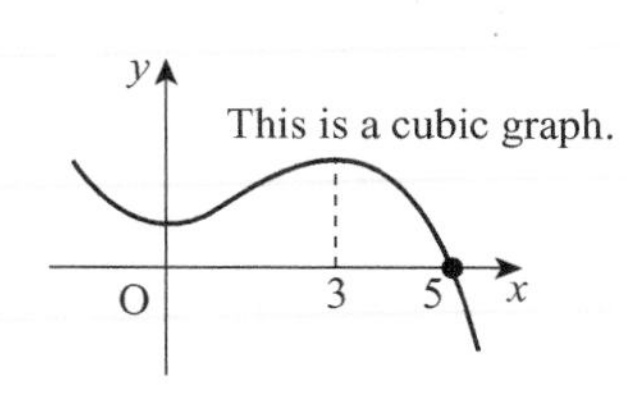

Further rules and techniques

Differentiating sine and cosine

The rules are:

$f(x)$	$f'(x)$
$\sin x$ $\cos x$	$\cos x$ $-\sin x$

These rules only hold if x is measured in radians

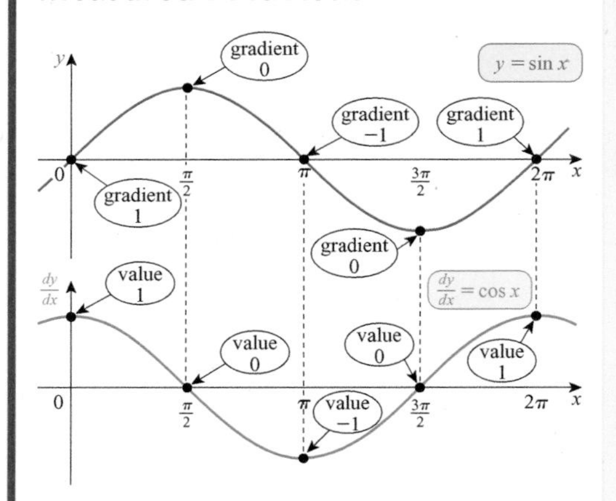

Example

Find the equation of the tangent to the graph $y = \cos x$ at the point where $x = \dfrac{\pi}{3}$

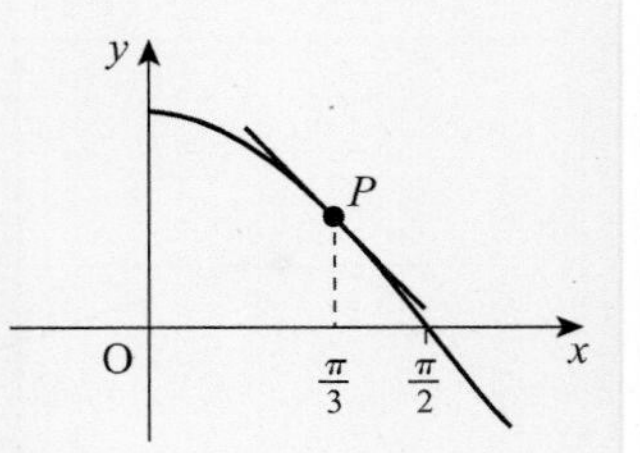

Solution

$y = \cos x$ so $\dfrac{dy}{dx} = -\sin x$

When $x = \dfrac{\pi}{3}$ $\dfrac{dy}{dx} = -\sin\dfrac{\pi}{3} = -\dfrac{\sqrt{3}}{2}$

and when $x = \dfrac{\pi}{3}$ $y = \cos\dfrac{\pi}{3} = \dfrac{1}{2}$

So the gradient of the tangent is $-\frac{\sqrt{3}}{2}$ and a point on the tangent is $P\left(\frac{\pi}{3}, \frac{1}{2}\right)$

The equation of the tangent is

$$y - \frac{1}{2} = -\frac{\sqrt{3}}{2}\left(x - \frac{\pi}{3}\right)$$

This gives $2y - 1 = -\sqrt{3}\left(x - \dfrac{\pi}{3}\right)$

or $\mathbf{2y + \sqrt{3}x = \dfrac{\sqrt{3}}{3}\pi + 1}$

The 'chain' rule

The 'chain' rule is used to extend the scope of the basic power rule and the two trig rules above:

The power rule: $y = x^n \Rightarrow \dfrac{dy}{dx} = nx^{n-1}$

The extended power rule: $y = (f(x))^n \Rightarrow \dfrac{dy}{dx} = n(f(x))^{n-1} \times f'(x)$

An example: $y = \sin^2 x = (\sin x)^2 \Rightarrow \dfrac{dy}{dx} = 2(\sin x)^1 \times \cos x = 2\sin x\cos x$

A trig rule: $y = \sin x \Rightarrow \frac{dy}{dx} = \cos x$

The extended trig rule: $y = \sin(f(x)) \Rightarrow \frac{dy}{dx} = \cos(f(x)) \times f'(x)$

An example: $y = \sin 3x \Rightarrow \frac{dy}{dx} = \cos 3x \times 3 = 3\cos 3x$

A trig rule: $y = \cos x \Rightarrow \frac{dy}{dx} = -\sin x$

The extended trig rule: $y = \cos(f(x)) \Rightarrow \frac{dy}{dx} = -\sin(f(x)) \times f'(x)$

An example: $y = \cos(x^2) \Rightarrow \frac{dy}{dx} = -\sin(x^2) \times 2x = -2x\sin(x^2)$

Working with the 'chain' rule

The general form of the 'chain' rule is: $y = g(h(x)) \Rightarrow \frac{dy}{dx} = g'(h(x)) \times h'(x)$

Example 1

Differentiate a) $\cos\left(3x - \frac{\pi}{2}\right)$ b) $\sqrt{x+1}$

Solution

a) $y = \cos\left(3x - \frac{\pi}{2}\right) \Rightarrow \frac{dy}{dx} = -\sin\left(3x - \frac{\pi}{2}\right) \times 3$

So $\frac{dy}{dx} = -3\sin\left(3x - \frac{\pi}{2}\right)$

b) $y = \sqrt{x+1} = (x+1)^{\frac{1}{2}} \Rightarrow \frac{dy}{dx} = \frac{1}{2}(x+1)^{-\frac{1}{2}} \times 1$

So $\frac{dy}{dx} = \frac{1}{2(x+1)^{\frac{1}{2}}} = \frac{1}{2\sqrt{(x+1)}}$

Example 2

Find $f'\left(\frac{3\pi}{4}\right)$ where $f(x) = \frac{1}{3}\sin^3 x$

Solution

$f(x) = \frac{1}{3}(\sin x)^3$

$\Rightarrow f'(x) = \frac{1}{3} \times 3(\sin x)^2 \times \cos x$

So $f'\left(\frac{3\pi}{4}\right) = \sin^2 \frac{3\pi}{4} \cos\frac{3\pi}{4}$

$= \left(\frac{1}{\sqrt{2}}\right)^2 \times \left(-\frac{1}{\sqrt{2}}\right) = -\frac{1}{2\sqrt{2}}$

Quick Test 24

1. Part of the graph $y = 2\sin^2 x$ is shown.
Find the gradient of the curve at the point P where $x = \frac{\pi}{4}$.

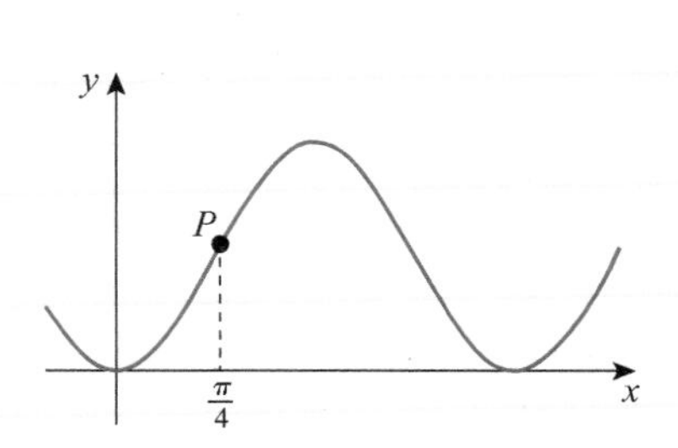

2. Differentiate a) $\sqrt{2-4x}$ b) $\frac{1}{\cos x}$

3. The function g is defined by $g(x) = \frac{1}{x}$ where $x \neq 0$.
Show that the graph $y = g(x)$ is always decreasing.

Rates of change

What is a rate of change?

Here is a distance/time graph:

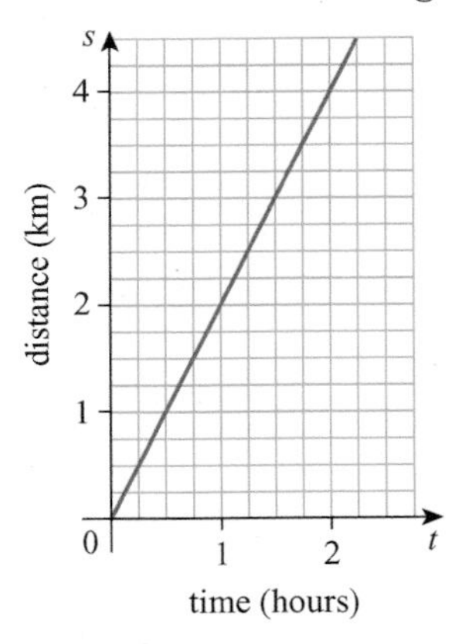

The formula for this graph is:

$s = 2t$

The distance changes at a rate of 2 km each hour.

You say that the rate of change of distance (s) with respect to time (t) is 2 km/hr.

In this case another name for this rate of change is speed.

The gradient of the graph measures the rate of change.

The speed is therefore given by: $\frac{ds}{dt} = 2$

In general for two variables x and y where the values of y depend on the values of x according to some rule $y = f(x)$ then the rate of change of y with respect to x is given by

$$\frac{dy}{dx} = f'(x)$$

and is shown by the gradient of the graph.

Notes:
- 'Displacement' and 'velocity' are used instead of 'distance' and 'speed' when direction matters.
- For a speed/time graph, the rate at which speed changes with respect to time is acceleration.

TOP TIP

When asked to find the rate of change of $f(x)$ you need to differentiate.

Working with the notation

Often the following letters are used:

s for displacement

v for velocity

a for acceleration

t for time

This gives:

$$\frac{ds}{dt} = v$$

The rate of change of displacement gives the velocity.

$$\frac{dv}{dt} = a$$

The rate of change of velocity gives the acceleration.

Example

An accelerating object has a displacement s metres after t seconds given by the formula: $s = t^2$. What is its velocity after 1·5 seconds?

Solution

$s = t^2$

$v = \frac{ds}{dt} = 2t$

When $t = 1{\cdot}5$

$v = 2 \times 1{\cdot}5 = 3$

The velocity is 3 m/s (gradient at indicated point).

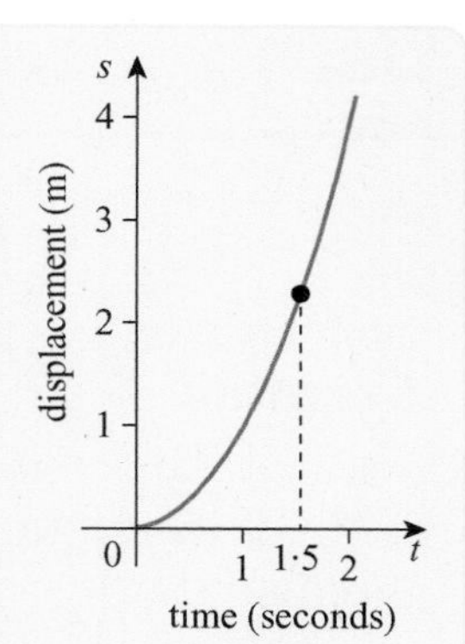

Solving rate of change problems

Example 1

The displacement, s cm, of a weight on a spring, t seconds after release is given by $s = 50t - 100t^2$. Find its velocity when released and after $\frac{1}{4}$ second.

Solution

$$v = \frac{ds}{dt} = 50 - 200t$$

At time of release $t = 0$ so
$v = 50 - 200 \times 0 =$ **50 m/s**

After $\frac{1}{4}$ second $t = \frac{1}{4}$ so

$$v = 50 - 200 \times \frac{1}{4} = \mathbf{0\ m/s}$$

(it has reached its greatest extent)

Example 2

The volume, V cm^3, of a spherical balloon with radius r cm is given by $V = \frac{4}{3}\pi r^3$. It is inflated.

Find the rate of change of V with respect to r when $r = 8$ cm.

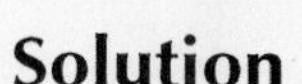

Solution

$$V = \frac{4}{3}\pi r^3$$

so $\frac{dV}{dr} = 3 \times \frac{4}{3}\pi r^2 = 4\pi r^2$ This is the 'rate of change formula'.

When $r = 8$, $\frac{dV}{dr} = 4 \times \pi \times 8^2 \doteqdot 804$ (to 3 sig. figs).

This means that when the balloon has radius 8 cm and if the rate at which the volume is changing were to remain the same, then for an increase of 1 cm in the radius the volume would increase by 804 cm^3.

Using integration

TOP TIP

The constant of integration is a crucial feature of these problems.

$\frac{ds}{dt} = v \Rightarrow s = \int v\,dt$ and also $\frac{dv}{dt} = a \Rightarrow v = \int a\,dt$

Example 3

The velocity v of a reference point on a rotating disc when measured from the edge of its container is given by $v = -5\sin 2t$ where t is the time in seconds from the disc starting to rotate.

When the disc started to rotate, the reference point was $s = 5{\cdot}5$ cm from the edge.

How far is it from the edge after 3 seconds?

Solution

$v = \frac{ds}{dt} = -5\sin 2t \Rightarrow s = \int -5\sin 2t\,dt = \frac{5}{2}\cos 2t + C$

When $t = 0$, $s = 5{\cdot}5$ (the start of rotation)

so $5{\cdot}5 = \frac{5}{2}\cos 0 + C = \frac{5}{2} + C \Rightarrow C = 3$

so $s = \frac{5}{2}\cos 2t + 3$ (the displacement formula)

When $t = 3$ $s = \frac{5}{2}\cos 6 + 3 \doteqdot 5{\cdot}4$ cm

Quick Test 25

1. The volume of a solid is given by $V(r) = \frac{2}{3}\pi r^3 + \pi r^2$.
   Find the rate of change of V cm^3 with respect to the radius r cm when $r = \frac{1}{2}$ cm.
2. A firework is launched. The height h metres of the firework t minutes after launch is given by the formula $h = 400t - 200t^2$.
   a) Find its speed at launch.
   b) Find its speed after 1 minute and explain your answer.
   c) Compare its speed after 2 minutes with its speed at launch and explain your result.

Basic integration rules and techniques

What is integration?

Integration reverses the process of Differentiation:

$y = f(x)$	$y = f(x) + C$
$\Downarrow$ differentiation	$\Uparrow$ integration
$\frac{dy}{dx} = f'(x)$	$\frac{dy}{dx} = f'(x)$

The integral sign $\int$ is used to indicate integration:

so $\int f'(x)\,dx = f(x) + C$

or $\int g(x)\,dx = G(x) + C$
where $G'(x) = g(x)$

C is called the **Constant of Integration**.

Note: $\int f(x)\,dx$ is read 'the integral of $f(x)$ with respect to x' and is called an **Indefinite Integral**.

Basic rules

$f(x)$	$\int f(x)\,dx$
x^n	$\frac{x^{n+1}}{n+1} + C \quad (n \neq -1)$
$g(x) \pm h(x)$	$\int g(x)\,dx \pm \int h(x)\,dx$
(Integrate each term of a sum or difference.)	
$ag(x)$	$a\int g(x)\,dx$
(When a term is multiplied by a constant then integrate as normal and multiply the result by the same constant.)	

Notes: Special cases

1. Integrating a constant: $\int k\,dx = kx + C$
2. $\int \frac{1}{x}\,dx = \int x^{-1}\,dx$.
   This integral does not follow the rule above. You do not need to know how to integrate x^{-1}.

Example

$\frac{dy}{dx} = 3x^2 - \frac{1}{x^2}$ Find y.

Solution

$y = \int (3x^2 - x^{-2})\,dx$

$= \frac{3x^3}{3} - \frac{x^{-1}}{(-1)} + C$

$= \boldsymbol{x^3 + \frac{1}{x} + C}$

TOP TIP

You should always check your integration answer by differentiating it. If it is correct you should get the expression you started with!

Example

Find $\int \left(\sqrt{x} - \frac{3}{\sqrt{x}}\right) dx$

Solution

First 'prepare' for integrating.

$$\int \left(x^{\frac{1}{2}} - \frac{3}{x^{\frac{1}{2}}}\right) dx = \int \left(x^{\frac{1}{2}} - 3x^{-\frac{1}{2}}\right) dx$$

3. A formula may involve a variable other than x, for example $f(t)$, and this may still be integrated:

   $\int f(t)\,dt$ (the integral with respect to t).

$$= \frac{x^{\frac{3}{2}}}{\frac{3}{2}} - \frac{3x^{\frac{1}{2}}}{\frac{1}{2}} + C$$

(Now double top and bottom of fractions.)

$$= \frac{2x^{\frac{3}{2}}}{3} - 6x^{\frac{1}{2}} + C$$

Integrating trig functions

The rules are:

$f(x)$	$\int f(x)\,dx$
$\cos x$	$\sin x + C$
$\sin x$	$-\cos x + C$

These rules only hold if x is measured in radians.

Example Find $\int \left(\frac{3\cos x}{2} - \frac{2\sin x}{3}\right) dx$

Solution

$$\int \left(\frac{3\cos x}{2} - \frac{2\sin x}{3}\right) dx$$

$$= \int \left(\frac{3}{2}\cos x - \frac{2}{3}\sin x\right) dx$$

$$= \frac{3}{2}\sin x - \frac{2}{3}(-\cos x) + C$$

$$= \frac{3}{2}\sin x + \frac{2}{3}\cos x + C$$

Solving differential equations

$\frac{dy}{dx} = f(x)$ is a **differential equation**. It has general solution $y = \int f(x)\,dx = F(x) + C$ where $F'(x) = f(x)$. If, in addition, a point (a, b) is known on the graph $y = F(x) + C$ then the value of the constant C can be found.

Example Point (9, 25) lies on the graph $y = f(x)$ where $f'(x) = \frac{2x-1}{\sqrt{x}}$. Find $f(x)$.

Solution $f(x) = \int \frac{2x-1}{x^{\frac{1}{2}}}dx = \int \frac{2x^1}{x^{\frac{1}{2}}} - \frac{1}{x^{\frac{1}{2}}}dx$ — Splitting the fraction

$= \int 2x^{\frac{1}{2}} - x^{-\frac{1}{2}}dx$ — Using the laws of indices

$= \frac{2x^{\frac{3}{2}}}{\frac{3}{2}} - \frac{x^{\frac{1}{2}}}{\frac{1}{2}} + C$ — Using the power rule: $\int x^n\,dx = \frac{x^{n+1}}{n+1} + C$

$= \frac{4x^{\frac{3}{2}}}{3} - \frac{2x^{\frac{1}{2}}}{1} + C$ — Doubling top and bottom of the fractions

So $f(x) = \frac{4}{3}\left(\sqrt{x}\right)^3 - 2\sqrt{x} + C$

Using the fact that (9, 25) lies on the graph $y = f(x)$ you know that $25 = f(9)$

Now $f(9) = \frac{4}{3}\left(\sqrt{9}\right)^3 - 2\sqrt{9} + C = 36 - 6 + C = 30 + C$

So $30 + C = 25 \Rightarrow C = -5$ giving $f(x) = \frac{4}{3}\left(\sqrt{x}\right)^3 - 2\sqrt{x} - 5$

Quick Test 26

1. Find: a) $\int\left(\frac{1}{x^3} - 2x^2\right)dx$ b) $\int\left(5x^2 + 5 - \frac{3}{2x^2}\right)dx$ c) $\int\left(\frac{2}{3\sqrt{x}} + \sqrt{x}\right)dx$

2. Find: a) $\int(5x^3 + x + \sin x)\,dx$ b) $\int(4x - \cos x)\,dx$

3. If $y = \frac{2}{3}$ when $x = 1$ solve the differential equation $\frac{dy}{dx} = \frac{3-x^2}{3x^2}$ by finding y in terms of x.

Definite integrals and special integrals

Definite integrals

$\int_a^b f(x)\,dx$ is called a **Definite Integral**.

a is the **lower limit** and b is the **upper limit** of the integral. This kind of integral has a particular value. Here are the steps to find this value:

Step 1 Integrate $f(x)$ as normal to get $F(x)$ but leave out the constant of integration.

Step 2 Calculate $F(b)$ using the upper limit $x = b$.

Step 3 Calculate $F(a)$ using the lower limit $x = a$.

Step 4 Calculate $F(b) - F(a)$. This is the required value.

Example 1

Find the **exact** value of $\int_1^3 \left(\frac{3}{\sqrt{x}} - x\right) dx$

Solution

$$\int_1^3 (3x^{-\frac{1}{2}} - x)\,dx = \left[\frac{3x^{\frac{1}{2}}}{\frac{1}{2}} - \frac{x^2}{2}\right]_1^3 = \left[6x^{\frac{1}{2}} - \frac{x^2}{2}\right]_1^3$$

$$= \left[6\sqrt{x} - \frac{x^2}{2}\right]_1^3$$

$$= \left(6\sqrt{3} - \frac{3^2}{2}\right) - \left(6\sqrt{1} - \frac{1^2}{2}\right)$$

$$= 6\sqrt{3} - \frac{9}{2} - 6 + \frac{1}{2}$$

$$= \mathbf{6\sqrt{3} - 10}$$

(Do not give a decimal approximation when an exact value is required.)

Example 2

Evaluate $\int_{\frac{\pi}{6}}^{\frac{\pi}{3}} 2\sin x - 3\cos x\,dx$ giving your answer as a surd in its simplest form.

Remember the exact values diagram:

Solution

$$\int_{\frac{\pi}{6}}^{\frac{\pi}{3}} 2\sin x - 3\cos x\,dx = [-2\cos x - 3\sin x]_{\frac{\pi}{6}}^{\frac{\pi}{3}}$$

$$= \left(-2\cos\tfrac{\pi}{3} - 3\sin\tfrac{\pi}{3}\right) - \left(-2\cos\tfrac{\pi}{6} - 3\sin\tfrac{\pi}{6}\right)$$

$$= \left(-2\times\frac{1}{2} - 3\times\frac{\sqrt{3}}{2}\right) - \left(-2\times\frac{\sqrt{3}}{2} - 3\times\frac{1}{2}\right)$$

$$= -1 - \frac{3\sqrt{3}}{2} + \sqrt{3} + \frac{3}{2}$$

$$= -\frac{2}{2} - \frac{3\sqrt{3}}{2} + \frac{2\sqrt{3}}{2} + \frac{3}{2}$$

$$= \frac{-2 - 3\sqrt{3} + 2\sqrt{3} + 3}{2} = \frac{1 - \sqrt{3}}{2}$$

TOP TIP

You should take great care with the negative signs in this sort of calculation, especially when subtracting the lower limit expression.

Special integrals

It is not possible to extend the power rule and trig rules for integration in the same way that you did for differentiation using the 'chain rule'.

Here are the limited extended rules:

$$\int (ax+b)^n\,dx = \frac{(ax+b)^{n+1}}{a(n+1)} + C$$

$(n \neq -1,\ a \neq 0)$

$$\int \cos(ax+b)\,dx = \frac{\sin(ax+b)}{a} + C$$

$(a \neq 0)$

$$\int \sin(ax+b)\,dx = -\frac{\cos(ax+b)}{a} + C$$

$(a \neq 0)$

Only a linear expression like $ax + b$ is allowed and you divide by a, the coefficient of x.

Example 1 Find $\int_{-1}^{1} \frac{4}{(5-3x)^2}\,dx$

Solution

$$\int_{-1}^{1} 4(5-3x)^{-2}dx = \left[\frac{4(5-3x)^{-1}}{-3\times(-1)}\right]_{-1}^{1}$$

$$= \left[\frac{4}{3(5-3x)}\right]_{-1}^{1} = \frac{4}{3(5-3\times 1)} - \frac{4}{3(5-3\times(-1))}$$

$$= \frac{4}{3\times 2} - \frac{4}{3\times 8} = \frac{4}{6} - \frac{1}{6} = \frac{3}{6} = \mathbf{\frac{1}{2}}$$

Example 2

Find a) $\int \sin 3x\,dx$ b) $\int 4\cos(2x - \frac{\pi}{6})\,dx$

Solution

a) $\int \sin 3x\,dx = \frac{-\cos 3x}{3} + C = -\frac{1}{3}\cos 3x + C$

b) $\int 4\cos(2x - \frac{\pi}{6})dx = \frac{4\sin(2x-\frac{\pi}{6})}{2} + C$

$= 2\sin(2x - \frac{\pi}{6}) + C$

Example 3

a) Use $\cos 2x = 2\cos^2 x - 1$ to write $\cos^2 x$ in terms of $\cos 2x$

b) Hence, find $\int \cos^2 x\,dx$

Solution

a) Rearranging gives

$\cos^2 x = \frac{1}{2}\cos 2x + \frac{1}{2}$

b) $\int \cos^2 x\,dx = \int \frac{1}{2}\cos 2x + \frac{1}{2}dx$

$= \frac{\sin 2x}{2\times 2} + \frac{1}{2}x + C$

$= \frac{1}{4}\sin 2x + \frac{1}{2}x + C$

TOP TIP

You are given this table of standard integrals in your exam:

$f(x)$	$\int f(x)x\,dx$
$\sin ax$	$-\frac{1}{a}\cos ax + C$
$\cos ax$	$\frac{1}{a}\sin ax + C$

Quick Test 27

1. Find the exact value of: $\int_1^4 \left(\sqrt{x} + \frac{1}{\sqrt{x}}\right) dx$

2. Find the exact value of $\int_0^{\frac{\pi}{4}} \frac{1}{2}\cos x - \sin x\,dx$

3. a) Use $\cos 2x = 1 - 2\sin^2 x$ to write $\sin^2 x$ in terms of $\cos 2x$

   b) Hence, find $\int \sin^2 x\,dx$

   c) Evaluate $\int_0^{\pi} \sin^2 x\,dx$

Check-up questions (Chapter 2)

2.1 Applying algebraic skills to solve equations

1. $g(x)$ is a cubic function. Here are some facts concerning $g(x)$:
   - The graph $y = g(x)$ crosses the x-axis at $(-2, 0)$
   - $x + 3$ is a factor of $g(x)$
   - If $g(x)$ is divided by $x - 1$ the remainder is zero

   Find the roots of the equation $g(x) = 0$
2. Myla draws the graph of $y = 2x^2 - x + a$ using an app on her phone.
   She notices that it has no x-axis intercepts.
   What is the range of values for a in this case?
3. A cubic function f is defined by $f(x) = x^3 - x^2 - 16x - 20$
   a) Factorise $f(x)$ fully.   b) Solve $f(x) = 0$

2.2 Applying trig skills to solve equations

1. Solve $2\cos 3x° = 1$ for $0 < x < 120$
2. Solve $\cos\theta° - 5\sin 2\theta° = 0$ for $0 \leq \theta \leq 90$
3. Carla was asked to write $4\sin\theta° - 3\cos\theta°$ in the form $k\sin(\theta - \alpha)°$
   She correctly calculated $k = 5$ and $\alpha = 36{\cdot}9$
   Hence she wrote $4\sin\theta° - 3\cos\theta° = 5\sin(\theta - 36{\cdot}9)°$
   Use her result to solve $4\sin 2\theta° - 3\cos 2\theta° = 2{\cdot}75$ for $0 \leq \theta \leq 90$

2.3 Applying calculus skills of differentiation

1. Differentiate $f(x)$ where $f(x) = -2\cos x$
2. Find $\frac{dy}{dx}$ where $y = \frac{5\sqrt{x}}{x} - 2\sqrt{x}$ $(x > 0)$
3. Part of the graph $y = \frac{1}{2}x^2 + 3x - 2$ is shown in the diagram.
   Find the equation of the tangent to the curve at the point P where $x = 2$

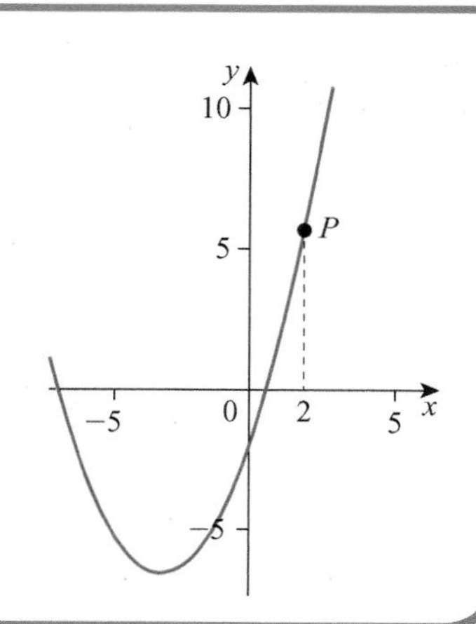

4. A stone is thrown into the air.

   The height h metres t seconds after it was thrown is given by the formula:

   $$h = 16t - 4t^2$$

   a) Find the velocity of the stone when it was thrown.

   b) Find the velocity of the stone after 2 seconds and explain your answer.

2.4 Applying calculus skills of integration

1. Find $\int \frac{1}{2}\cos x \, dx$
2. For the function g it is known that $g'(x) = (5 + x)^{-2}$

   Find $g(x)$
3. Find $\int \left(\frac{1}{x^{\frac{1}{2}}} - 2x^{\frac{1}{2}} \right) dx$ where $x > 0$
4. Evaluate the definite integral $\int_{-2}^{2} (x + 2)^3 \, dx$

Sample end-of-course exam questions (Chapter 2)

Non-calculator

1. The functions f, g and h are defined on the set of Real numbers by:

   $f(x) = 2x^3 + x - 3$

   $g(x) = -3x^2 + x - 2$

   $h(x) = 2x^3 + 3x^2 - 1$

   a) i) Show that $x + 1$ is a factor of $h(x)$.

   ii) Hence, factorise $h(x)$ fully.

   iii) Solve $h(x) = 0$.

   b) The two curves $y = f(x)$ and $y = g(x)$ share a common tangent at point T. Find the coordinates of T.

2. Find the value of $\int_{-1}^{1} \frac{6x^3 - x}{3x^3}\, dx$.

3. Find $\frac{dy}{dx}$ given that $y = \frac{1}{\sin^2 x}$.

4. Differentiate $\frac{2x+6}{\sqrt{x}}$ with respect to x.

5. The diagram shows a sketch of part of a cubic graph $y = f(x)$ with stationary points $(0, a)$ and (b, c). Sketch the graph of $y = f'(x)$.

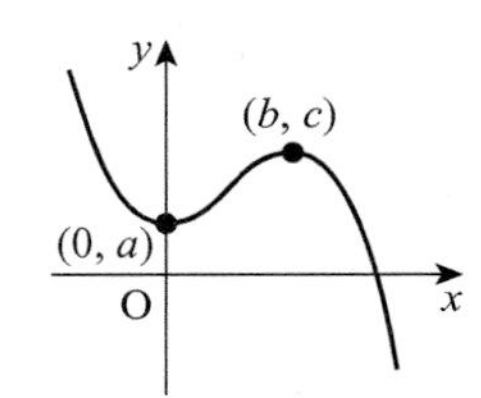

6. a) Show that $(\sin A + \cos B)^2 + (\cos A + \sin B)^2 = 2 + 2\sin(A + B)$

   b) Hence, if $(\sin A + \cos B)^2 + (\cos A + \sin B)^2 = 3$, find two possible values for angle $(A + B)$ between 0 and 2π.

7. Part of the graphs of $y = 2\sqrt{x+1}$ and $2y - 3x = 6$ are shown in the diagram. A tangent to the curve is drawn parallel to the given straight line.

   Find the x-coordinate of the point of contact of this tangent to the curve.

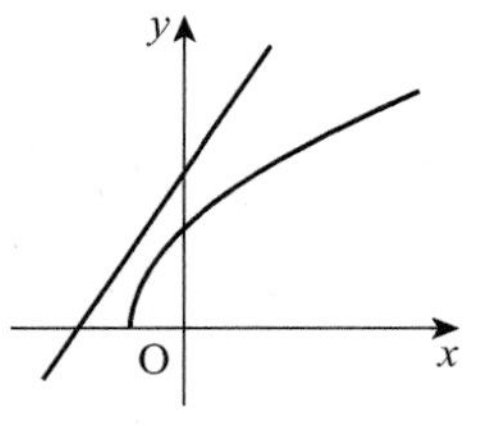

Calculator allowed

1. $f(x) = \cos x° - 5\sin x°$
   a) Express $f(x)$ in the form $k\cos(x + \alpha)°$ where $k > 0$ and $0 \leq \alpha \leq 360$.
   b) Hence, solve $f(x) = 1$ for $0 \leq x < 360$.
   c) The graph $y = f(x)$ cuts the x-axis at the point $(a, 0)$ where $180 < a < 270$. Find the value of a.
2. a) Show that $\cos^2 x° - \cos 2x° = 1 - \cos^2 x°$.
   b) Hence, solve the equation $3\cos^2 x° - 3\cos 2x° = 8\cos x°$ in the interval $0 \leq x < 90$.
3. The roots of the equation $(kx + 2)(x + 3) = 8$ are equal. Find the values of k.
4. A preliminary partial sketch of the curve with equation $y = 3 + 2x^2 - x^4$ is shown in the diagram.

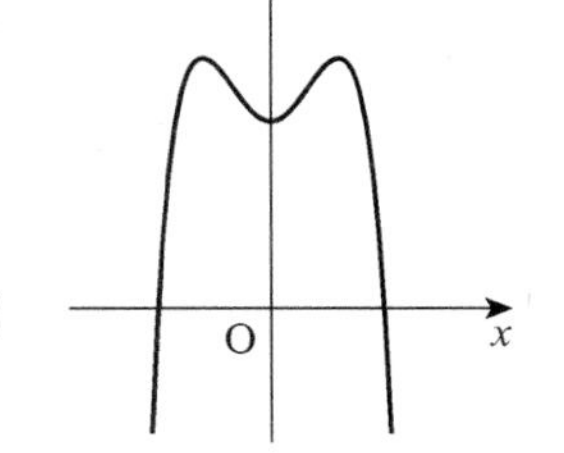

   a) Find the coordinates of the stationary points on the curve.
   b) Confirm the information in the sketch by determining the nature of the stationary points.
5. The function f is defined by $f(\theta) = 3\cos\theta° - \sin\theta°$.
   a) Show that $f(\theta)$ can be expressed in the form $f(\theta) = k\cos(\theta + \alpha)°$ where $k > 0$ and $0 \leq \alpha \leq 360$ and determine the values of k and α.
   b) Hence, find the maximum and minimum values of $f(\theta)$ and the corresponding values of θ, where θ lies in the interval $0 \leq \theta < 360$.
   c) Write down the minimum value of $\sqrt{10}(3\cos\theta° - \sin\theta°) + 10$.
6. The function $f(x) = x^4 + 4x^3 + 5x^2 + 14x + 24$

$$= (x + a)(x + b)(x^2 - x + 4)$$

   a) Find the values of a and b.
   b) Hence, show that the equation $f(x) = 0$ has only two real solutions.
7. Given that $5 \times 3^{\alpha} = 2$ where $\alpha = \cos^2 x - \sin^2 x$, calculate the smallest possible positive value of x.

Chapter 3

Gradient and the straight line revisited

What is gradient?

Gradient is a **number** that measures the slope of a line. Divide the **vertical distance** by the **horizontal distance**.

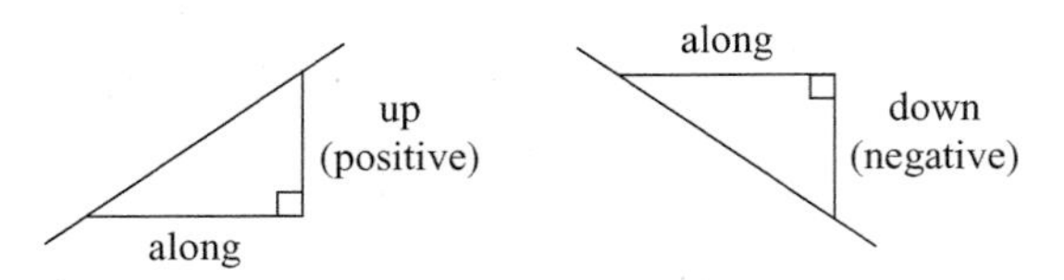

$$\textbf{gradient} = \frac{\textbf{distance up or down}}{\textbf{distance along}}$$

Here is some notation to do with gradient:

m	unknown value of a gradient
m_{AB}	the gradient of the line AB
$m_{\perp}$	the gradient of a perpendicular line
m_1, m_2	the values of two different gradients

The gradient formula

If you know the two pairs of coordinates for two points A and B, here is how to find m_{AB}:

$B(x_2, y_2)$

$A(x_1, y_1)$

$$m_{AB} = \frac{y_2 - y_1}{x_2 - x_1}$$

y-coordinate difference

x-coordinate difference

Notes:

1. $x_1 \neq x_2$. If $x_1 = x_2$ you get a 'vertical' line which has **no** gradient (gradient is undefined).
2. $\frac{y_1 - y_2}{x_1 - x_2}$ gives the same result.

you can swap **both** top and bottom order but not just one.

Example Find the gradient of AB where A and B have coordinates $(-1, 1)$ and $(5, -2)$.

Solution $m_{AB} = \frac{1-(-2)}{-1-5} = \frac{3}{-6} = -\frac{1}{2}$ Note that $\frac{-2-1}{5-(-1)} = \frac{-3}{6} = -\frac{1}{2}$ gives the same result.

The equation of a straight line I

Here is a typical equation of a straight line:

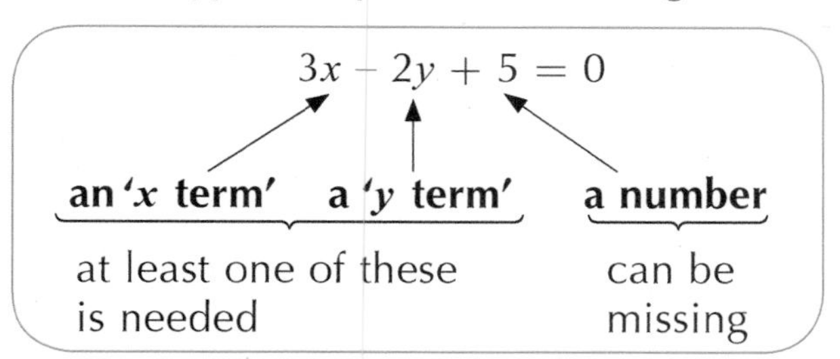

Unless the line is parallel to the y-axis it is always possible to rearrange the equation into the form:

$$y = mx + c$$

gradient y-intercept is $(0, c)$

Example

Find the gradient and y-intercept of the line with equation $3x - 2y + 5 = 0$

Solution

Rearrange the equation:

$$2y = 3x + 5$$

$$\Rightarrow \quad y = \tfrac{3}{2}x + \tfrac{5}{2}$$

So the gradient is $\frac{3}{2}$

and the y-intercept is $(0, \frac{5}{2})$

Special cases

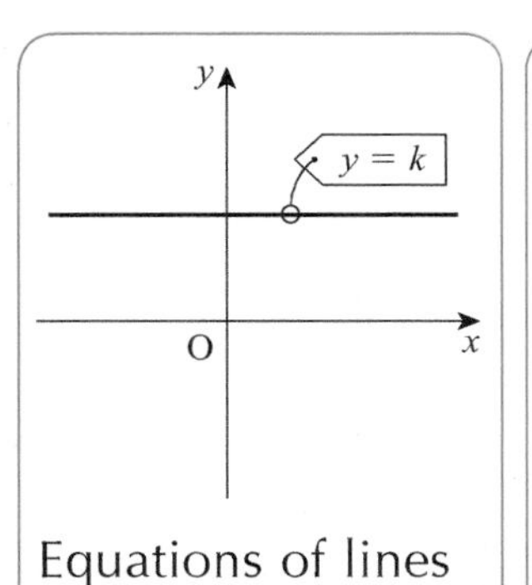

Equations of lines parallel to x-axis are of the form **y = 'a number'**

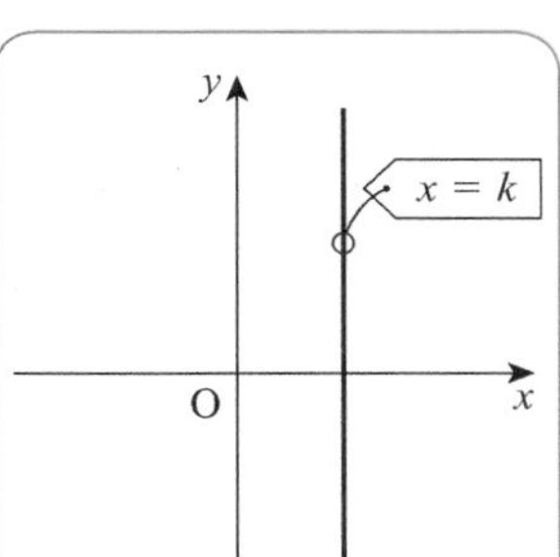

Equations of lines parallel to y-axis are of the form **x = 'a number'**

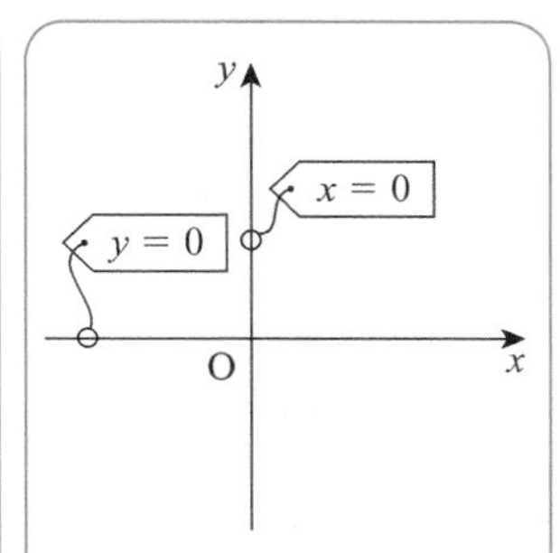

Equation of the x-axis is $\boldsymbol{y = 0}$
Equation of the y-axis is $\boldsymbol{x = 0}$

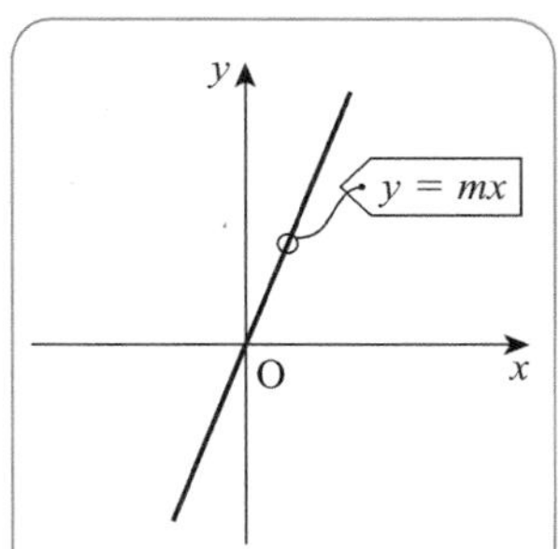

All lines passing through the origin (apart from the y-axis) have equations of the form $\boldsymbol{y = mx}$

Quick Test 28

1. Find the gradient m_{PQ} where $P(-1, 4)$ and $Q(-4, 3)$.

2. a) Find the gradient and y-intercept of the line with equation $3x - 2y = 4$.
   b) Find the equation of the straight line with the same y-axis intercept but with gradient $\frac{2}{3}$.

3. Find the equation of the line through $(-2, -1)$ and $(-2, 7)$.

Chapter 3

Working with the gradient

Gradient and angles

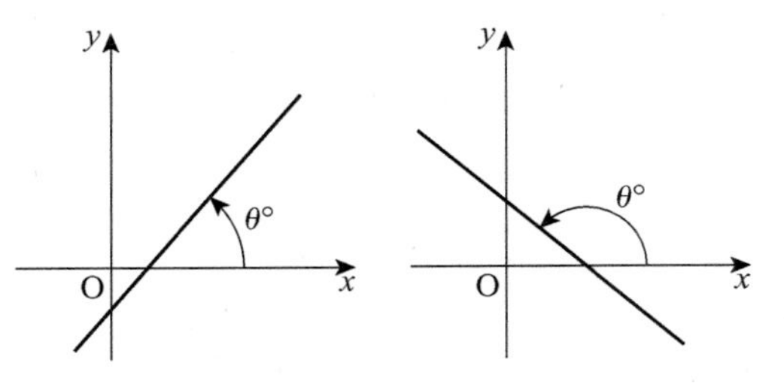

tan $\theta°$ is positive for $0 < \theta < 90$

tan $\theta°$ is negative for $90 < \theta < 180$

Assuming the scales are the same on both axes then...

tan θ = the gradient of the line (m)

Example

Find the angle that a line with gradient $-\frac{1}{3}$ makes with the positive direction of the x-axis.

Solution

Suppose the angle is $\theta°$, then...

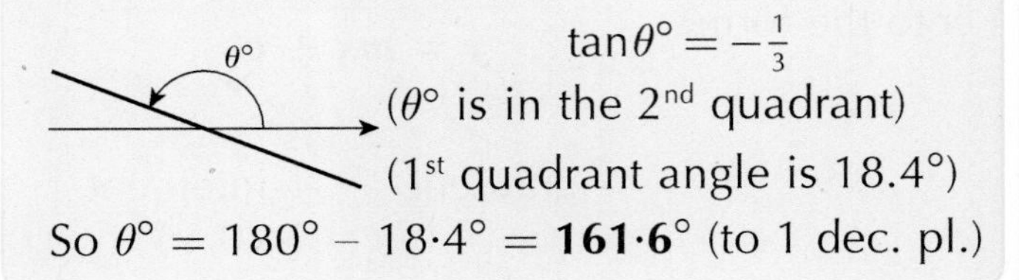

$\tan\theta° = -\frac{1}{3}$

($\theta°$ is in the 2nd quadrant)

(1st quadrant angle is 18.4°)

So $\theta° = 180° - 18{\cdot}4° =$ **161·6°** (to 1 dec. pl.)

Parallel lines and collinear points

$m_1 = m_2$

If one is true ↕ then so is the other

The lines are parallel

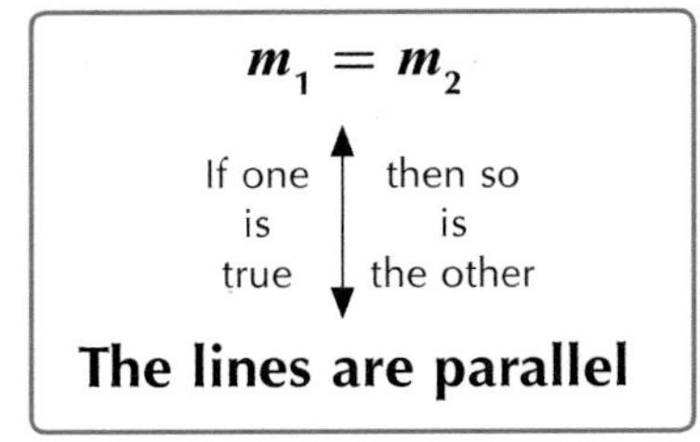

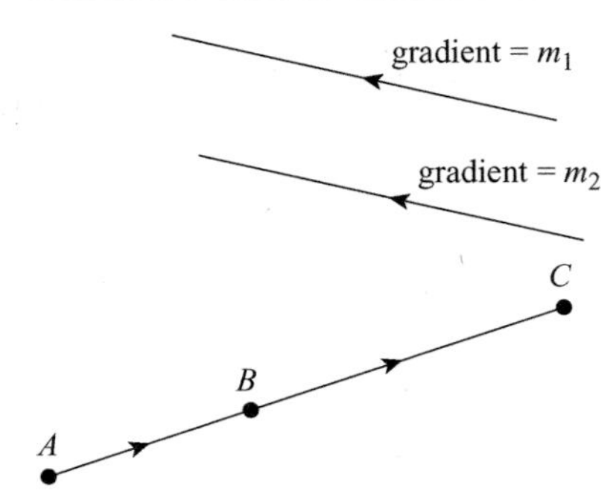

Example

Show that the points $A(-1, -5)$, $B(1, -4)$ and $C(7, -1)$ are collinear.

Solution

$$m_{AB} = \frac{-4-(-5)}{1-(-1)} = \frac{1}{2} \text{ and}$$

$$m_{BC} = \frac{-1-(-4)}{7-1} = \frac{1}{2}$$

So AB and BC are parallel.

Since they share a common point B, they are collinear.

Perpendicular lines

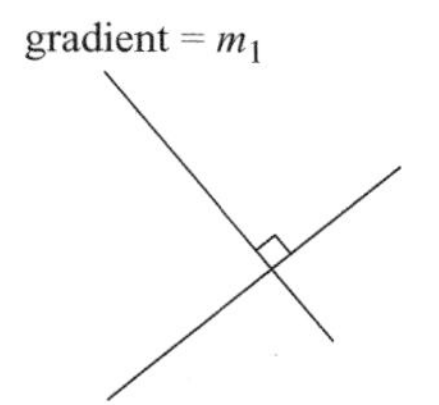

gradient = m_2
(m_1, m_2 are both non-zero)

$$m_1 \times m_2 = -1$$

If one is true ↕ then so is the other

The lines are perpendicular

$$m = \frac{a}{b} \Rightarrow m_\perp = -\frac{b}{a} \quad (a \neq 0)$$

$$m = a \Rightarrow m_\perp = -\frac{1}{a} \quad (a \neq 0)$$

Remember: $m_\perp$ means the gradient of a perpendicular line.

TOP TIP

To find a perpendicular gradient you: 'change the sign and invert'
$m = \frac{2}{3}$ becomes $m_\perp = -\frac{3}{2}$

Example

A triangle ABC has vertices $A(-3, -1)$, $B(-1, 2)$ and $C(5, -2)$. Show that it is right-angled.

Solution

$$m_{AB} = \frac{-1-2}{-3-(-1)} = \frac{-3}{-2} = \frac{3}{2}$$

$$m_{BC} = \frac{2-(-2)}{-1-5} = \frac{4}{-6} = -\frac{2}{3}$$

Since $m_{AB} \times m_{BC} = \frac{3}{2} \times \left(-\frac{2}{3}\right) = -1$

then $AB \perp BC$ (AB is perpendicular to BC) and so ΔABC is right-angled at B.

The equation of a straight line II

If you know:

Fact 1 the gradient: m

Fact 2 a point on the line: (a, b)

then the equation of the straight line is:

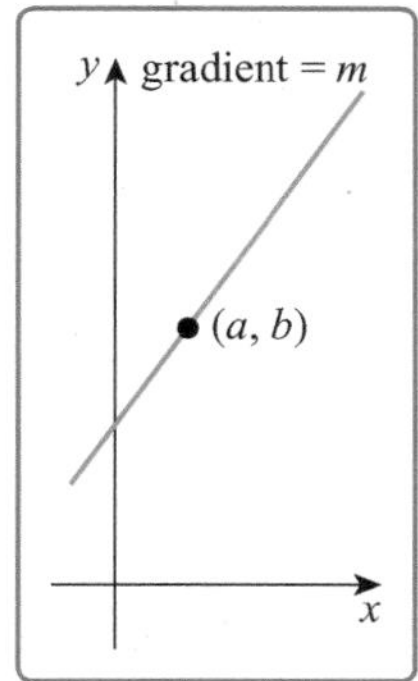

$$y - b = m(x - a)$$

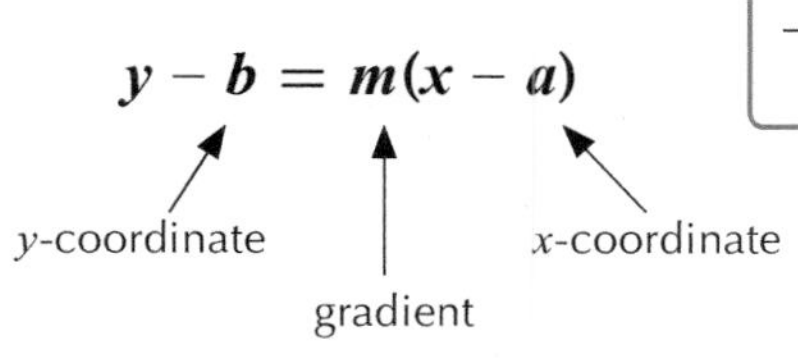

Example

Find the equation of the line passing through (2, –3) with gradient $\frac{1}{2}$.

Solution

Use $y - b = m(x - a)$

Fact 1 gradient: $\frac{1}{2}$

Fact 2 a point on line: (2, –3)

So equation is $y - (-3) = \frac{1}{2}(x - 2)$

$$y + 3 = \frac{1}{2}(x - 2)$$

$$2y + 6 = x - 2$$

$$\mathbf{2y - x = -8}$$

(or $2y = x - 8$)

Get rid of this fraction by doubling

Example

Find the equation of the line passing through $A(-1, 3)$ and $B(4, -2)$.

Solution

Find the gradient:

$$m_{AB} = \frac{3-(-2)}{-1-4} = \frac{5}{-5} = -1$$

Use $y - b = m(x - a)$

Fact 1 gradient: -1

Fact 2 a point on line: $(-1, 3)$

So equation is $y - 3 = -1(x - (-1))$

$\Rightarrow \quad y - 3 = -(x + 1)$

$\Rightarrow \quad y - 3 = -x - 1$

giving $\mathbf{y + x = 2}$

Points of intersection

To find where a line crosses the axes:

to find:		
to find:	**x-axis intercept** ⟷	**Set $y = 0$ in the equation**
to find:	**y-axis intercept** ⟷	**Set $x = 0$ in the equation**

To find where two lines intersect:

equation of 1st line → **solve these simultaneously**
equation of 2nd line →

Example

Find the point of intersection of the lines $3y = 2x + 4$ and $3x = 7 - 2y$

Solution

After rearranging equations

$$\left.\begin{aligned} 3y - 2x &= 4 \\ 2y + 3x &= 7 \end{aligned}\right\} \begin{aligned} &\times 3 \rightarrow 9y - 6x = 12 \\ &\times 2 \rightarrow 4y + 6x = 14 \end{aligned}$$

Add: $13y = 26$

$y = 2$

Put $y = 2$ in $2y + 3x = 7 \Rightarrow 4 + 3x = 7$

So $3x = 3$ giving $x = 1$

(1, 2) is the point of intersection.

TOP TIP

To prove perpendicularity: Use m_1 and m_2 for the two gradients and then calculate $m_1 \times m_2$ showing it equals -1.

Quick Test 29

1. For the line AB where $A(-3, -5)$ and $B(5, -1)$ find

a) the gradient b) the equation

c) the angle it makes with the positive direction of the x-axis

2. Show that the lines $y + 2x + 1 = 0$ and $2y - x - 3 = 0$ are perpendicular and find their point of intersection.

Problem solving using the gradient

The distance formula

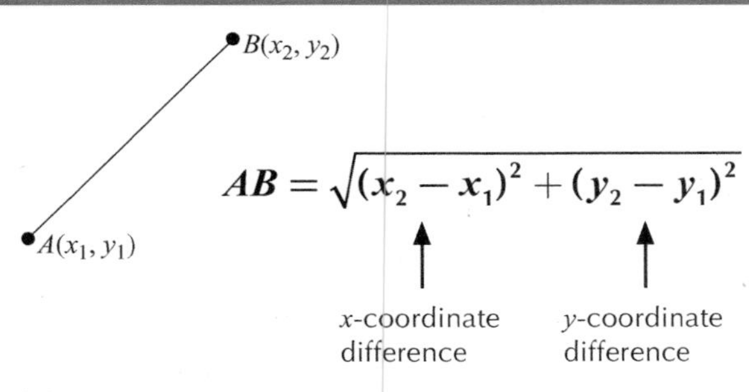

$$AB = \sqrt{(x_2 - x_1)^2 + (y_2 - y_1)^2}$$

Note:

$AB^2 = (x_2 - x_1)^2 + (y_2 - y_1)^2$

if you wish to avoid the square root.

Example

Show that triangle ABC with vertices $A(1, 2)$, $B(3, 0)$ and $C(-1, -2)$ is isosceles.

Solution

$$AC = \sqrt{(1-(-1))^2 + (2-(-2))^2} = \sqrt{2^2 + 4^2}$$

$$= \sqrt{4 + 16} = \sqrt{20}$$

$$BC = \sqrt{(3-(-1))^2 + (0-(-2))^2} = \sqrt{4^2 + 2^2}$$

$$= \sqrt{16 + 4} = \sqrt{20}$$

So $AC = BC$ and the triangle is isosceles.

The midpoint formula

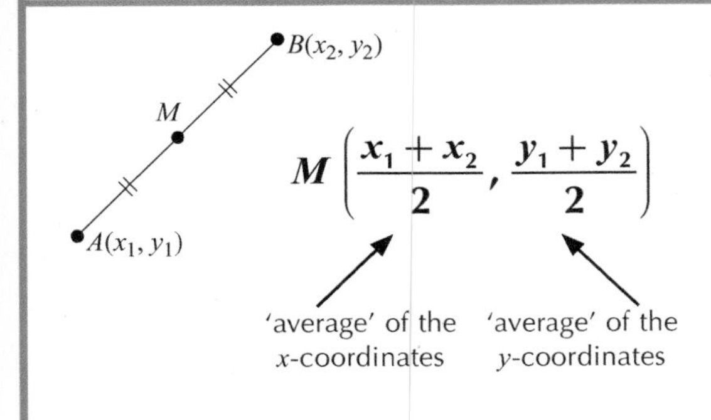

$$M\left(\frac{x_1 + x_2}{2}, \frac{y_1 + y_2}{2}\right)$$

Example

Find the coordinates of M, the midpoint of CD, where C is the point $(-1, 5)$ and D is $(-5, 2)$.

Solution

$$M\left(\frac{-1+(-5)}{2}, \frac{5+2}{2}\right)$$

$$= M\left(\frac{-6}{2}, \frac{7}{2}\right) = M\left(-3, \frac{7}{2}\right)$$

Perpendicular bisectors

The **perpendicular bisector** of line AB is a line with these two properties:

Property 1 It is perpendicular to AB

Property 2 It passes through the midpoint of AB

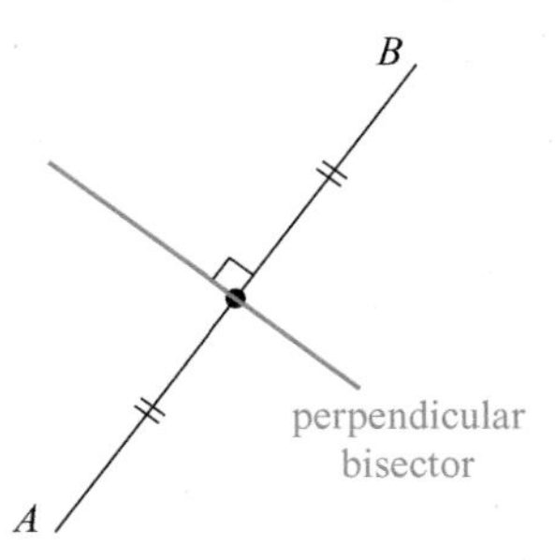

Example

Find the equation of the perpendicular bisector of the line joining $P(2, 3)$ and $Q(10, 1)$.

Solution

The midpoint of PQ is $M\left(\frac{2+10}{2}, \frac{3+1}{2}\right) = M(6, 2)$

also $m_{PQ} = \frac{1-3}{10-2} = \frac{-2}{8} = -\frac{1}{4}$

$\Rightarrow m_{\perp} = 4$ (perpendicular gradient)

Using '$y - b = m(x - a)$' the equation is:

$y - 2 = 4(x - 6)$

$\Rightarrow y - 2 = 4x - 24$

giving $y - 4x = -22$

Special lines in triangles

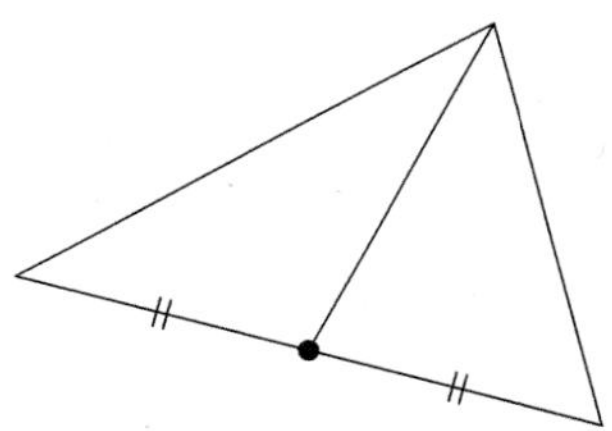

A **median** joins a vertex to the midpoint of the opposite side.

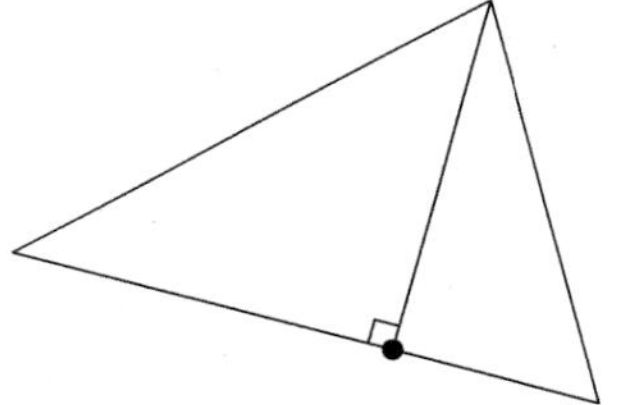

An **altitude** is a line through a vertex perpendicular to the opposite side.

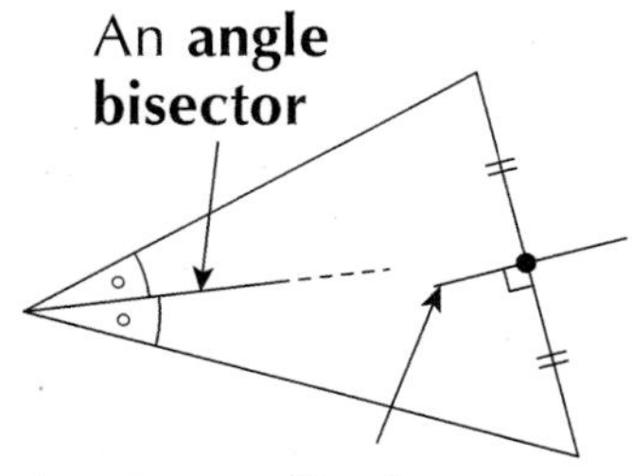

A **perpendicular bisector** of a side.

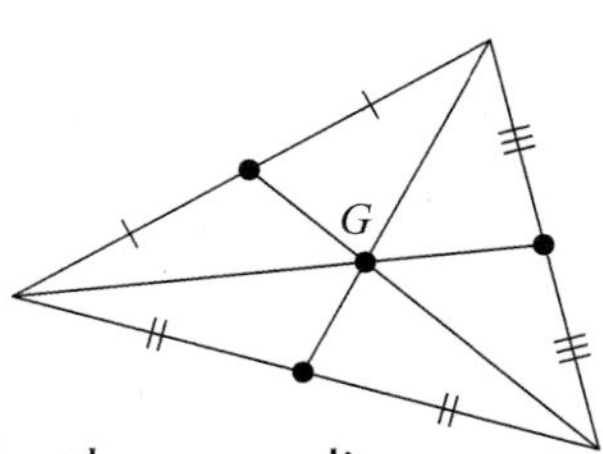

The three medians are **concurrent**. They meet at G, the **centroid** of the triangle.

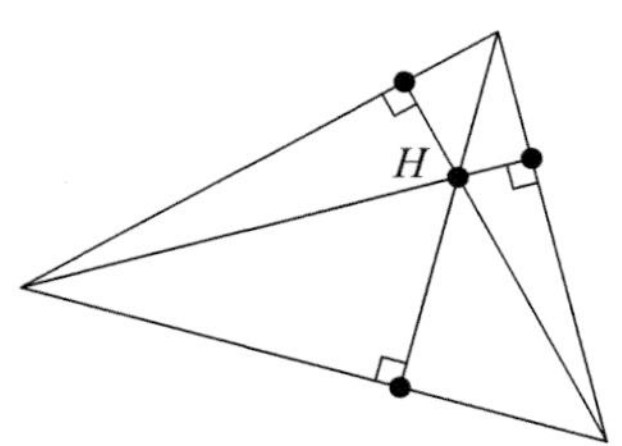

The three altitudes are also concurrent (meeting at H, the **orthocentre**).

Notes:
1. The centroid G divides each median in the ratio 2 : 1 (vertex to midpoint).
2. 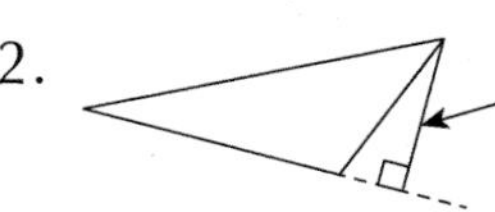 Altitudes can lie 'outside' the triangle.
3. The three angle bisectors are concurrent as are the three perpendicular bisectors of the sides.

TOP TIP

Lines that are **concurrent** meet in one point called the point of concurrency.

Example

A triangle has vertices $P(-2, 3)$, $Q(6, -1)$ and $R(-4, -5)$. Find the coordinates of G, the point of intersection of the medians PS and RT.

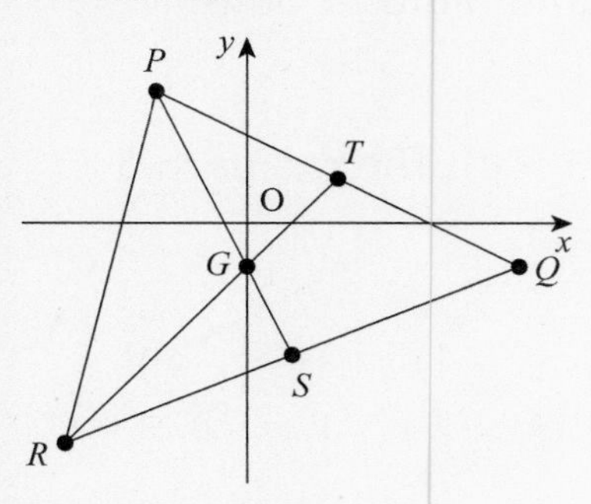

Solution

$S\left(\frac{-4+6}{2}, \frac{-5+(-1)}{2}\right) = S(1, -3)$

also $m_{PS} = \left(\frac{-3-3}{1-(-2)}\right) = \frac{-6}{3} = -2$

Equation of PS: $y - (-3) = -2(x - 1)$

$\Rightarrow y + 3 = -2x + 2 \Rightarrow y + 2x = -1$

Similar calculations lead to: $T(2,1)$, $m_{RT} = 1$

Equation of RT: $y - x = -1$

Solve: $\begin{cases} y + 2x = -1 \\ y - x = -1 \end{cases}$

subtract: $3x = 0 \Rightarrow x = 0$

Put $x = 0$ in $y - x = -1 \Rightarrow y = -1$

So $G(0, -1)$

TOP TIP

Always work with axes diagrams to make sure your answers make sense, e.g. that gradients look right and points of intersection are where they are supposed to be!

Quick Test 30

A triangle has vertices $A(-1, 2)$, $B(3, 4)$ and $C(3, -2)$.

1. Find the equation of the perpendicular bisector of
   a) side AB b) side AC.
2. Find the point of intersection G of these two perpendicular bisectors.
3. Show that in triangle ABC all three perpendicular bisectors are concurrent.

The circle equation

Centre (a, b) radius r

The equation of a circle centre (a, b) with radius r is

$$(x - a)^2 + (y - b)^2 = r^2$$

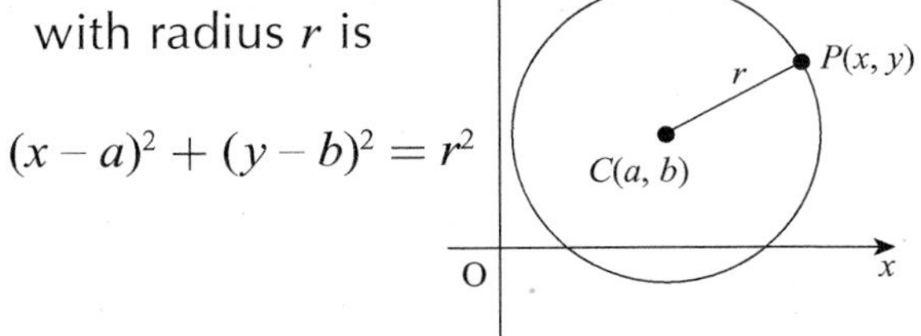

Note: For a circle centre the origin $a = 0$ and $b = 0$ the equation reduces to: $x^2 + y^2 = r^2$

Example

Find the equation of the circle with centre (1, −4) and radius 5 units.

Solution

The centre is (1, −4). The radius = 5

The equation is $(x - 1)^2 + (y - (-4))^2 = 5^2$

$\Rightarrow (x - 1)^2 + (y + 4)^2 = 25$

TOP TIP

Make sure your equation starts: x^2+y^2 ...
Or this process will not work!

The general equation

In general any equation of a circle can be rearranged to look like:

$$x^2 + y^2 + ax + by + c = 0$$

The centre is $\left(-\frac{a}{2}, -\frac{b}{2}\right)$

The radius $= \sqrt{\left(-\frac{a}{2}\right)^2 + \left(-\frac{b}{2}\right)^2 - c}$

Note: The number under the square root sign has to be positive otherwise the equation does not represent a circle.

The process you follow to find the centre and the radius from the general equation is:

- Identify the x and y coefficients a and b and the constant term c.
- Change the sign and halve the values of a and b to get the coordinates of the centre.
- Square each coordinate of the centre and add then subtract the constant term c. Now take the square root of the result.

Example

For what range of values of k does $x^2 + y^2 - 2x + 6y + k = 0$ represent a circle?

Solution

Centre is (1, –3)

Radius $= \sqrt{1^2 + (-3)^2 - k} = \sqrt{10 - k}$

So $10 - k > 0$, i.e. $10 > k$ or $\mathbf{k < 10}$ (or there's no circle!)

Example

Find the centre and radius of the circle $x^2 + y^2 - 2x + 3y - 3 = 0$

Solution

$$x^2 + y^2 - 2x + 3y - 3 = 0$$

Centre is $\left(1, -\frac{3}{2}\right)$

Radius $= \sqrt{1^2 + (-\frac{3}{2})^2 - (-3)}$

$\sqrt{1 + \frac{9}{4} + 3} = \sqrt{\frac{25}{4}} = \frac{5}{2}$

Completing squares

It is possible to 'complete squares' to find the radius and centre from the general equation:

$$x^2 + y^2 - 4x + 5y + 1 = 0$$

$$x^2 - 4x + y^2 + 5y = -1$$

$$(x - 2)^2 + (y + \tfrac{5}{2})^2 = -1 + 4 + \tfrac{25}{4} = \tfrac{37}{4}$$

$x^2 - 4x + 4$ $\quad$ $y^2 + 5y + \frac{25}{4}$

The centre is $\left(2, -\frac{5}{2}\right)$ and the radius $= \sqrt{\frac{37}{4}} = \frac{\sqrt{37}}{2}$

Rearrange to get x terms and y terms together.

Complete the squares. You will have introduced extra terms (in purple) on the left and need to add them on the right to balance the equation.

Now compare with:

$$(x - a)^2 + (y - b)^2 = r^2$$

Centre: (a, b) $\qquad$ Radius $= r$

Quick Test 31

1. Find the equation of the circle with centre (–3, 7) and radius $\sqrt{3}$
2. Find the centre and the radius of each circle:

   a) $\left(x - \frac{1}{2}\right)^2 + \left(y + \frac{5}{2}\right)^2 = 5$ $\qquad$ b) $x^2 + y^2 - x - y - \frac{1}{2} = 0$
3. A circle with radius 2 units touches the x-axis and touches the y-axis. Give the four possible equations for the circle.
4. Show that the circle $x^2 + y^2 - 4x + 6y + 9 = 0$ touches the y-axis but does not intersect the x-axis.

Lines and circles

How many points of intersection?

To find points of intersection of a line and a circle, you solve their equations simultaneously. This will result in a quadratic equation. Three distinct situations can arise…

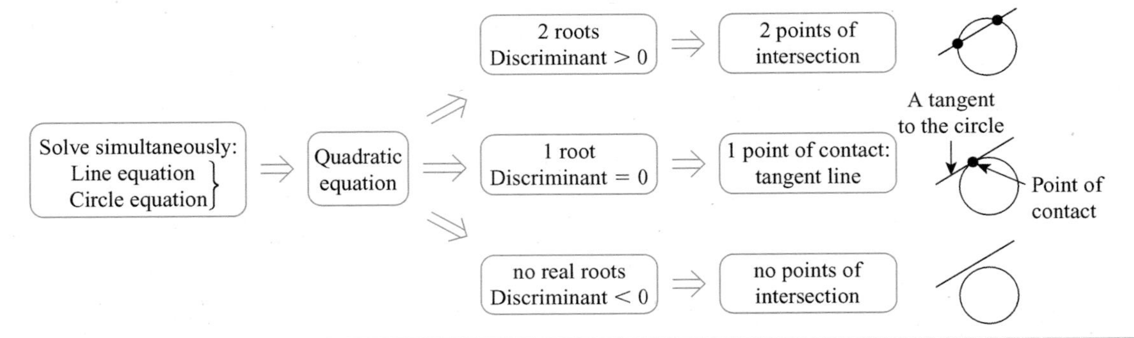

Showing a line is a tangent

You need to show that when you solve the line equation and the circle equation simultaneously, you only get one solution. You then can use this one value of x to find the point of contact.

Example

Show that $y = 2x - 10$ is a tangent to the circle $x^2 + y^2 - 4x + 2y = 0$ and find the point of contact.

Solution

For the points of intersection, solve:

$$\left.\begin{array}{l} y = 2x - 10 \\ x^2 + y^2 - 4x + 2y = 0 \end{array}\right\}$$

Substitute $y = 2x - 10$ in the circle equation.

This gives… $x^2 + (2x - 10)^2 - 4x + 2(2x - 10) = 0$

$x^2 + 4x^2 - 40x + 100 - 4x + 4x - 20 = 0$

$5x^2 - 40x + 80 = 0 \Rightarrow 5(x^2 - 8x + 16) = 0 \Rightarrow 5(x - 4)(x - 4) = 0$

Since there is only one solution, $x = 4$, the line $y = 2x - 10$ is a tangent to the circle. Now let $x = 4$ in $y = 2x - 10 \Rightarrow y = 2 \times 4 - 10 = -2$. The point of contact is $(4, -2)$.

Tangent at a given point

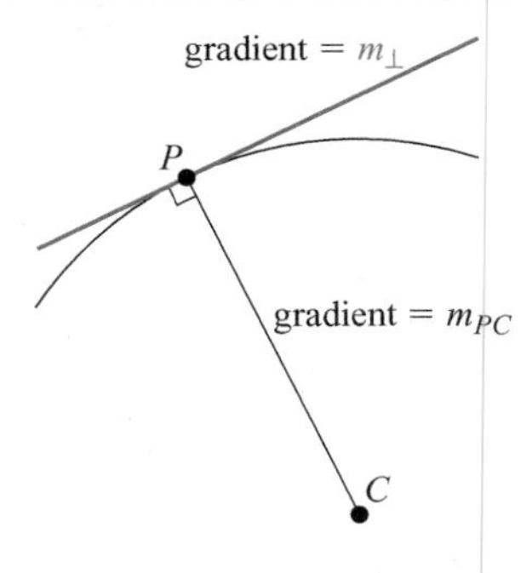

The tangent at P is perpendicular to the radius CP from the centre C to the point of contact P.

So $m_\perp \times m_{pc} = -1$

Example

Find the equation of the tangent at the point $P(-4, 4)$ on the circle $(x + 2)^2 + y^2 = 20$.

Solution

The centre is $C(-2, 0)$ so $m_{cp} = \frac{4-0}{-4-(-2)} = \frac{4}{-2} = -2$

$\Rightarrow m_\perp = \frac{1}{2}.$ This is the gradient of the tangent.

A point on the tangent is $(-4, 4)$.

The equation is $y - 4 = \frac{1}{2}(x - (-4))$

$\Rightarrow 2y - 8 = x + 4 \Rightarrow 2y - x = 12$

Using the discriminant to find a tangent

A tangent has **one** point of intersection with a circle. By imposing a 'one root only' condition, in some problems you can find the equation of an unknown tangent. This usually involves setting the discriminant of a quadratic equation to zero.

Example

$y = mx$ is a tangent to the circle
$(x + 2)^2 + y^2 = 3$
Find the possible values of m

Solution

The points of intersection can be found by solving:

$$\left.\begin{array}{l} y = mx \\ (x + 2)^2 + y^2 = 3 \end{array}\right\} \Rightarrow \begin{array}{l} (x+2)^2 + (mx)^2 = 3 \\ x^2 + 4x + 4 + m^2x^2 = 3 \\ (1 + m^2)x^2 + 4x + 1 = 0 \end{array}$$

This is a quadratic equation in x...
Discriminant $= 4^2 - 4(1 + m^2) \times 1$
('$b^2 - 4ac$') $= 16 - 4 - 4m^2$
$= 12 - 4m^2$

For the line to be a tangent, the quadratic equation must have 1 root (or equal roots) so...
Discriminant = 0 giving $12 - 4m^2 = 0$

$\Rightarrow m^2 = 3 \Rightarrow m = \sqrt{3}$ or $-\sqrt{3}$

Check with a diagram. The circle has centre $(-2, 0)$ and radius = $\sqrt{3}$
The two possible tangents are shown.

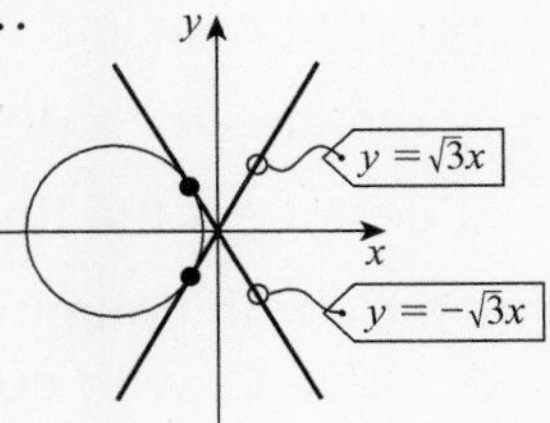

Diametrically opposite points

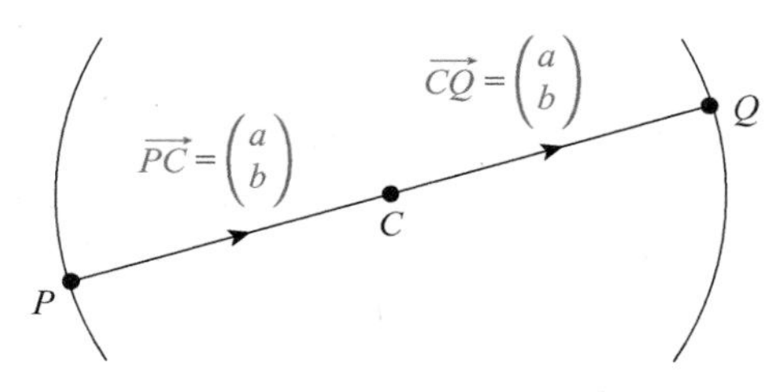

PQ is a diameter. C is the centre.

Q is **diametrically opposite** P.

This means: $\overrightarrow{PC} = \overrightarrow{CQ}$

If you know two of the points then you can find the third point:

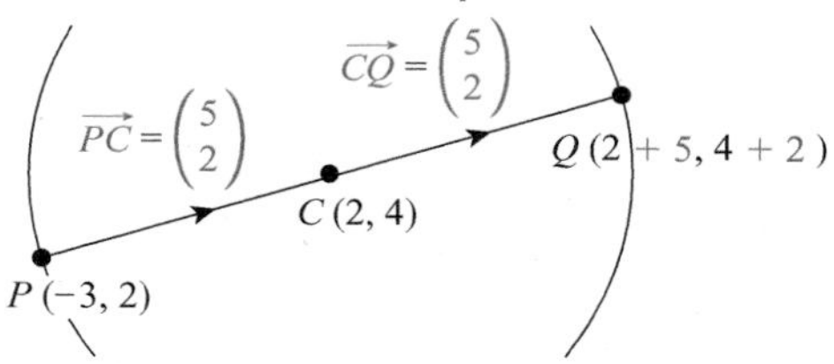

From C to Q you move 5 right and 2 up the same as from P to C. So Q is the point (7, 6).

TOP TIP

It is sensible to sketch the circles and lines before trying to solve the problem. It helps with your strategy!

Quick Test 32

1. Show that the line $y - x = 5$ is a tangent to the circle $x^2 + y^2 + 6x + 4y + 5 = 0$ and find the coordinates of the point of contact.
2. Find the equation of the tangent to the circle $x^2 + y^2 - 16x + 35 = 0$ at the point (3, –4).
3. Find the possible values of k if the line $y = x - k$ is a tangent to the circle $(x - 1)^2 + y^2 = 2$.
4. $A(-1, 0)$ lies on circle C_1 with equation $(x - 2)^2 + (y - 1)^2 = 10$. B is diametrically opposite A. Find the equation of circle C_2 with centre B and with the same radius as circle C_1.

Problem solving with circles

Are the circles touching?

D : distance between the two centres
R : radius of large circle
r : radius of small circle

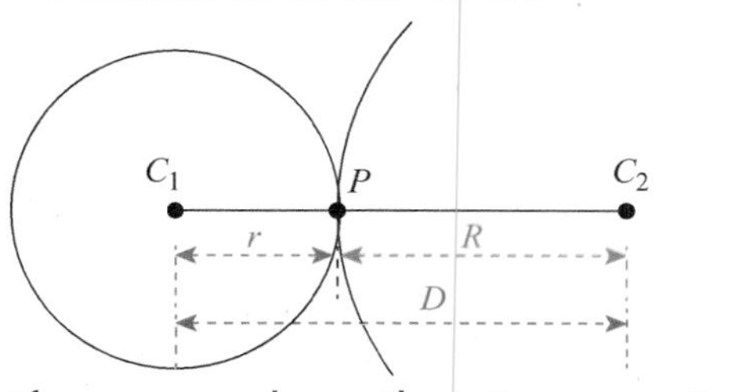

If you can show that $D = r + R$ then you have shown the two circles touch externally.

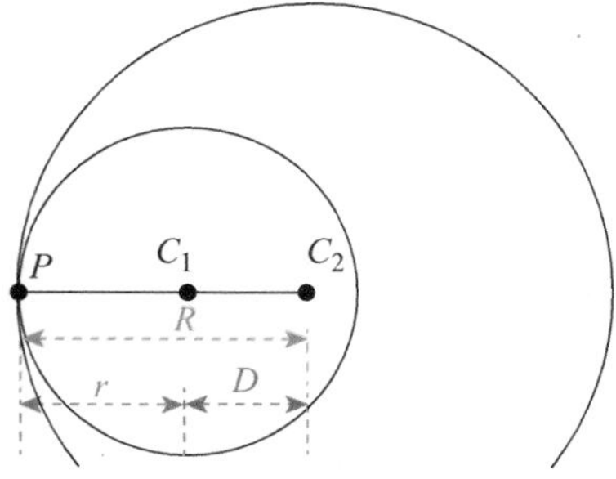

If you can show that $D = R - r$ then you have shown the two circles touch internally.

Notes:
- If neither of these two equations is true then the circles do not touch (they could intersect in two points or miss each other).
- The point of contact P can be found as the point which divides C_1C_2 in the ratio $r : R$ in the case of externally touching circles.

Example

Two circles have equations $(x-3)^2 + (y-1)^2 = 5$ and $(x+3)^2 + (y+2)^2 = 20$. Show that the circles touch and find the coordinates of the point of contact.

Solution

Circle 1: Centre is C_1 (3, 1) Radius $r = \sqrt{5}$

Circle 2: Centre is C_2 (−3, −2) Radius $R = \sqrt{20} = \sqrt{4 \times 5} = \sqrt{4} \times \sqrt{5} = 2\sqrt{5}$

$C_1C_2 = \sqrt{(-3-3)^2 + (-2-1)^2} = \sqrt{36+9} = \sqrt{45} = \sqrt{9 \times 5} = \sqrt{9} \times \sqrt{5} = 3\sqrt{5}$

$r + R = \sqrt{5} + 2\sqrt{5} = 3\sqrt{5} = C_1C_2$

So the two circles touch.

The point of contact P divides C_1C_2 in the ratio 1 : 2

$\overrightarrow{PC_2} = 2\overrightarrow{C_1P} \Rightarrow \mathbf{c}_2 - \mathbf{p} = 2(\mathbf{p} - \mathbf{c}_1)$

$\Rightarrow 2\mathbf{p} + \mathbf{p} = 2\mathbf{c}_1 + \mathbf{c}_2 \Rightarrow 3\mathbf{p} = 2\mathbf{c}_1 + \mathbf{c}_2$

So $3\mathbf{p} = 2\begin{pmatrix}3\\1\end{pmatrix} + \begin{pmatrix}-3\\-2\end{pmatrix} = \begin{pmatrix}3\\0\end{pmatrix}$ giving $\mathbf{p} = \frac{1}{3}\begin{pmatrix}3\\0\end{pmatrix} = \begin{pmatrix}1\\0\end{pmatrix}$

Thus $P(1, 0)$

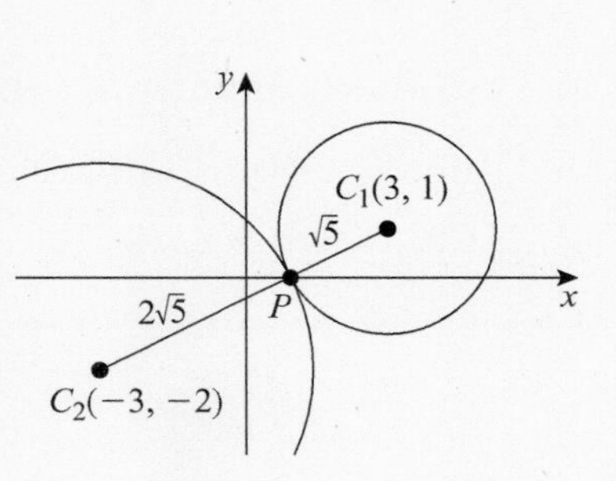

What's the locus?

A **locus** (plural **loci**) is the path traced out by a moving point. Interesting loci are the ones that have restrictions imposed on the way the point moves.

Example

Find the equation of the locus of a point that moves so that it stays at a fixed distance of 3 units from the point (5, –3).

Solution

The point travels in a circle with radius 3 units and centre (5, –3) so its equation is $(x - 5)^2 + (y + 3)^2 = 9$.

Where's the centre?

Suppose you know three points A, B and C on a circle:

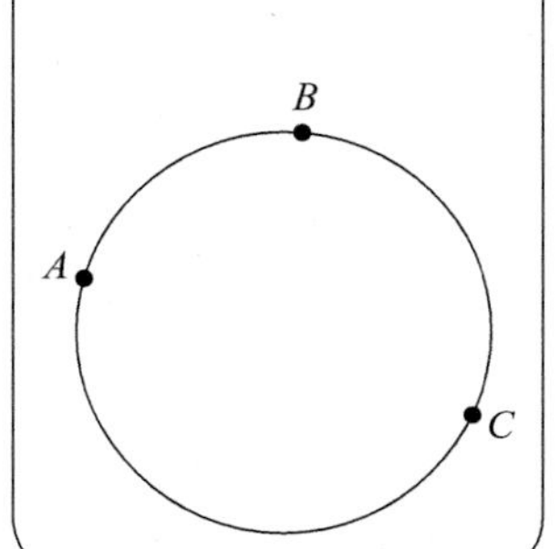

The centre lies somewhere on the perpendicular bisector of chord AB:

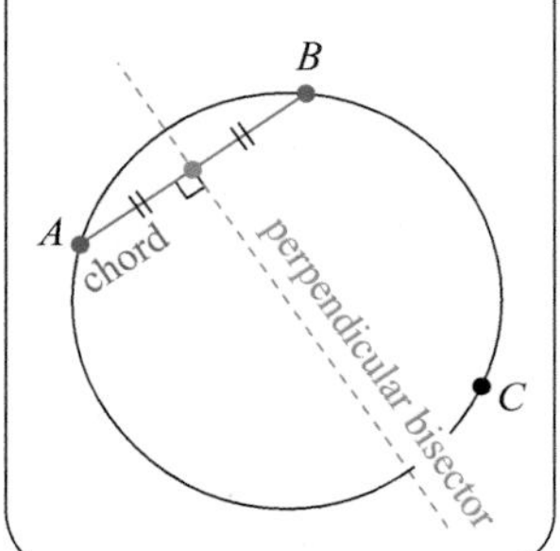

It also lies somewhere on the perpendicular bisector of chord BC:

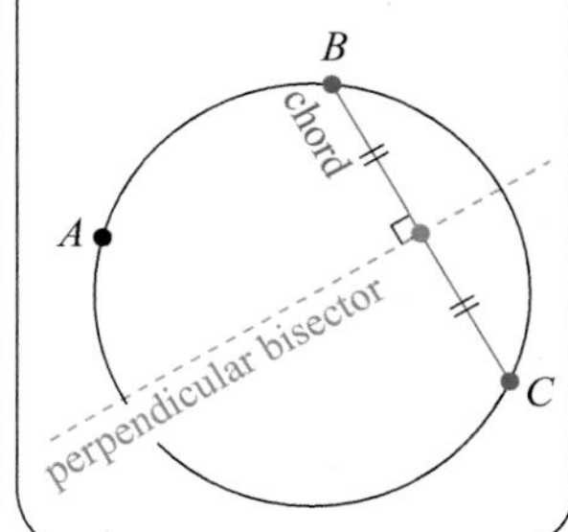

So the centre is at the intersection of the two perpendicular bisectors:

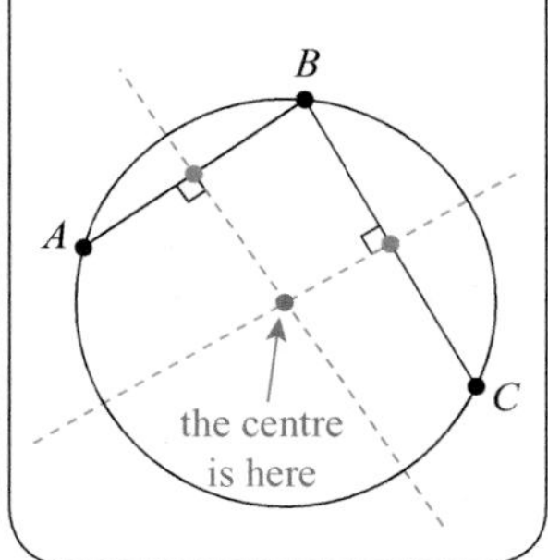

Example

Points $P(-3, 4)$, $Q(5, 4)$ and $R(9, -2)$ lie on a circle with centre C and radius r units.

a) Find the equation of the perpendicular bisector of chord QR.

b) Find the coordinates of the centre C.

c) Write down the value of r and hence find the equation of the circle.

Solution

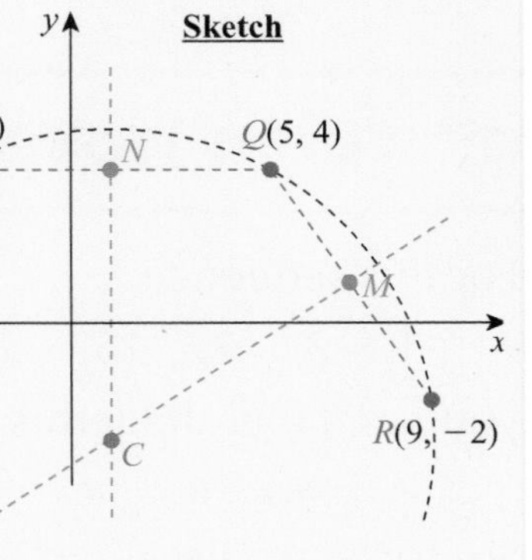

a) The midpoint of QR is $M\left(\frac{5+9}{2}, \frac{4+(-2)}{2}\right) = M(7,1)$

$m_{QR} = \frac{-2-4}{9-5} = \frac{-6}{4} = -\frac{3}{2}$ $m_{\perp} = \frac{2}{3}$ (perpendicular gradient)

So the equation of the perpendicular bisector is:

$y - 1 = \frac{2}{3}(x-7) \Rightarrow 3y - 3 = 2x - 14 \Rightarrow 3y = 2x - 11$

b) PQ is parallel to the x-axis so the midpoint of PQ is $N\left(\frac{-3+5}{2}, 4\right) = N(1,4)$

The perpendicular bisector of PQ is $x = 1$

Solve $\left.\begin{array}{r} x = 1 \\ 3y = 2x - 11 \end{array}\right\} \Rightarrow 3y = 2\times1 - 11 \Rightarrow 3y = -9 \Rightarrow y = -3$

The centre is $C\ (1, -3)$

c) The radius $= CR = \sqrt{(9-1)^2 + (-2-(-3))^2} = \sqrt{8^2 + 1^2} = \sqrt{65}$

With centre $C(1, -3)$ and radius $= \sqrt{65}$ the equation of the circle is $(x-1)^2 + (y+3)^2 = 65$

Note: Check that the coordinates of each point A, B and C satisfy this equation.

TOP TIP

In the exam, always try to check equations of lines or circles that you get by substituting the values of coordinates of points that are supposed to lie on them!

Quick Test 33

1. Two circles have equations $x^2 + y^2 = 2$ and $x^2 + y^2 + 8x - 8y + 14 = 0$. Show that the circles touch and find the coordinates of the point of contact.
2. Find the equation of the locus of a point which stays $\sqrt{3}$ units from the point $(-2, -1)$.
3. Find the centre of the circle that passes through the points $A(-1, -1)$, $B(2, -1)$ and $C(2, 3)$.

Chapter 3

Notation and calculation

n^{th} term notation

For this sequence:

1, 3, 7, 15, 31, ...

you can label the terms:

u_1, u_2, u_3, u_4, u_5, ... u_{n-1}, u_n, u_{n+1}, ...

So, for example, $u_3 = 7$ and $u_5 = 31$

u_{n-1} is the term immediately before u_n and u_{n+1} is the term immediately after u_n in the sequence.

u_n is the label attached to the nth term of the sequence

TOP TIP

Sometimes u_0 is used for the 1st term in a sequence.

n^{th} term formulae

Let's investigate the sequence 1, 3, 7, ... more closely:

u_1, u_2, u_3, u_4, ... u_n

1, 3, 7, 15 ...

$2^1 - 1$, $2^2 - 1$, $2^3 - 1$, $2^4 - 1$ ... $2^n - 1$

So $u_n = 2^n - 1$

This is an example of a nth term formula.

Example

Find the 10th term in the sequence with n^{th} term $u_n = 3n^2 - 2$

Solution

Substitute $n = 10$ in the formula:

$u_{10} = 3 \times 10^2 - 2 = \mathbf{298}$

Recurrence relations

TOP TIP

The same recurrence relation can give different sequences if you use different 1st terms.

Here's another way to get the sequence 1, 3, 7, ...

1 → 3 → 7 → 15 ...
(double add 1 at each step)

So $u_1 \to u_2$ (double add 1) gives $u_2 = 2u_1 + 1$

$u_2 \to u_3$ (double add 1) gives $u_3 = 2u_2 + 1$

Example

Find a recurrence relation and an nth term formula for the sequence 1, 3, 9, 27, ...

Solution

The 'build-up' rule is 'multiply by 3' so the recurrence relation is

$$\boldsymbol{u_{n+1} = 3u_n} \text{ with } u_1 = 1$$

(where u_n is the nth term)

To find the nth term formula, compare the terms with the powers of 3:

and in general

$u_n \quad u_{n+1}$ (double, add 1) gives $u_{n+1} = 2u_n + 1$

This is a **recurrence relation**: it recurs as you build up the sequence.

u_1	u_2	u_3	u_4
1	3	9	27
3^0	3^1	3^2	3^3

one less each time

This gives the formula:

$$u_n = 3^{n-1}$$

Linear recurrence relations

A **linear recurrence** relation is one with the form:

$$u_{n+1} = mu_n + c$$

m is called the **multiplier**

c is the constant term

A sequence generated by this type of recurrence relation is built up in this manner:

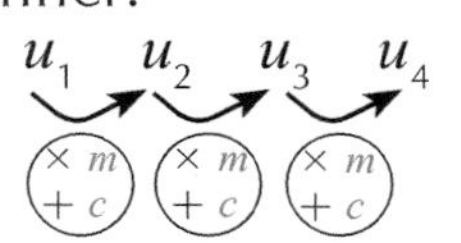

So $u_2 = mu_1 + c$

$u_3 = mu_2 + c$

$u_4 = mu_3 + c$

and so on.

Example

A sequence is defined by $u_{n+1} = au_n + b$ where a and b are constants. The first three terms are $u_1 = 20$, $u_2 = 15$ and $u_3 = 12{\cdot}5$.

Find the recurrence relation and hence calculate u_5

Solution

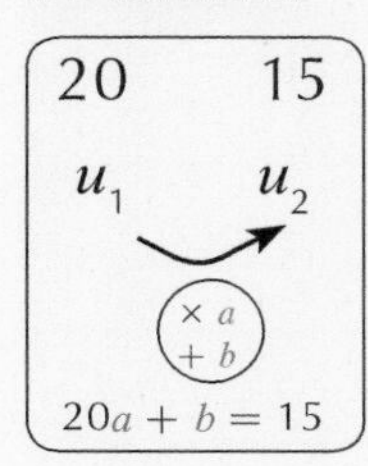

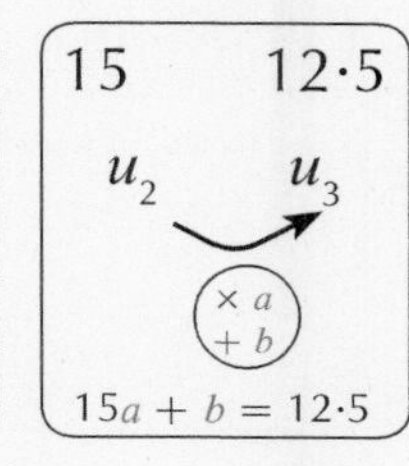

Solve: $20a + b = 15$, $15a + b = 12{\cdot}5$

Subtract: $5a = 2{\cdot}5$

$\Rightarrow a = 0{\cdot}5$

Put $a = 0{\cdot}5$ in $20a + b = 15$

so $20 \times 0{\cdot}5 + b = 15$

$\Rightarrow 10 + b = 15 \Rightarrow b = 5$

The recurrence relation is $u_{n+1} = 0.5u_n + 5$

So $u_4 = 0{\cdot}5u_3 + 5 = 0{\cdot}5 \times 12{\cdot}5 + 5 = 11{\cdot}25$

which gives $u_5 = 0{\cdot}5 \times 11{\cdot}25 + 5 = 10{\cdot}625$

Quick Test 34

1. A recurrence relation is defined by $u_{n+1} = \frac{2}{3}u_n - 1$. If $u_1 = 27$, calculate u_3.
2. Evaluate u_6 if $u_n = 2^{5-n} + \frac{1}{2}$
3. The first three terms of a sequence generated by $u_{n+1} = pu_n + q$ are 2, 5 and 14. Calculate the values of the constants p and q.

Limits and context problems

Conditions for a limit to exist

Sequences generated by the recurrence relation $u_{n+1} = mu_n + c$ behave in remarkably different ways as the sequence continues. The different types of behaviour depend entirely on the value of the multiplier m. These graphs are typical and show the values of u_n as n increases:

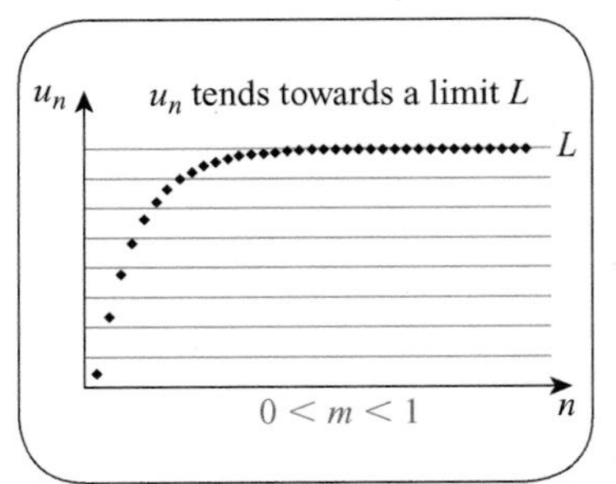

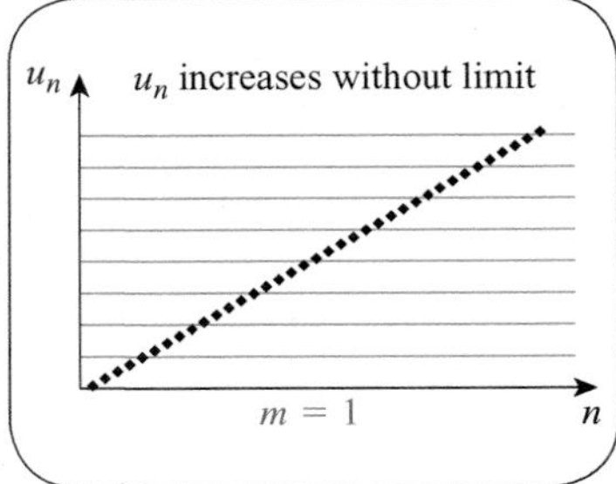

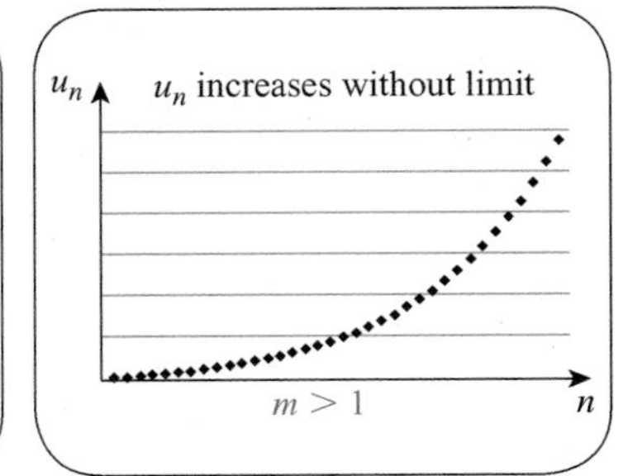

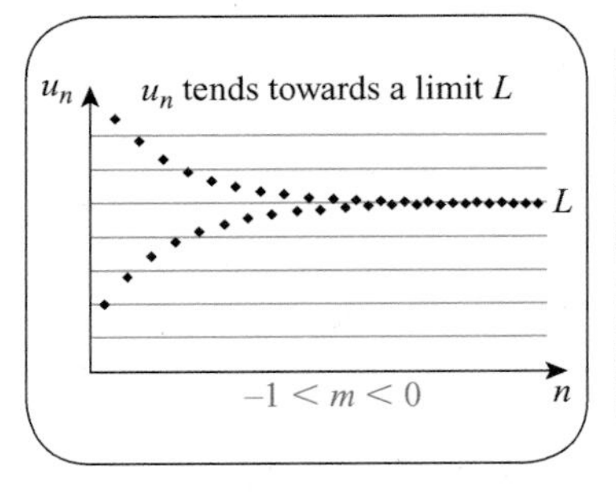

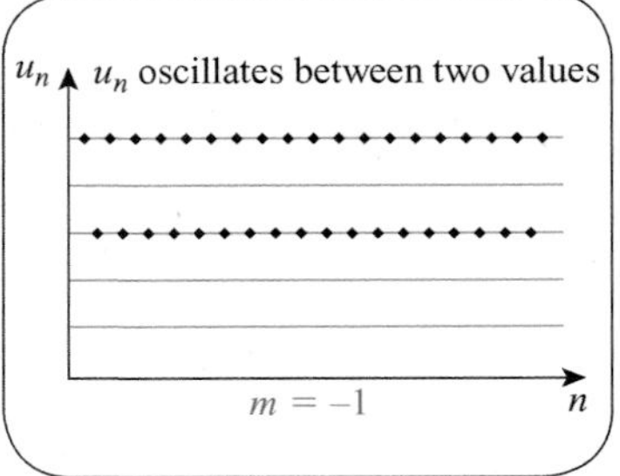

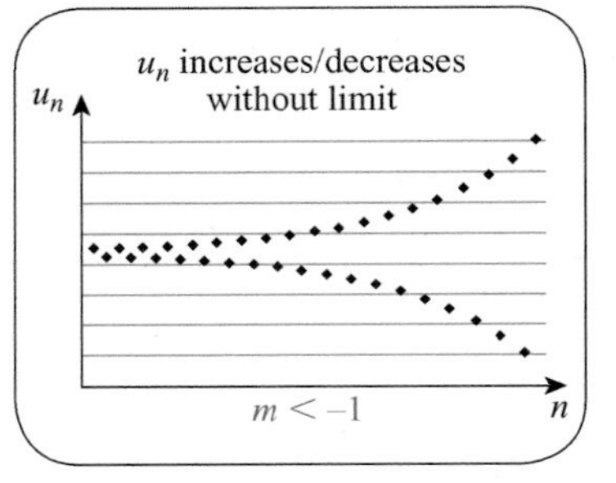

In two cases $0 < m < 1$ and $-1 < m < 0$ as n gets larger and larger, the values of u_n get closer and closer to a limiting value L.

For a **limit to exist** the multiplier m must lie between –1 and 1 ($-1 < m < 1$).

Finding the limit *L*

For the recurrence relation

$$u_{n+1} = mu_n + c$$

if $-1 < m < 1$ (the multiplier lies between –1 and 1) then any sequence of values generated by this relation will eventually 'level out' at some limiting value L.

Putting this value L into the recurrence relation will give L again for the next term:

$$L = mL + c \qquad \text{so} \qquad L - mL = c \qquad (1 - m)L = c$$

$$L = \frac{c}{1-m} \quad (-1 < m < 1)$$

This is the **algebraic method** for calculating the limit.

Example

A sequence is defined by the recurrence relation $u_{n+1} = 0{\cdot}7u_n + 14$ with 1st term u_1

Explain why this sequence has a limit as n tends to infinity. Find the **exact** value of this limit.

Solution

The multiplier 0·7 lies between −1 and 1 and so **a limit exists.**

Let the limit be L, then $L = 0{\cdot}7L + 14$

so $0{\cdot}3L = 14$, so $L = \frac{14}{0{\cdot}3}$, giving $L = \frac{140}{3} = \mathbf{46\frac{2}{3}}$

TOP TIP

When finding a limit L remember always to show first that $-1 < m < 1$ before you proceed.

Problems in context

Context problems that are solved using recurrence relations have many features in common. The following series of steps follows the course of a typical problem:

Step 1 To solve the problem you will probably need to do a recurring calculation to produce a sequence of values. Try to calculate the first few values (e.g. 10 tonnes of pollutant, 1 week later 12·5 tonnes, etc).

Step 2 If u_n is the nth term in this sequence of values then state clearly what meaning u_n has in the given context (e.g. u_n is number of tonnes of pollutant after n weeks).

Step 3 Describe the recurring calculation using u_{n+1} and u_n (e.g. $u_{n+1} = 0{\cdot}8u_n + 2$).

Step 4 Use this recurrence relation to solve the problem (e.g. calculate a limit, etc).

Note: Be very careful when finding the multiplier:

A reduction of 70% from u_n to u_{n+1} involves a multiplier of 0·3 since 30% remains of u_n.

Example

Two different types of water-purifying machines are in use.

Type A removes 35% of pollutants each day and is used in a tank that receives 15 litres of new pollutant at the end of each day. Type B daily removes 60% of pollutants but is operating in a tank where 25 litres of new pollutant are dumped after each day's Operation.

In the long run, which tank contains less pollutant?

Solution

After n days let the amount of pollutant in tank A be A_n litres and in tank B be B_n litres. The recurrence relations that model this situation are:

$A_{n+1} = 0{\cdot}65A_n + 15$ and
$B_{n+1} = 0{\cdot}4B_n + 25$

In both cases the multipliers, namely 0·65 and 0·4, lie between −1 and 1 and so a limit exists in each case. Let the limit for tank A be L litres and for tank B be M litres, then:

$L = 0{\cdot}65L + 15$

$\Rightarrow L - 0{\cdot}65L = 15$

$\Rightarrow 0{\cdot}35L = 15$

$\Rightarrow L = \dfrac{15}{0{\cdot}35}$

$\doteqdot 42{\cdot}9$

$M = 0{\cdot}4M + 25$

$\Rightarrow M - 0{\cdot}4M = 25$

$\Rightarrow 0{\cdot}6M = 25$

$\Rightarrow M = \dfrac{25}{0{\cdot}6}$

$\doteqdot 41{\cdot}7$

In the long run tank B will contain 41·7 litres (to 3 sig. figs) of pollutant, approximately 1·2 litres less than the amount of pollutant in tank A.

Quick Test 35

1. For the sequence defined by each recurrence relation, explain whether or not there is a limit as n tends to infinity. If there is a limit then find its **exact** value.

   a) $u_{n+1} = 0{\cdot}4u_n + 1$ b) $u_{n+1} = \dfrac{u_n}{4} + 5$ c) $7u_{n+1} = 2u_n + 8$

2. The body destroys 70% of a drug in a day, so a daily 28 unit injection is then given. The initial injection was 50 units.

   a) At no time should the body contain more than 50 units of the drug. Is this course of treatment safe?

   b) Is it safe to increase the daily injections to 36 units?

Applications of differentiation

Intervals

In context problems, the variable (x or t etc) is often restricted to a limited set of values.

For example, the cardboard box shown might have its depth, x cm, restricted to no less than 40 cm and no more than 60 cm.

So the interval of acceptable values is $40 \leq x \leq 60$

Other types of intervals are, for example: $t > 10$ or $v \leq 2{\cdot}3$

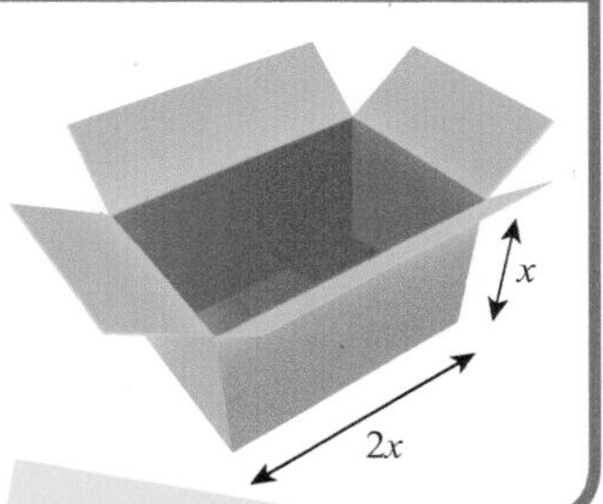

TOP TIP

You should try to sketch the graph of the function to see if your answers make sense. For Paper 2 questions this is a good use of a graphing calculator.

Max/min values on an interval

For a graph $y = f(x)$ if the values of x are resticted to an interval $a \leq x \leq b$ then the maximum and minimum values of f will be found at the stationary points or at the end points of the interval:

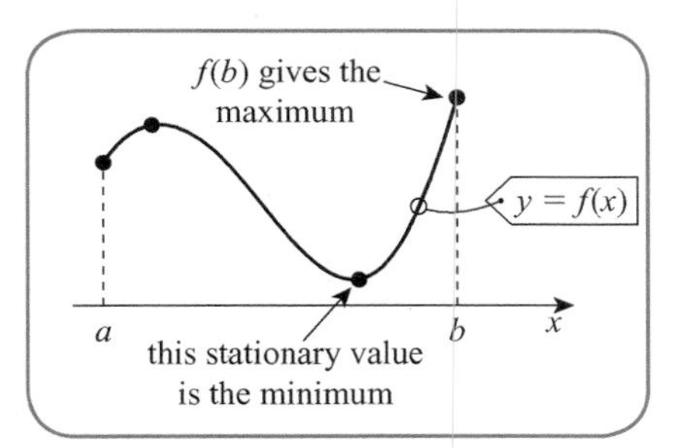

But look what happens for a different interval for the same graph:

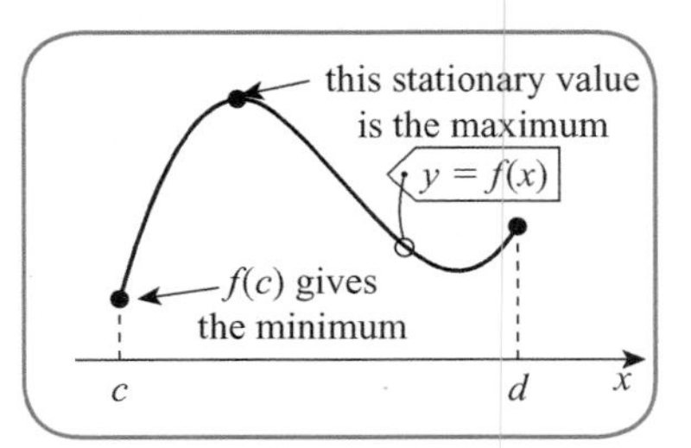

Note: The actual maximum or minimum value can be found by substituting the corresponding x-value into the formula $f(x)$.

Example

Find the maximum and minimum values of $f(x) = x^3 - 3x + 2$ on the interval $-2 \leq x \leq 3$

Solution

$f(x) = x^3 - 3x + 2$

$\Rightarrow f'(x) = 3x^2 - 3$

For stationary points set $f'(x) = 0$

So $3x^2 - 3 = 0$

$\Rightarrow 3(x^2 - 1) = 0$

$\Rightarrow 3(x - 1)(x + 1) = 0$

$\Rightarrow x - 1 = 0$ or $x + 1 = 0$

$\Rightarrow x = 1$ or $x = -1$

The stationary values are given by:

$f(1) = 1^3 - 3\times1 + 2 = 0$

$f(-1) = (-1)^3 - 3\times(-1) + 2 = -1 + 3 + 2 = 4$

The end points of the interval give values:

$f(-2) = (-2)^3 - 3 \times (-2) + 2 = -8 + 6 + 2 = 0$

$f(3) = 3^3 - 3 \times 3 + 2 = 27 - 9 + 2 = 20$

The maximum value is 20 (when $x = 3$)
The minimum value is 0 (when $x = 1$ or -2)

Optimisation problems

The following example shows how finding stationary points can help to solve an **optimisation** problem:

Example

The diagram shows a poster with width x metres.
The total area of the poster is required to be $2\,m^2$.
The print area leaves margins 0·1 m at the top and bottom edges and 0·05 m margins at the left and right edges as shown.
Find the dimensions for the poster to maximise the print area.

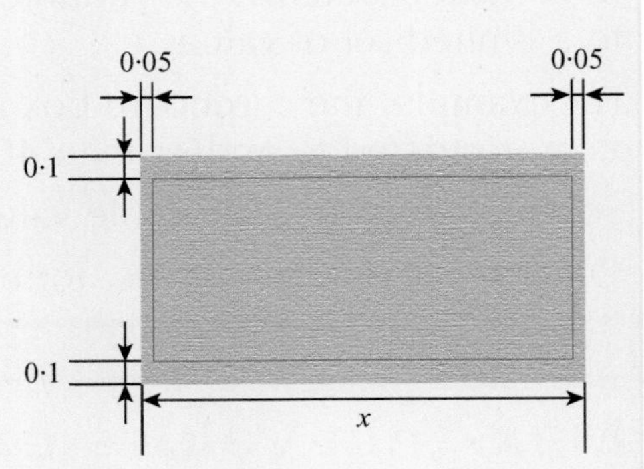

Solution

width $\times x = 2 \Rightarrow$ width $= \frac{2}{x}$

so print area $A(x) = (x - 2\times0{\cdot}05)\left(\frac{2}{x} - 2\times0{\cdot}1\right) = (x - 0{\cdot}1)\left(\frac{2}{x} - 0{\cdot}2\right)$

$$= 2 - 0{\cdot}2x - \frac{0{\cdot}2}{x} + 0{\cdot}02 = 2{\cdot}02 - 0{\cdot}2x - \frac{0{\cdot}2}{x} = 2{\cdot}02 - 0{\cdot}2x - 0{\cdot}2x^{-1}$$

For stationary points set $A'(x) = 0$

This gives $A'(x) = -0{\cdot}2 + 0{\cdot}2x^{-2} = -0{\cdot}2 + \frac{0{\cdot}2}{x^2} = 0 \Rightarrow \frac{0{\cdot}2}{x^2} = 0{\cdot}2 \Rightarrow x^2 = 1$

Since $x > 0$ (x measures a length) $x = 1$ is the only solution.

You now show this gives a maximum value:

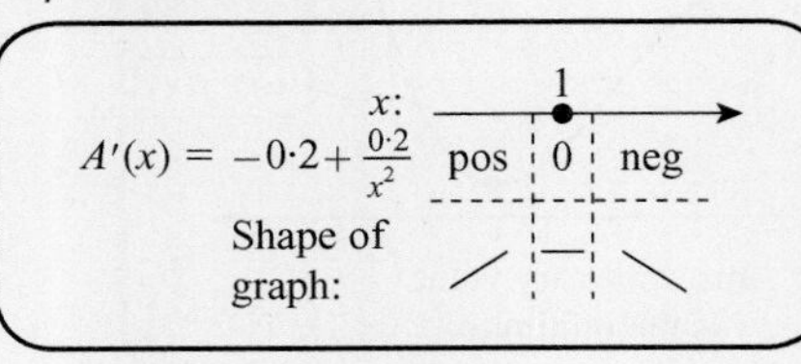

So $x = 1$ gives a maximum print area with:

length $= x = 1$ m and width $= \frac{2}{x} = \frac{2}{1} = 2$ m.
The required dimensions are 1 m × 2 m.

Quick Test 36

1. Find the maximum and minimum values of $f(x) = x^4 - 2x^2 - 3$ on the interval $-0{\cdot}5 \leq x \leq 1{\cdot}5$

2. A cuboid is to be cut from a long wedge of wood as shown in the diagram. The wedge is in the shape of a prism with a right-triangular end with base 20 cm and height 10 cm. When cut, the cuboid has to have dimensions x cm × h cm × $6x$ cm.

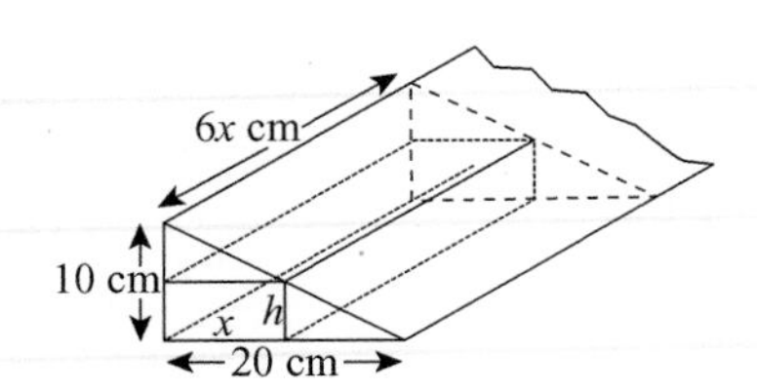

a) Show that $h = 10 - \frac{1}{2}x$ (hint: use similar triangles).

b) Show that the volume, $V\,cm^3$, of the cuboid is given by $V(x) = 60x^2 - 3x^3$

c) Hence, find the dimensions of the cuboid with the greatest volume that can be cut from the wedge.

Applications of integration

The area under a curve

above the x-axis

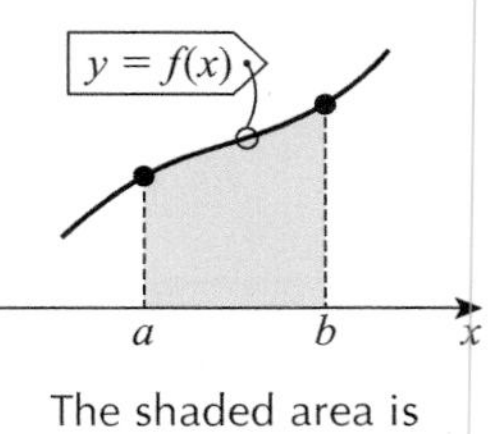

The shaded area is given by the integral

$\int_a^b f(x)dx$

below the x-axis

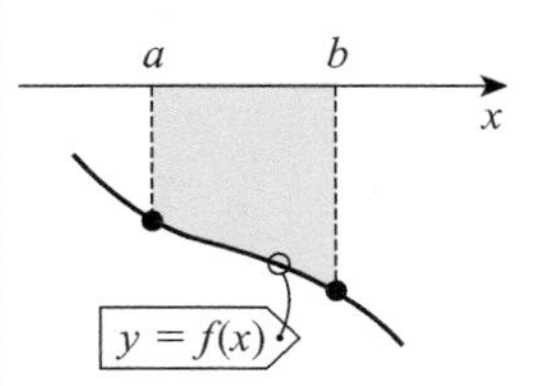

The shaded area is given by the integral

$-\int_a^b f(x)dx$

above and below the x-axis

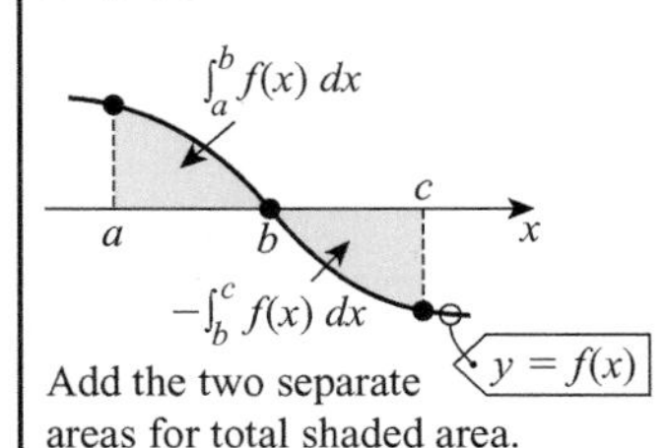

Add the two separate areas for total shaded area.

Often the x-axis intercepts are used as the limits for the integral:

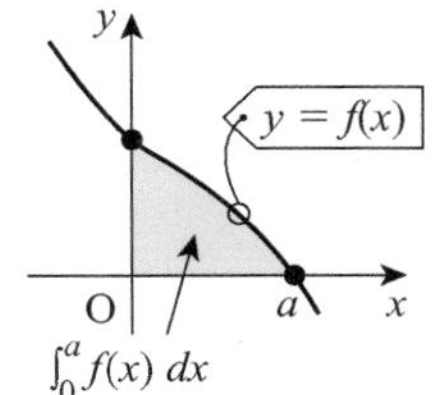

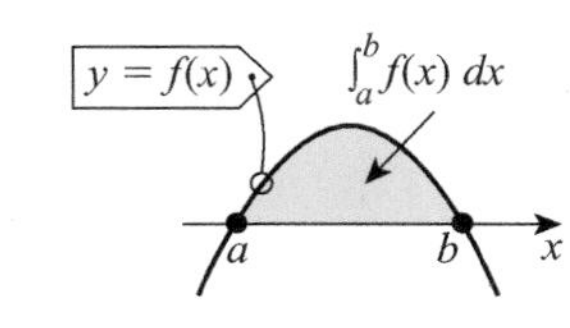

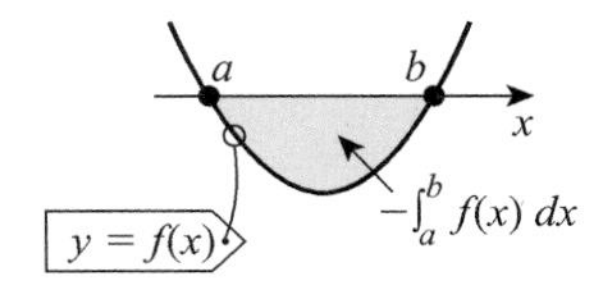

Example 1

Find the area enclosed by $y = x^2 + x - 2$ and the x-axis.

Solution

For x-axis intercepts set $y = 0$

So solve $x^2 + x - 2 = 0$

$\Rightarrow (x + 2)(x - 1) = 0$

$\Rightarrow x = -2$ or $x = 1$

Now draw a sketch:

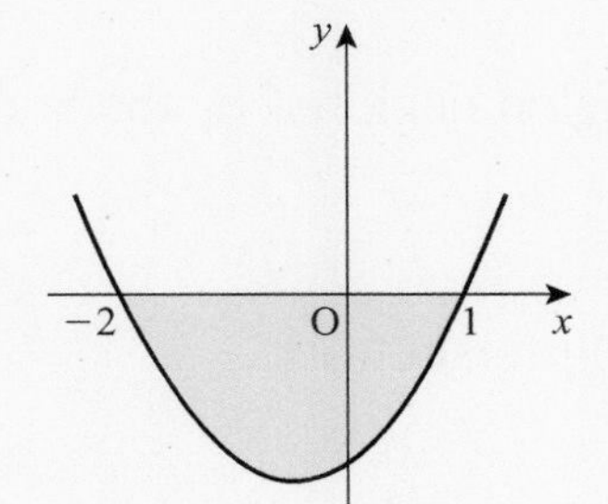

Calculating the integral gives:

$$\int_{-2}^{1}\left(x^2 + x - 2\right)dx = \left[\frac{x^3}{3} + \frac{x^2}{2} - 2x\right]_{-2}^{1} = \left(\frac{1^3}{3} + \frac{1^2}{2} - 2\times 1\right) - \left(\frac{(-2)^3}{3} + \frac{(-2)^2}{2} - 2\times(-2)\right)$$

$$= \frac{1}{3} + \frac{1}{2} - 2 - \left(-\frac{8}{3} + 2 + 4\right) = -\frac{9}{2} = -4\frac{1}{2}$$

Since the area is below the x-axis, to get the area change this value from negative to positive. So required area is **$4\frac{1}{2}$ unit^2**.

Note:

The unit of area used is a square of 1 unit along each axis. You can sometimes estimate whether your answer makes sense!

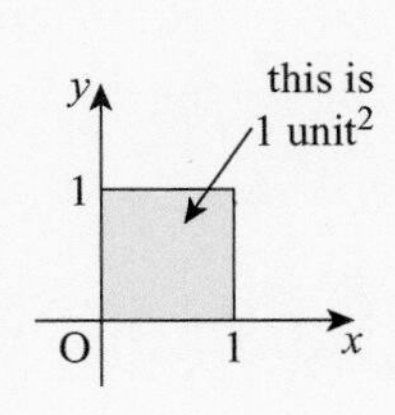

Example 2

Find the shaded area.

Solution

The area is given by...

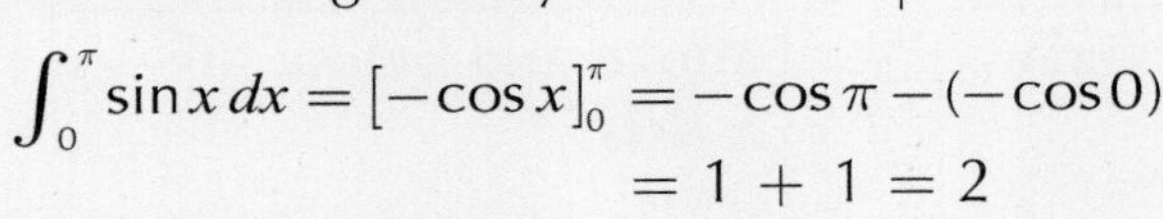

$$\int_0^{\pi} \sin x\,dx = [-\cos x]_0^{\pi} = -\cos\pi - (-\cos 0)$$
$$= 1 + 1 = 2$$

The shaded area = **2 unit²**.

TOP TIP

Try to estimate your answer using a unit 'square' on your sketch: 1 unit along each axis. Depending on scales they may look like a non-square rectangle!

The area between curves

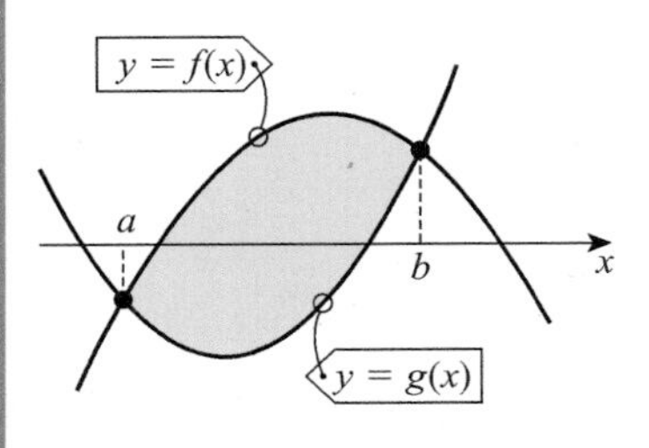

The shaded area is given by:

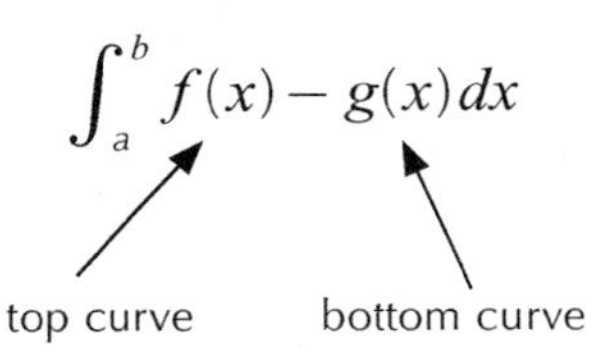

$$\int_a^b f(x) - g(x)\,dx$$

$x = a$ and $x = b$ are where the curves intersect.

Notes:

1. Simplify $f(x) - g(x)$ **first** before integrating... it's usually easier!
2. This result holds wherever the region is relative to the x-axis... above or below or even crossing the x-axis... use the same result.

Example 1

Find the area of the region enclosed by the line $y = x + 1$ and the parabola $y = 10 + 7x - 3x^2$

Solution

To find the points of intersection:

Solve:

$$\left.\begin{aligned} y &= x + 1 \\ y &= 10 + 7x - 3x^2 \end{aligned}\right\}$$

$x + 1 = 10 + 7x - 3x^2$

$3x^2 - 6x - 9 = 0$

$3(x + 1)(x - 3) = 0$

$x = -1$ or $x = 3$

Make a sketch:

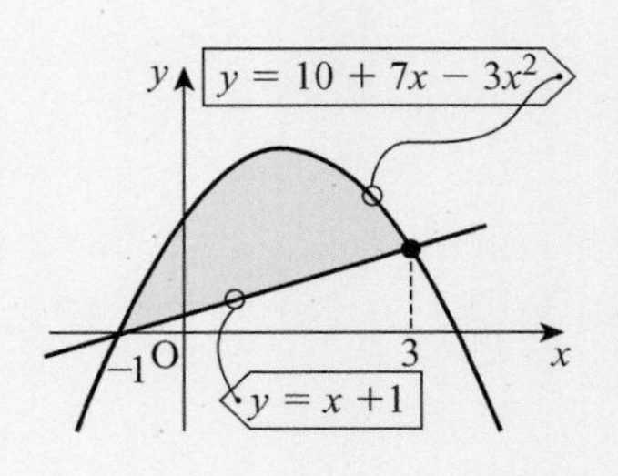

Shaded area

top graph bottom graph

$$\int_{-1}^{3} (10 + 7x - 3x^2) - (x + 1)\,dx$$

$$\int_{-1}^{3} (10 + 7x - 3x^2 - x - 1)\,dx$$

$$\int_{-1}^{3}(9+6x-3x^2)\,dx=\left[9x+\frac{6x^2}{2}-\frac{3x^3}{3}\right]_{-1}^{3}$$

$$\left[9x+3x^2-x^3\right]_{-1}^{3}$$

$= (9\times3 + 3\times3^2 - 3^3) - (9 \times (-1) + 3 \times (-1)^2 - (-1)^3)$

$= 27 + 27 - 27 + 9 - 3 - 1 =$ **32 unit²**

Example 2

Calculate the area enclosed by the curves $y = \sin x$ and $y = \sin 2x$ in the range $0 \le x \le \frac{\pi}{2}$ (shaded area in the diagram).

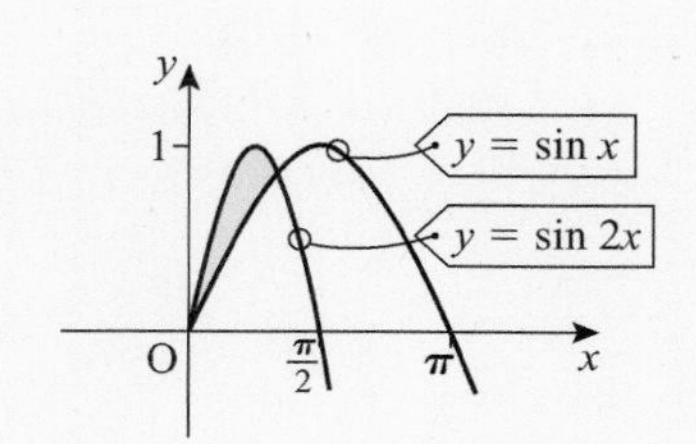

Solution

For intersection points: solve: $\left.\begin{matrix} y = \sin 2x \\ y = \sin x \end{matrix}\right\}$

$\sin 2x = \sin x \Rightarrow 2\sin x \cos x - \sin x = 0$

$\sin x(2\cos x - 1) = 0 \Rightarrow \sin x = 0$ or $\cos x = \frac{1}{2}$

So $x = 0, \pi, 2\pi\ldots$ or $x = \frac{\pi}{3}, \frac{5\pi}{3}, \ldots$

The intersection you use is $x = \frac{\pi}{3}$:

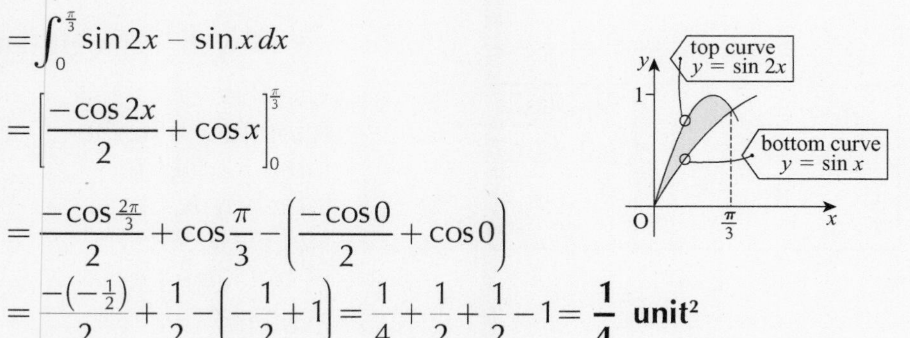

$$\text{Required area} = \int_0^{\frac{\pi}{3}} \sin 2x - \sin x\,dx$$

$$= \left[\frac{-\cos 2x}{2} + \cos x\right]_0^{\frac{\pi}{3}}$$

$$= \frac{-\cos\frac{2\pi}{3}}{2} + \cos\frac{\pi}{3} - \left(\frac{-\cos 0}{2} + \cos 0\right)$$

$$= \frac{-(-\frac{1}{2})}{2} + \frac{1}{2} - \left(-\frac{1}{2} + 1\right) = \frac{1}{4} + \frac{1}{2} + \frac{1}{2} - 1 = \mathbf{\frac{1}{4}}\ \textbf{unit}^2$$

Quick Test 37

1. Calculate the shaded area for each diagram:

a)

y = cos x

O, π/6, π/2, x, y

b)

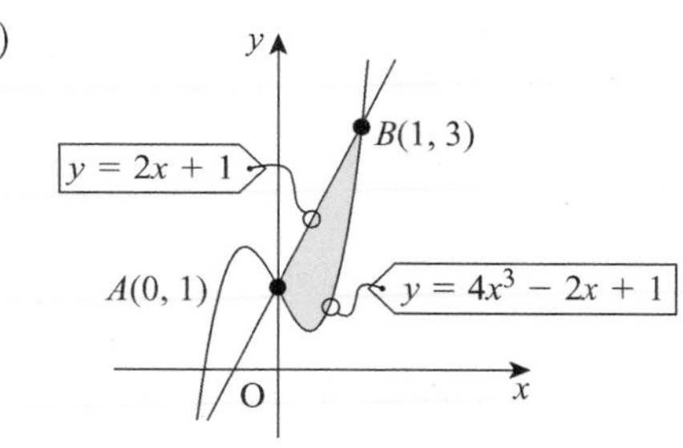

2. Calculate the area enclosed by the line and the curve:

$y = \sqrt{x}$

$y = \frac{1}{2}x$

Check-up questions (Chapter 3)

3.1 Applying algebraic skills to rectilinear shapes, to circles and to sequences

1. Find the equation of the straight line parallel to $y - 5x = 2$ which passes through the point $(-1, 3)$.
2. $PQRS$ is a kite
   Diagonal PR has equation $y = -\frac{1}{3}x + 2$
   Vertex S is the point $(0,-3)$

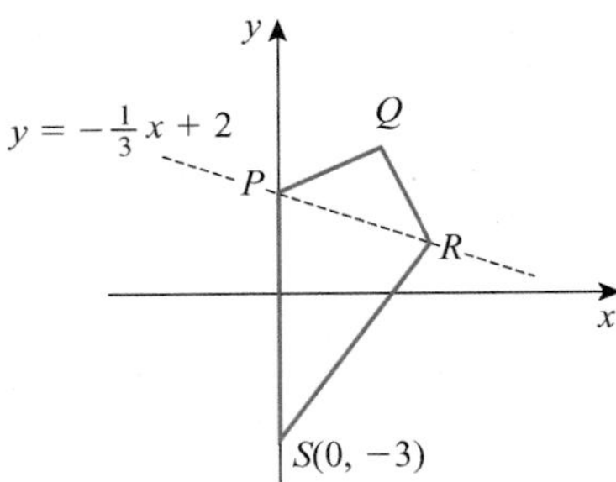

   Find the equation of QS, the other diagonal of the kite.
3. Calculate the obtuse angle that line $y = \frac{1}{2}x + 2$ makes with the x-axis.
4. Most road surfaces have a camber. This means the surface slopes down from the centre of the road to allow rain to flow off easily.

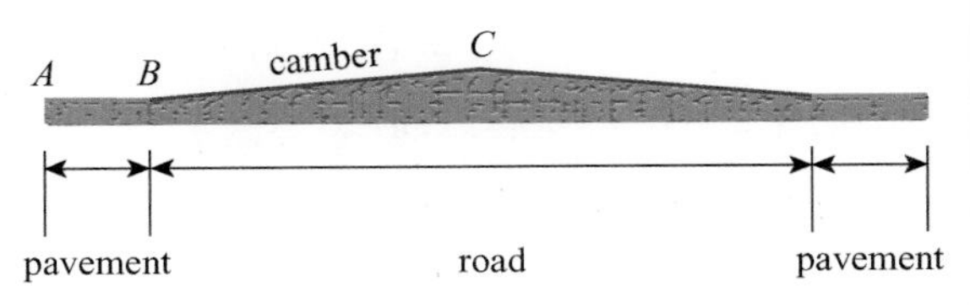

   The diagram shows the cross-section of a road with a straight-line camber BC.

Type of surface	Gradient of camber BC
Concrete	$0{\cdot}01 < m \leq 0.02$
Gravel	$m > 0{\cdot}02$

   Use the information in the table to decide which type of surface is shown in this diagram:

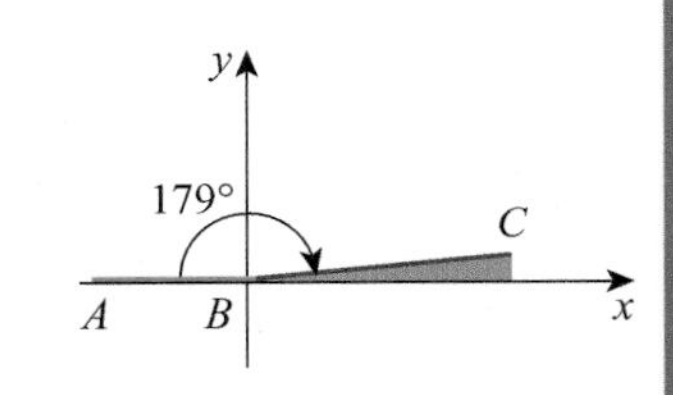

   Explain your answer fully.
5. Two congruent circles are shown. One touches the x-axis and the other touches the y-axis and passes through the point $(12, 0)$. Both circles have centres on the axes.
   Find the equation of the circle touching the x-axis.

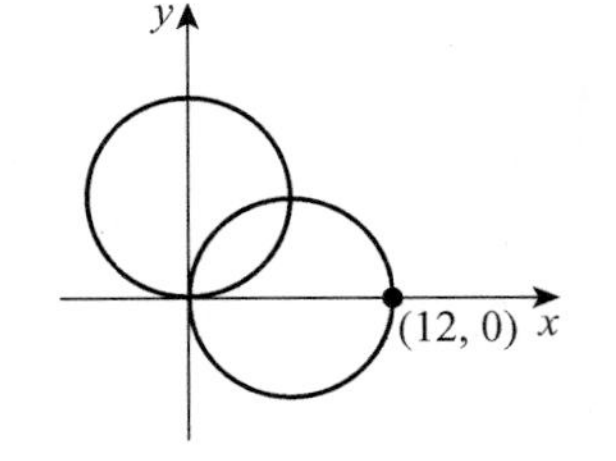

6. Ailsa says she can prove algebraically that $y = x - 1$ is a tangent to the circle with equation $(x - 1)^2 + (y + 4)^2 = 16$
   Can what she says be true? Show the reasoning for your answer.
7. The recurrence relation $u_{n+1} = au_n + b$ generates the sequence $u_1 = -2$, $u_2 = 1$ and $u_3 = 7$
   Find the values of a and b and hence calculate u_5.

8. Only 30% of a drug remains after a day in a patient's body.

   The patient is put on a course of daily 28 unit injections of the drug. The initial injection on the first day of the course was 50 units.

   Set up a recurrence relation showing the amount of drug in the patient's body immediately after an injection. Decide whether this course of injections is safe in the long run for the patient if the danger level is more than 50 units of the drug in the body.

3.2 Applying calculus skills to optimisation and area

1. The average manufacturing cost, £C, of a tablet screen protector depends on x, the number of thousands of the item that are produced per week.

   The formula is:
   $$C(x) = 3x^4 - 4x^3 + 5$$
   Find the value of x that minimises the cost C.

2. Part of the graph $y = x(16 - x^2)$ is shown in the diagram. Calculate the shaded area.

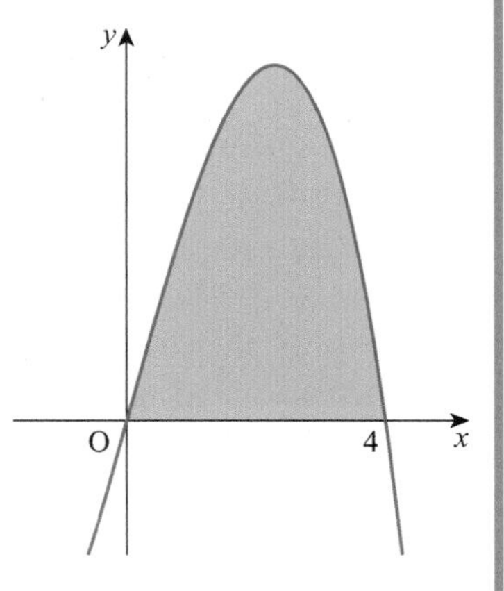

3. The diagram shows the line $y = 5 - x$ and the curve $y = x^2 - 3x + 5$.

   The two graphs intersect where $x = 0$ and $x = 2$.

   Find the shaded area enclosed by the graphs.

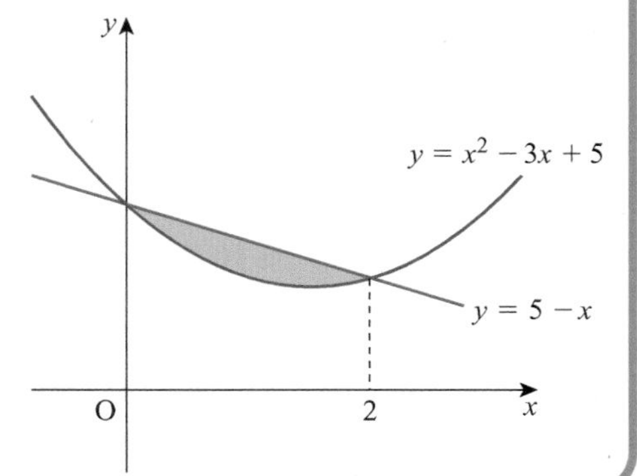

Chapter 3

Sample end-of-course exam questions (Chapter 3)

Non-calculator

1. $A(3, -1)$, $B(-1, 3)$ and $C(-2, 0)$ are the vertices of triangle ABC as shown in the diagram. M is the midpoint of AB. Find the equation of the line through M perpendicular to BC.

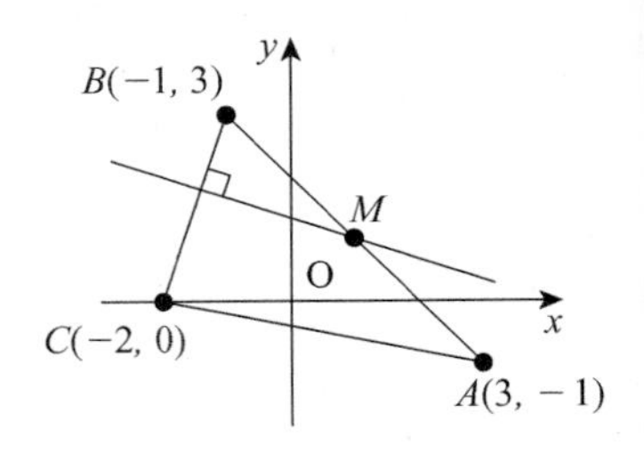

2. A sequence is defined by the recurrence relation $u_{n+1} = au_n + b$ with first term $u_1 = 5$.
   a) State the condition for this sequence to have a limit as n tends to infinity.
   b) If $u_2 = 8$ and $u_3 = 9{\cdot}5$ calculate the values of a and b.
   c) Find the exact value of the limit of this sequence as n tends to infinity.

3. The diagram shows the design stage for a tambourine with four jingles. The line of centres of the tambourine (large circle) and the upper and lower jingles (small circles) is parallel to the y-axis. The centres of all four jingles lie on the circumference of the tambourine. The tambourine touches the x-axis and two of the jingles touch the y-axis as shown.

   If the equation of the tambourine (large circle) is $x^2 + y^2 - 2x - 6y + 1 = 0$ find the equation of the upper jingle (small circle).

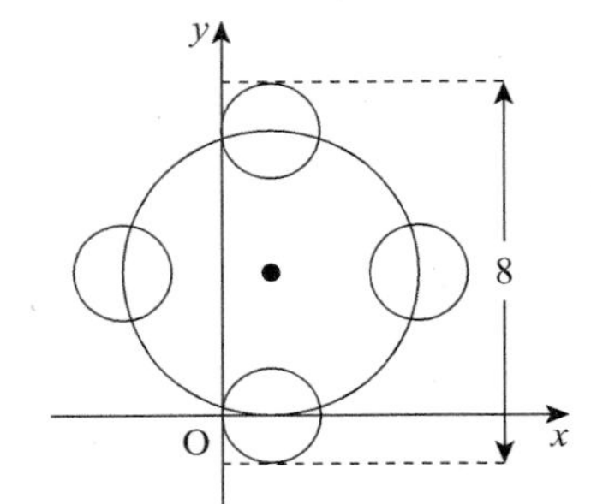

4. The diagram shows part of the graph $y = \sin x$ Find the shaded area shown.

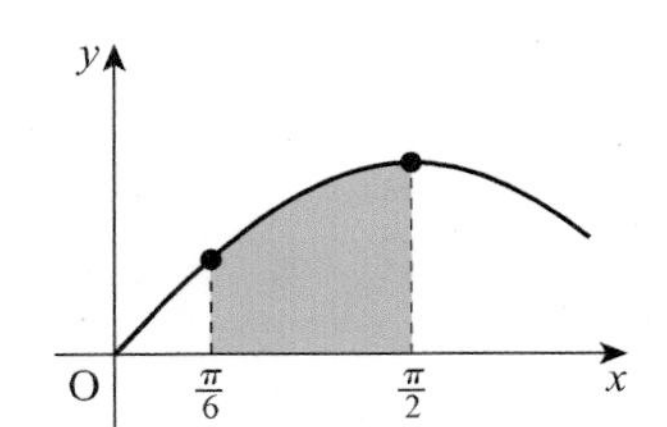

5. $A(1, 2)$, $B(-3, -1)$ and $C(4, -2)$ are the vertices of triangle ABC as shown in the diagram.
   a) Show that triangle ABC is isosceles.
   b) Side AB makes an angle θ with the positive direction of the x-axis. Find the exact value of $\tan\,\theta$.

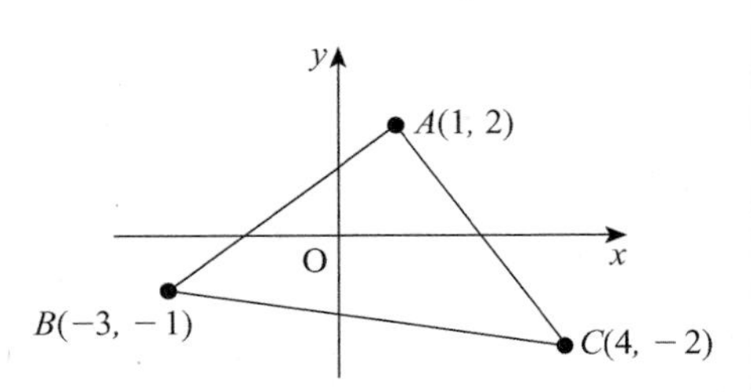

Calculator allowed

1. $PQRS$ is a rhombus. P, Q and R have coordinates $P(-2, -1)$, $Q(-1, 4)$ and $R(4, 5)$. Find the equation of SR.

2. The diagram shows the cross-section of a wall of a hillside irrigation trench that has been modelled by the curve $y = x^3 - 2x + 1$.

   The surface of the hillside has been modelled by QP, the tangent to the curve at $P(1, 0)$.

   a) Find the equation of the tangent QP.

   b) Find the coordinates of Q.

   c) Calculate the area of the shaded cross-section of the wall of the trench.

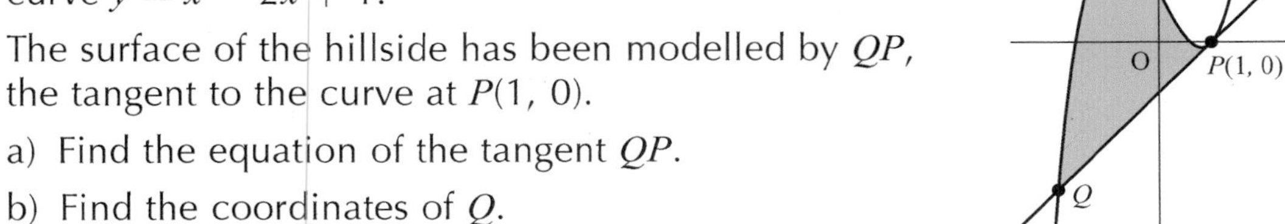

3. A metal component is in the form of a prism with cross-sectional area bounded by the curves with equations $y = 4\sqrt{x}$ and $y = 5 - x^2$ as shown in the diagram (all measurements are in centimetres). 30,000 of these components are to be produced.

   What total volume of metal will be required if 11% of this total is added to allow for wastage in the casting process?

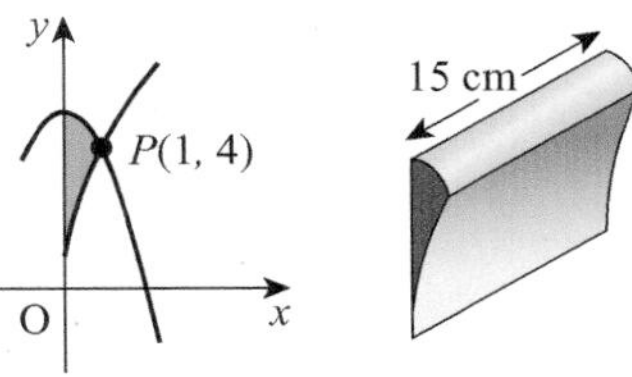

4. The diagram shows the circle with equation $x^2 + y^2 + 4x + 2y - 15 = 0$. A line with gradient 2 passes through C, the centre of the circle, and intersects the circle at points A and B as shown in the diagram.

   a) Find the equation of the line AB and hence show that A lies on the y-axis.

   b) Find the equation of the tangent to the circle at the point B.

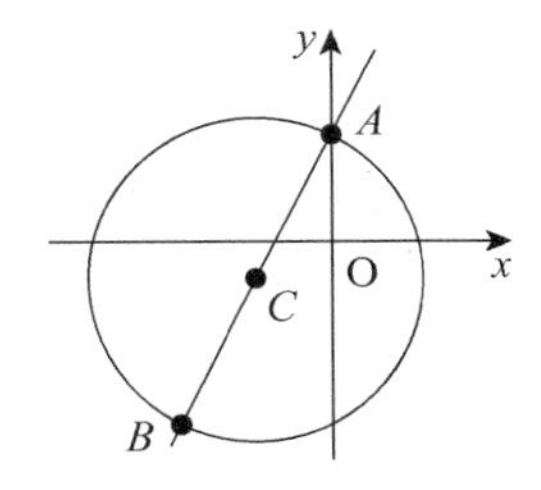

5. The diagram shows two identically sized circles that have line AB as a common tangent with T as the point of contact. The equation of AB is $x = -2$.

   a) One of the circles has equation $x^2 + y^2 - 2x + 2y - 7 = 0$.
      Find the equation of the other circle.

   b) A third circle is to be added to the figure in such a way that the other two circles lie inside it. The third circle has equation $x^2 + y^2 + 4x + 2y + c = 0$. Find the range of possible values of c.

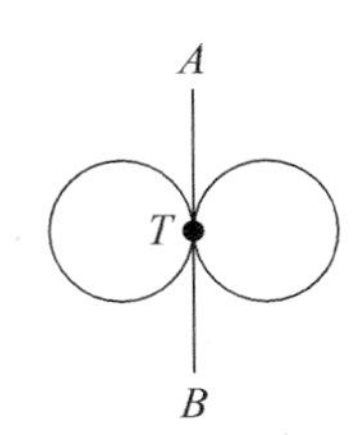

Quick Test answers

Quick Test 1

1. a) $x \in \mathbb{R}, x \neq -2$ b) $x \leq 10, x \in \mathbb{R}$
2. a) $b = 9$ b) 19

Quick Test 2

1. a) $(3x - 1)^2$ b) $3x^2 - 1$ c) $9x - 4$ d) x^4
2. a) $f^{-1}(x) = 2x - 2$ b) $f^{-1}(x) = 7 - x$ c) $f^{-1}(x) = x^2 + 1$

Quick Test 3

1. a) $2(x - 1)^2 + 4, a = 2, b = -1, c = 4$ b) $3(x - 1)^2 - 1, a = 3, b = -1, c = -1$
2. $k = \frac{1}{4}, m = 4, n = 2$

Quick Test 4

1.

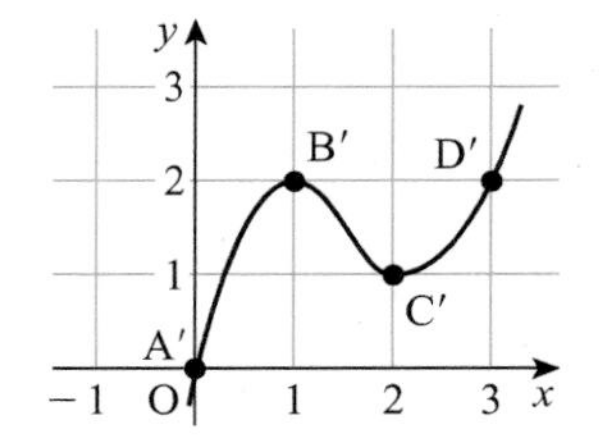

2.

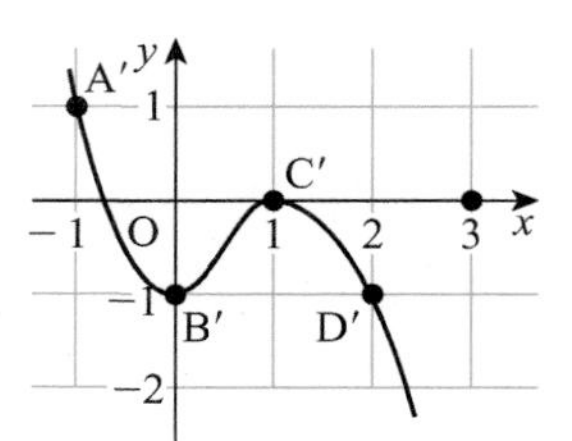

Quick Test 5

1. a) $0 < a < 1$ b) $a = \frac{1}{3}$
2. £311,000 (to the nearest £1000)

Quick Test 6

1. a)

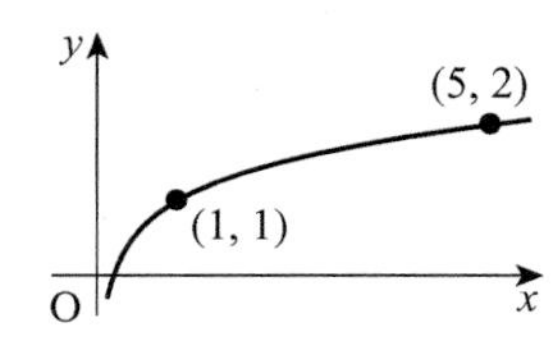

b)

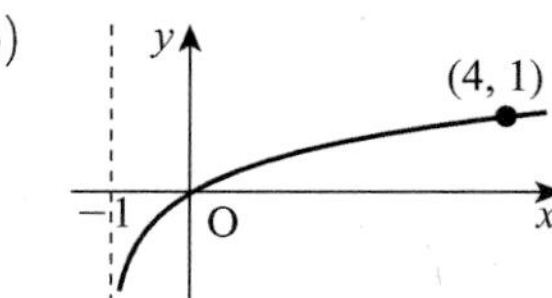

c)

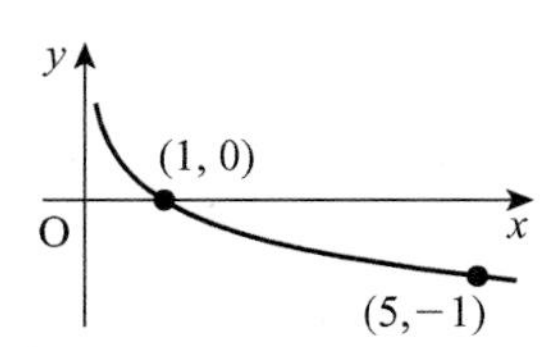

d)

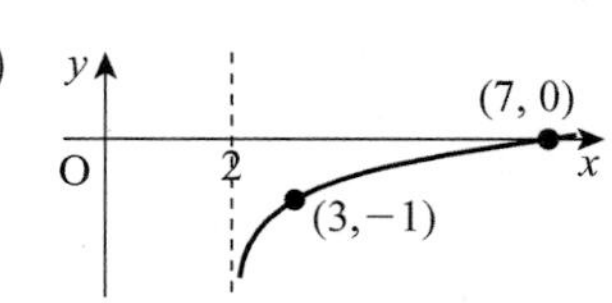

2. a) $\log_{10} 92 = x$ b) $y = \log_e 4$ c) $10^4 = t$ d) $e^B = A$

Quick Test 7

1. 2
2. $\log_a x$
3. a) $x = 1{\cdot}16$ b) $x = 0{\cdot}83$
4. $a = \frac{1}{2}$ $b = 3$

Quick Test 8

1. $\frac{3}{2}$ 2. $0{\cdot}64$ 3. $0, 2\pi$ 4. a) $315°$ b) $\frac{\pi}{180} \times 40 \doteqdot 0{\cdot}698$

Quick Test 9

1. $a = 4, b = 2$ 2. a)

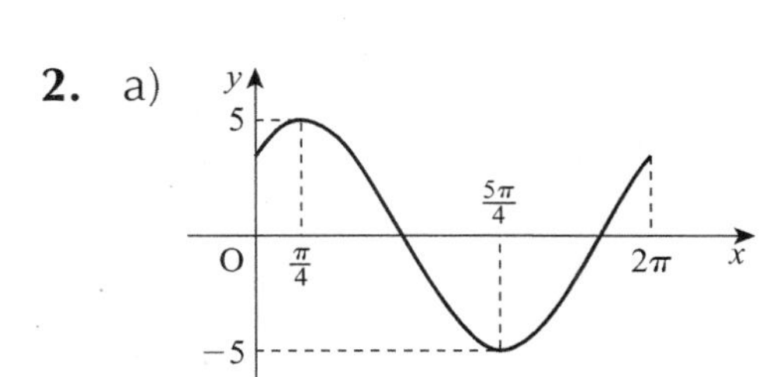

b) $\left(\frac{\pi}{4}, 5\right), \left(\frac{5\pi}{4}, -5\right)$

Quick Test 10

1. $\frac{4}{\sin x°} = \frac{5}{\sin\frac{1}{2}(180 - x)°} \Rightarrow \frac{4}{\sin x°} = \frac{5}{\sin(90 - \frac{1}{2}x)°} \Rightarrow \frac{4}{\sin x°} = \frac{5}{\cos\frac{1}{2}x°} \Rightarrow \cos\frac{1}{2}x° = \frac{5}{4}\sin x°$

Quick Test 11

1. $(\cos\theta + \sin\theta)(\cos\theta - \sin\theta) = \cos^2\theta - \sin^2\theta = \cos 2\theta$
2. $\frac{3}{5}$
3. $\cos(x + y) = \cos x \cos y - \sin x \sin y = \frac{8}{10} \times \frac{10}{5\sqrt{5}} - \frac{6}{10} \times \frac{5}{5\sqrt{5}} = \frac{8 - 3}{5\sqrt{5}} = \frac{5}{5\sqrt{5}} = \frac{1}{\sqrt{5}}$

Quick Test 12

1. $\sqrt{10}\cos(x + 71{\cdot}6)°$ 2. $2\sin(x + \frac{7\pi}{6})$

Quick Test 13

1. a) $\begin{pmatrix} 2 \\ -4 \\ 3 \end{pmatrix}$ b) 9

Quick Test 14

1. a) (i) $\begin{pmatrix} 4 \\ 5 \\ -4 \end{pmatrix}$ (ii) $\begin{pmatrix} 4 \\ 5 \\ -4 \end{pmatrix}$ b) parallelogram
2. $\overrightarrow{BC} = 3\overrightarrow{AB}$ so $AB \parallel BC$ with B a shared point; B divides AC in the ratio $1:3$

Quick Test 15

1. $33{\cdot}1°$ or $0{\cdot}577$ 2. $m = -2$ or 3 3. $\begin{pmatrix} -\frac{1}{3} \\ -\frac{2}{3} \\ \frac{2}{3} \end{pmatrix}$

Quick Test 16

1. a) $|b| = |c| = \sqrt{2}$ b) 2
2. a) $M(0, 3, 2)$, $N(5, 2, 0)$ b) $5i - j - 2k$
3. a) $F_4 = \begin{pmatrix} -3 \\ -3 \\ 2 \end{pmatrix}$ b) $\sqrt{22}$

Quick Test 17

1. $x^2 - 11x + 51 = 3x + 2 \Rightarrow x^2 - 14x + 49 = 0 \Rightarrow$ Discriminant $= (-14)^2 - 4 \times 1 \times 49 = 0$ $\Rightarrow$ one solution $\Rightarrow$ line is a tangent
2. $-\frac{3}{2} \leq k \leq \frac{3}{2}$
3. $-4 < x < 5$
4. $12x^2 + x - 1 = 0$
5. $\left(-\frac{1}{2}, -\frac{3}{4}\right), (3, 22)$

Quick Test 18

1. x^{-1}
2. $f(3) = 0$ and $f(-2) = 0$
3. $4x^2 - 8x + 4$; rem $= -1$
4. $\frac{85}{81}$

Quick Test 19

1. a) $f(1) = 0$ b) $f(x) = (x - 1)(2x^3 - x^2 - 2x + 1)$ c) $2\times(-1)^3 - (-1)^2 - 2\times(-1) + 1 = 0$ d) $f(x) = (x - 1)^2 (x + 1)(2x - 1)$
2. $x = -\frac{1}{2}$, $x = \frac{2}{3}$ and $x = 1$

Quick Test 20

1. a) 126·9, 306·9 b) 3·48, 5·94
2. a) $\frac{\pi}{6}, \frac{7\pi}{6}$ b) $\frac{2\pi}{3}, \frac{4\pi}{3}$
3. $\left(\frac{\pi}{6}, -\frac{1}{2}\right), \left(\frac{\pi}{2}, -\frac{1}{2}\right)$
4. a) 60, 300 b) $\frac{7\pi}{6}, \frac{3\pi}{2}, \frac{11\pi}{6}$
5. a) $g(x) = 2\sin(x - 30)°$ b) $x = 53·6$ or $x = 186·4$

Quick Test 21

1. a) $3x^{-4}$ b) $-x^{-\frac{3}{2}}$ c) $\frac{1}{2}x^{-\frac{3}{2}}$
2. a) $\frac{3}{2}x^{-\frac{1}{2}} + 2x^{-2}$ b) $x^{-\frac{1}{2}} - \frac{1}{2}x^{-\frac{3}{2}}$
3. a) $\frac{1}{4}$ b) $\frac{13}{3}$
4. only $\left(-1, \frac{1}{2}\right)$

Quick Test 22

1. a) at A and C b) A: $3y + 3x = 5$, B: $y + 2x = 1$, C: $3y + 3x = 1$
2. $(-1, -1)$ maximum, $(0, -2)$ minimum

Quick Test 23

1.

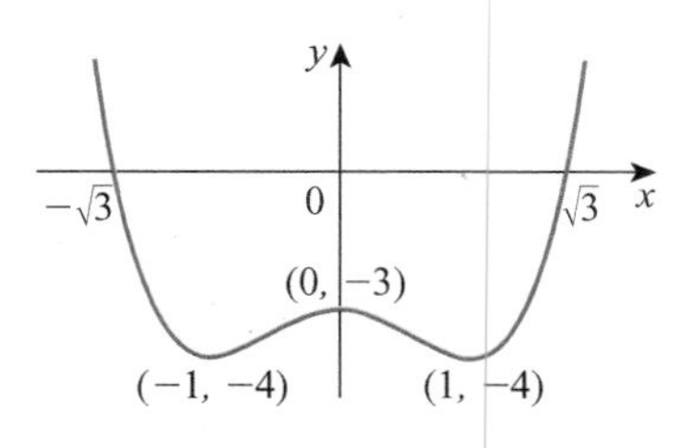

2.

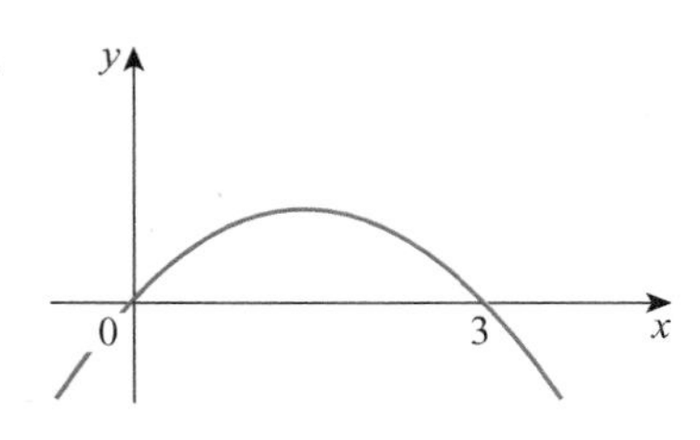

Quick Test 24

1. 2
2. a) $-\dfrac{2}{\sqrt{2-4x}}$ b) $\dfrac{\sin x}{\cos^2 x}$
3. $g'(x) = -\dfrac{1}{x^2}$ For $x \neq 0$: $\dfrac{1}{x^2} > 0$ so $-\dfrac{1}{x^2} < 0 \Rightarrow g'(x) < 0 \Rightarrow$ graph is decreasing

Quick Test 25

1. $\dfrac{3\pi}{2}$
2. a) 400 m/min b) 0 m/min: max height reached
   c) –400 m/min: same as at start but in opposite direction (fallen back to ground)

Quick Test 26

1. a) $-\dfrac{1}{2x^2} - \dfrac{2}{3}x^3 + C$ b) $\dfrac{5}{3}x^3 + 5x - \dfrac{3}{2x} + C$ c) $\dfrac{4}{3}x^{\frac{1}{2}} + \dfrac{2}{3}x^{\frac{3}{2}} + C$
2. a) $\dfrac{5}{4}x^4 + \dfrac{1}{2}x^2 - \cos x + C$ b) $2x^2 - \sin x + C$
3. $y = -\dfrac{1}{x} - \dfrac{x}{3} + 2$

Quick Test 27

1. $\dfrac{20}{3}$
2. $\dfrac{3\sqrt{2}-4}{4}$
3. a) $\sin^2 x = \dfrac{1}{2} - \dfrac{1}{2}\cos 2x$ b) $\dfrac{1}{2}x - \dfrac{1}{4}\sin 2x + C$ c) $\dfrac{\pi}{2}$

Quick Test 28

1. $\dfrac{1}{3}$
2. a) $\dfrac{3}{2}$; $(0, -2)$ b) $y = \dfrac{2}{3}x - 2$
3. $x = -2$

Quick Test 29

1. a) $\frac{1}{2}$ b) $2y - x = -7$ c) $26{\cdot}6°$
2. $m_1 = -2$ and $m_2 = \frac{1}{2} \Rightarrow m_1 \times m_2 = -1 \Rightarrow$ lines are perpendicular. Intersection point is $(-1, 1)$

Quick Test 30

1. a) $y + 2x = 5$ b) $y - x = -1$ 2. $G(2, 1)$
3. The 3rd equation is $y = 1$ and this passes through G also

Quick Test 31

1. $(x + 3)^2 + (y - 7)^2 = 3$ 2. a) $\left(\frac{1}{2}, -\frac{5}{2}\right)$; $\sqrt{5}$ b) $\left(\frac{1}{2}, \frac{1}{2}\right)$; 1
3. $(x - 2)^2 + (y - 2)^2 = 4$; $(x + 2)^2 + (y - 2)^2 = 4$; $(x + 2)^2 + (y + 2)^2 = 4$; $(x - 2)^2 + (y + 2)^2 = 4$
4. y-axis: set $x = 0 \Rightarrow y^2 + 6y + 9 = 0 \Rightarrow y = -3$ one solution so circle touches
   x-axis: set $y = 0 \Rightarrow x^2 - 4x + 9 = 0 \Rightarrow$ discriminant $< 0 \Rightarrow$ no solutions $\Rightarrow$ no intercepts

Quick Test 32

1. Solving simultaneously gives one root $x = -5 \Rightarrow$ line is a tangent; contact point: $(-5, 0)$
2. $4y + 5x = -1$ 3. $k = -1$ or 3 4. $(x - 5)^2 + (y - 2)^2 = 10$

Quick Test 33

1. Centres $(0, 0)$ and $(-4, 4)$ are $4\sqrt{2}$ units apart. The radii $\sqrt{2}$ and $3\sqrt{2}$ sum to $4\sqrt{2} \Rightarrow$ circles touch. Point of contact is $(-1, 1)$
2. $(x + 2)^2 + (y + 1)^2 = 3$ 3. $\left(\frac{1}{2}, 1\right)$

Quick Test 34

1. $\frac{31}{3}$ 2. 1 3. $p = 3, q = -1$

Quick Test 35

1. a) $m = 0{\cdot}4 \Rightarrow -1 < m < 1 \Rightarrow$ a limit L exists: $L = \frac{5}{3}$

   b) $m = \frac{1}{4} \Rightarrow -1 < m < 1 \Rightarrow$ a limit L exists: $L = \frac{20}{3}$

   c) $m = \frac{2}{7} \Rightarrow -1 < m < 1 \Rightarrow$ a limit L exists: $L = \frac{8}{5}$

2. Let u_n be amount of drug in body immediately after an injection $\Rightarrow u_{n+1} = 0{\cdot}3u_n + 28$ with $u_1 = 50$
   a) $m = 0{\cdot}3 \Rightarrow -1 < m < 1 \Rightarrow$ a limit L exists: $L = 40 \Rightarrow$ course is safe
   b) $u_1 = 50$; $u_2 = 51$ so this is not safe

Quick Test 36

1. max: $-2{\cdot}4375$ at $x = 1{\cdot}5$ min: -4 at $x = 1$

2. a) By similar triangles: $\frac{h}{10} = \frac{20 - x}{20} \Rightarrow h = 10 - \frac{1}{2}x$

b) $V(x) = x \times \left(10 - \frac{1}{2}x\right) \times 6x = 60x^2 - 3x^3$

c) Dimensions are: $\frac{40}{3}$ cm $\times$ $\frac{10}{3}$ cm $\times$ 80 cm

Quick Test 37

1. a) $\frac{1}{2}$ unit^2 b) 1 unit^2 **2.** $\frac{4}{3}$ unit^2

Check-up solutions (Chapter 1)

1.1 Applying algebraic skills to logarithims and exponentials

1. $\log_2 \dfrac{3ab}{3b} = \log_2 a$ 2. $\log_b a^3 \times a^2 = \log_b a^5 = 5\log_b a$
3. Equivalent statement is: $2^1 = y - 3 \Rightarrow y = 5$

1.2 Applying trig skills to manipulating expressions

1. $\left.\begin{array}{l} k\cos a = 3 \\ k\sin a = 1 \end{array}\right\} \Rightarrow \tan a = \frac{1}{3} \Rightarrow a \doteqdot 18{\cdot}4^\circ$ also $k = \sqrt{3^2 + 1^2} = \sqrt{10}$ giving $\sqrt{10}\cos(x - 18{\cdot}4)^\circ$
2. $1 - 2\sin x + \sin^2 x + 2\sin x = 1 + \sin^2 x = 1 + (1 - \cos^2 x) = 2 - \cos^2 x$
3. $\sin a \cos b + \cos a \sin b = \dfrac{9}{15} \times \dfrac{5}{13} + \dfrac{12}{15} \times \dfrac{12}{13} = \dfrac{189}{195} = \dfrac{63}{65}$

1.3 Applying algebra and trig skills to functions

1. 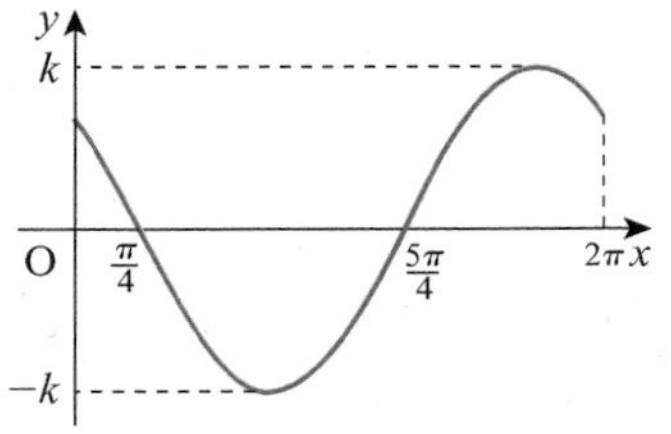

2. 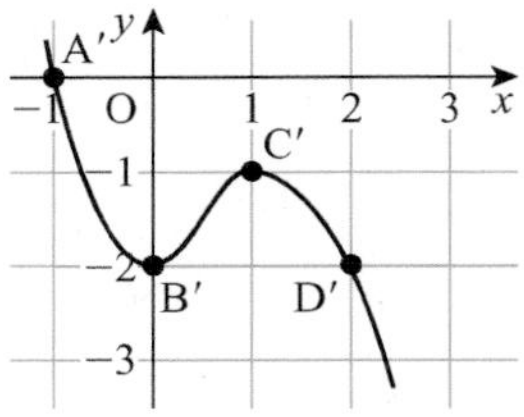

3. $b = \frac{1}{2}a + 3 \Rightarrow a = 2b - 6$ so $f^{-1}(x) = 2x - 6$ 4. $k = \frac{1}{2}$, $a = 2$, $b = -1$
5. $n = 3$, $m = 5$ 6. a) $f\left(-\sqrt{x}\right) = 2 - \left(-\sqrt{x}\right) = 2 + \sqrt{x}$ b) $x \geq 0$, $x \in \mathbb{R}$

1.4 Applying geometric skills to vectors

1. a) $\overrightarrow{QR} = 3\overrightarrow{PQ}$, therefore collinear since Q is a shared point
b) Since $\overrightarrow{QR} = 3\overrightarrow{PQ}$ pipe 1 is three times as long as pipe 2
2. $\overrightarrow{AT} = 2\overrightarrow{TB} \Rightarrow \boldsymbol{t} - \boldsymbol{a} = 2(\boldsymbol{b} - \boldsymbol{t}) \Rightarrow 3\boldsymbol{t} = 2\boldsymbol{b} + \boldsymbol{a} = 2\begin{pmatrix} 7 \\ 3 \\ 1 \end{pmatrix} + \begin{pmatrix} 1 \\ -3 \\ 4 \end{pmatrix} = \begin{pmatrix} 15 \\ 3 \\ 6 \end{pmatrix} \Rightarrow \boldsymbol{t} = \begin{pmatrix} 5 \\ 1 \\ 2 \end{pmatrix}$
so $T(5, 1, 2)$
3. $\overrightarrow{DB} = \overrightarrow{DA} + \overrightarrow{AB} = -\begin{pmatrix} -4 \\ 2 \\ 3 \end{pmatrix} + \begin{pmatrix} -8 \\ 0 \\ 0 \end{pmatrix} = \begin{pmatrix} -4 \\ -2 \\ -3 \end{pmatrix}$
4. 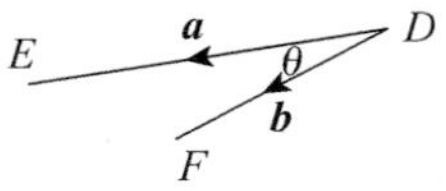

$\boldsymbol{a}.\boldsymbol{b} = 62$, $|\boldsymbol{a}| = \sqrt{70}$, $|\boldsymbol{b}| = \sqrt{66}$, $\dfrac{\boldsymbol{a}.\boldsymbol{b}}{|\boldsymbol{a}| \times |\boldsymbol{b}|} = 0{\cdot}912 \Rightarrow \theta \doteqdot 24{\cdot}2^\circ$

Check-up solutions (Chapter 2)

2.1 Applying algebraic skills to solve equations

1. Factors are $(x + 2)$, $(x + 3)$ and $(x - 1)$ so the roots are $x = -2$, $x = -3$ and $x = 1$
2. $2x^2 - x + a = 0$ has no solutions so discriminant $= (-1)^2 - 4 \times 2 \times a < 0$
   $\Rightarrow 1 - 8a < 0 \Rightarrow a > \frac{1}{8}$
3. $$\begin{array}{r|rrrr} -2 & 1 & -1 & -16 & -20 \\ & & -2 & 6 & 20 \\ \hline & 1 & -3 & -10 & 0 \end{array}$$
   so $f(x) = (x+2)(x^2 - 3x - 10)$
   $= (x+2)(x+2)(x-5)$
   $= (x+2)^2(x-5)$

2.2 Applying trig skills to solve equations

1. $\cos 3x = \frac{1}{2} \Rightarrow 3x = 60$ or 300 or ... $\Rightarrow x = 20$ or 100 or ...
   The only solutions in the range $0 < x \leq 120$ are $x = 20, 100$
2. $\cos\theta° - 5 \times 2\sin\theta° \cos\theta° = 0 \Rightarrow \cos\theta°(1 - 10\sin\theta°) = 0$
   $$\begin{cases} \Rightarrow \cos\theta° = 0 \text{ giving } \theta = 90 \\ \Rightarrow \sin\theta° = \frac{1}{10} \text{ giving } \theta \doteqdot 5{\cdot}7 \text{ (1st quadrant only)} \end{cases}$$
3. $\sin(\theta - 36{\cdot}9)° = \frac{2{\cdot}75}{5} = 0{\cdot}55 \Rightarrow \theta - 36{\cdot}9 \doteqdot 33{\cdot}4 \Rightarrow \theta \doteqdot 70{\cdot}3$ (1st quadrant only)

2.3 Applying calculus skills of differentiation

1. $f'(x) = -2 \times (-\sin x) = 2\sin x$
2. $y = \frac{5x^{\frac{1}{2}}}{x^1} - 2x^{\frac{1}{2}} = 5x^{-\frac{1}{2}} - 2x^{\frac{1}{2}} \Rightarrow \frac{dy}{dx} = -\frac{5}{2}x^{-\frac{3}{2}} - x^{-\frac{1}{2}}$
3. $\frac{dy}{dx} = x + 3$ so when $x = 2$: $\frac{dy}{dx} = 5$. Also when $x = 2$: $y = \frac{1}{2} \times 2^2 + 3 \times 2 - 2 = 6$
   So a point on the tangent is $(2, 6)$ and the gradient $= 5$:
   equation is $y - 6 = 5(x - 2) \Rightarrow y = 5x - 4$
4. a) $v = \frac{dh}{dt} = 16 - 8t$ so when $t = 0$ then $v = 16$ m/sec
   b) when $t = 2$ then $v = 16 - 8 \times 2 = 0$: it has reached its maximum height

2.4 Applying calculus skills of integration

1. $\frac{1}{2}\sin x + C$
2. $g(x) = \int (5 + x)^{-2}\, dx = \frac{(5+x)^{-1}}{-1} + C = -\frac{1}{5+x} + C$
3. $\int x^{-\frac{1}{2}} - 2x^{\frac{1}{2}}\, dx = \frac{x^{\frac{1}{2}}}{\frac{1}{2}} - \frac{2x^{\frac{3}{2}}}{\frac{3}{2}} + C = 2x^{\frac{1}{2}} - \frac{4}{3}x^{\frac{3}{2}} + C$
4. $\left[\frac{(x+2)^4}{4}\right]_{-2}^{2} = \frac{(2+2)^4}{4} - \frac{(-2+2)^4}{4} = \frac{4^4}{4} - 0 = 4^3 = 64$

Check-up solutions (Chapter 3)

3.1 Applying algebraic skills to rectilinear shapes, to circles and to sequences

1. gradient $= 5$ so the equation is $y - 3 = 5(x + 1) \Rightarrow y = 5x + 8$
2. $m_{PR} = -\frac{1}{3} \Rightarrow m_{\perp} = 3$ so $m_{QS} = 3$ and a point on QS is $(0, -3)$.
   The equation is $y + 3 = 3(x - 0) \Rightarrow y = 3x - 3$
3. The acute angle $= \tan^{-1}\frac{1}{2} \doteq 26{\cdot}6^{\circ}$ so the obtuse angle $= 180° - 26{\cdot}6° = 153{\cdot}4°$
4. BC makes a $1°$ angle with the x-axis $\Rightarrow m = \tan 1° = 0{\cdot}017... \Rightarrow 0{\cdot}01 < m \leq 0{\cdot}02$
   $\Rightarrow$ it's concrete
5. radius $= 6$ and centre is $(0, 6) \Rightarrow$ equation is $(x - 0)^2 + (y - 6)^2 = 6^2 \Rightarrow x^2 + (y - 6)^2 = 36$
6. Solving the equations simultaneously gives:
   $(x - 1)^2 + ((x - 1) + 4)^2 = 16 \Rightarrow x^2 - 2x + 1 + x^2 + 6x + 9 = 16 \Rightarrow 2x^2 + 4x - 6 = 0$
   $\Rightarrow x^2 + 2x - 3 = 0 \Rightarrow (x + 3)(x - 1) = 0 \Rightarrow x = -3$ or $x = 1$
   Since there are two solutions, the line cannot be a tangent so what she said was false.
7. $1 = a \times (-2) + b$ and $7 = a \times 1 + b$ so solve: $\left.\begin{matrix} -2a + b = 1 \\ a + b = 7 \end{matrix}\right\} \Rightarrow a = 2,\ b = 5$

   So $u_{n+1} = 2u_n + 5$. The sequence is $-2, 1, 7, 19, 43, \ldots$ So $u_5 = 43$.
8. Let u_n be the amount of the drug in the patient's body immediately after the nth injection.
   Then $u_{n+1} = 0{\cdot}3u_n + 28$ with $u_1 = 50$.
   The multiplier $m = 0{\cdot}3$ with $-1 < m < 1$ so a limit L exists.
   So $L = 0{\cdot}3L + 28 \Rightarrow 0{\cdot}7L = 28 \Rightarrow L = 40$ which is safe being less than the danger level of 50 units.

3.2 Applying calculus skills to optimisation and area

1. $C'(x) = 12x^3 - 12x^2 = 12x^2(x - 1)$. For stationary values set $C'(x) = 0 \Rightarrow x = 0$ or $x = 1$
   Since $x > 0$ only consider $x = 1$. Nature table gives:

x:		1	
$C'(x) = 12x^2(x - 1)$:	neg	0	pos
Shape of graph:	↘	—	↗

   So $x = 1$ gives a minimum value for C
2. $\int_0^4 16x - x^3\,dx = \left[8x^2 - \frac{x^4}{4}\right]_0^4 = (8 \times 4^2 - \frac{4^4}{4}) - (0 - 0) = 128 - 64 = 64$ unit^2
3. Shaded area
   $= \int_0^2 (5 - x) - (x^2 - 3x + 5)\,dx = \int_0^2 2x - x^2\,dx = \left[x^2 - \frac{x^3}{3}\right]_0^2 = 2^2 - \frac{2^3}{3} = \frac{4}{3}$ unit^2

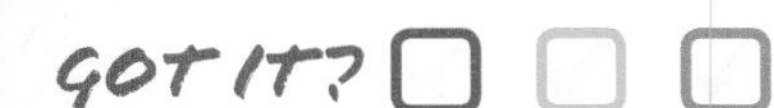

Chapter 1 sample exam question solutions

Non-calculator

1. $h\left(3\times\frac{\pi}{12}\right)=h\left(\frac{\pi}{4}\right)=\sin\left(2\times\frac{\pi}{4}\right)=\sin\frac{\pi}{2}=1$

2. $x=\frac{\pi}{6}$ so $\sin x=\frac{1}{2}$ and $\cos x=\frac{\sqrt{3}}{2}$

$\Rightarrow \sin 2x = 2\sin x\cos x = 2\times\frac{1}{2}\times\frac{\sqrt{3}}{2}=\frac{\sqrt{3}}{2}$

3. a) Since AP : PC = 2 : 1

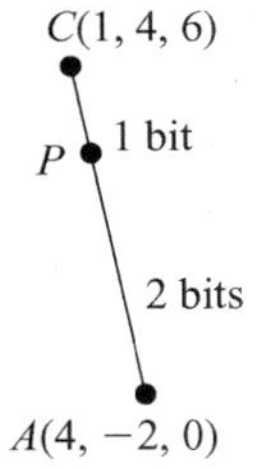

$\overrightarrow{AP}=2\overrightarrow{PC}$

$\Rightarrow \boldsymbol{p}-\boldsymbol{a}=2(\boldsymbol{c}-\boldsymbol{p})$

$\Rightarrow \boldsymbol{p}-\boldsymbol{a}=2\boldsymbol{c}-2\boldsymbol{p}$

$\Rightarrow \boldsymbol{p}+2\boldsymbol{p}=2\boldsymbol{c}+\boldsymbol{a}$

$\Rightarrow 3\boldsymbol{p}=2\boldsymbol{c}+\boldsymbol{a}$

So

$$3\boldsymbol{p}=2\begin{pmatrix}1\\4\\6\end{pmatrix}+\begin{pmatrix}4\\-2\\0\end{pmatrix}=\begin{pmatrix}6\\6\\12\end{pmatrix}$$

$$\Rightarrow \boldsymbol{p}=\frac{1}{3}\begin{pmatrix}6\\6\\12\end{pmatrix}=\begin{pmatrix}2\\2\\4\end{pmatrix}$$

So P (2, 2, 4)

b) $$\overrightarrow{BP}=\boldsymbol{p}-\boldsymbol{b}=\begin{pmatrix}2\\2\\4\end{pmatrix}-\begin{pmatrix}3\\5\\0\end{pmatrix}=\begin{pmatrix}-1\\-3\\4\end{pmatrix}$$

So $\overrightarrow{BP}=-\boldsymbol{i}-3\boldsymbol{j}+4\boldsymbol{k}$

4. $2(\sqrt{3}\cos x-\sin x)=k\cos(x+a)$

$\Rightarrow 2\sqrt{3}\cos x-2\sin x$

$=k\cos x\cos a-k\sin x\sin a$

So $k\cos a=2\sqrt{3}$, $k\sin a=2$ — Since cos a and sin a are positive, a is in the 1st quadrant.

$\frac{k\sin a}{k\cos a}=\frac{2}{2\sqrt{3}}$

$\Rightarrow \frac{\sin a}{\cos a}=\frac{1}{\sqrt{3}}$

$\Rightarrow \tan a=\frac{1}{\sqrt{3}}$

$\Rightarrow a=\frac{\pi}{6}$

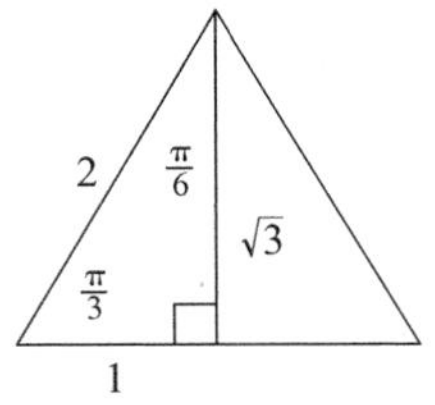

Also

$(k\sin a)^2+(k\cos a)^2=2^2+(2\sqrt{3})^2$

$\Rightarrow k^2\sin^2 a+k^2\cos^2 a=4+12$

$\Rightarrow k^2(\sin^2 a+\cos^2 a)=16$

$\Rightarrow k^2\times 1=16\Rightarrow k^2=16$

$\Rightarrow k=4\ (k>0)$

So $2(\sqrt{3}\cos x-\sin x)$

$=4\cos\left(x+\frac{\pi}{6}\right)$

5. a) $$\overrightarrow{AB}=\boldsymbol{b}-\boldsymbol{a}=\begin{pmatrix}k\\k\\0\end{pmatrix}-\begin{pmatrix}1\\-2\\-k\end{pmatrix}=\begin{pmatrix}k-1\\k+2\\k\end{pmatrix}$$

$$\overrightarrow{AC}=\boldsymbol{c}-\boldsymbol{a}=\begin{pmatrix}4\\-3\\3-k\end{pmatrix}-\begin{pmatrix}1\\-2\\-k\end{pmatrix}=\begin{pmatrix}3\\-1\\3\end{pmatrix}$$

since $\overrightarrow{AB}$ and $\overrightarrow{AC}$ are perpendicular

then $\overrightarrow{AB}.\overrightarrow{AC}=0\Rightarrow \begin{pmatrix}k-1\\k+2\\k\end{pmatrix}\cdot\begin{pmatrix}3\\-1\\3\end{pmatrix}=0$

$\Rightarrow 3(k-1)-1(k+2)+3k=0$

$\Rightarrow 3k-3-k-2+3k=0$

$\Rightarrow 5k-5=0$

$\Rightarrow 5k=5\Rightarrow k=1$

b) $A(1, -2, -1)$, $C(4, -3, 2)$ and $D(13, -6, 11)$

$$\overrightarrow{AC} = \begin{pmatrix} 3 \\ -1 \\ 3 \end{pmatrix} \text{ from part } (a) \text{ above}$$

$$\overrightarrow{CD} = \mathbf{d} - \mathbf{c} = \begin{pmatrix} 13 \\ -6 \\ 11 \end{pmatrix} - \begin{pmatrix} 4 \\ -3 \\ 2 \end{pmatrix}$$

$$= \begin{pmatrix} 9 \\ -3 \\ 9 \end{pmatrix}$$

So $\overrightarrow{CD} = 3\overrightarrow{AC}$ so $\overrightarrow{CD}$ and $\overrightarrow{AC}$ are parallel and since C is a shared point then A, C and D are collinear

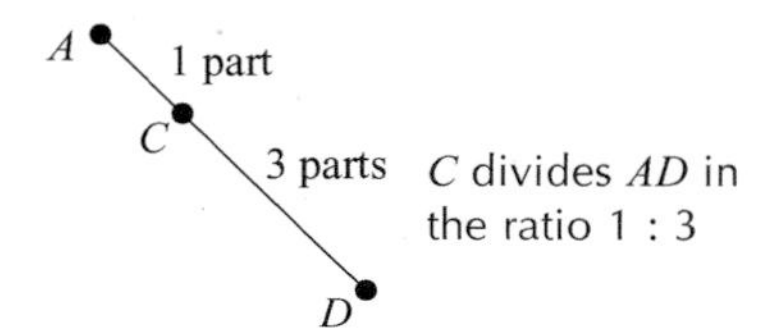

6. $a = 2$ (number of cycles in interval $0 \le x \le 2\pi$)
$b = -1$ (graph $y = 2\sin 2x$ moved down 1 unit)

7. a) $\log_{\sqrt{a}} b = 2c$

$$\Rightarrow \left(\sqrt{a}\right)^{2c} = b$$

$$\Rightarrow (a^{\frac{1}{2}})^{2c} = b \Rightarrow a^{\frac{1}{2}\times 2c} = b$$

$$\Rightarrow a^c = b \Rightarrow \log_a b = c$$

b) By part (a) above, $\log_{\sqrt{a}} b$ has twice the value of $\log_a b$ ($2c$ is twice c).
Since $\log_5 7$ is equal to $\log_{\sqrt{25}} 7$ it therefore has twice the value of $\log_{25} 7$ so:
$$\log_5 7 - \log_{25} 7 = 2\log_{25} 7 - \log_{25} 7$$
$$= \log_{25} 7 \text{ as required.}$$

Calculator allowed

1. a) $\dfrac{2x^2 - 7x + 6}{x^2 - 4}$

$$= \frac{(2x-3)(x-2)}{(x-2)(x+2)} = \frac{2x-3}{x+2}$$

b) $\log_3(2x^2 - 7x + 6) - \log_3(x^2 - 4) = 2$

$$\Rightarrow \log_3\left(\frac{2x^2 - 7x + 6}{x^2 - 4}\right) = 2$$

$$\Rightarrow \log_3\left(\frac{2x-3}{x+2}\right) = 2$$

$$\Rightarrow \frac{2x-3}{x+2} = 3^2$$

$$\Rightarrow 2x - 3 = 9(x+2)$$

$$\Rightarrow 2x - 3 = 9x + 18$$

$$\Rightarrow -3 - 18 = 9x - 2x$$

$$\Rightarrow -21 = 7x$$

$$\Rightarrow x = \frac{-21}{7} = -3$$

2. a) $\overrightarrow{CA} = \mathbf{a} - \mathbf{c} = \begin{pmatrix} 3 \\ 2 \\ 5 \end{pmatrix} - \begin{pmatrix} 6 \\ 4 \\ 3 \end{pmatrix} = \begin{pmatrix} -3 \\ -2 \\ 2 \end{pmatrix}$

$$\overrightarrow{CB} = \mathbf{b} - \mathbf{c} = \begin{pmatrix} 6 \\ 0 \\ 3 \end{pmatrix} - \begin{pmatrix} 6 \\ 4 \\ 3 \end{pmatrix} = \begin{pmatrix} 0 \\ -4 \\ 0 \end{pmatrix}$$

b) Use $\cos\theta° = \dfrac{\mathbf{p}\cdot\mathbf{q}}{|\mathbf{p}||\mathbf{q}|}$

with $\mathbf{p} = \begin{pmatrix} -3 \\ -2 \\ 2 \end{pmatrix}$

A, P, θ°, C, q, B

and $\mathbf{q} = \begin{pmatrix} 0 \\ -4 \\ 0 \end{pmatrix}$

$$\mathbf{p}.\mathbf{q} = -3 \times 0 + (-2) \times (-4) + 2 \times 0$$
$$= 8$$

2. b) (continued)

$$|p| = \sqrt{(-3)^2 + (-2)^2 + 2^2}$$
$$= \sqrt{9+4+4} = \sqrt{17}$$
$$|q| = \sqrt{0^2 + (-4)^2 + 0^2} = \sqrt{16} = 4$$

So $\cos\theta^\circ = \dfrac{8}{\sqrt{17}\times 4}$

$$\Rightarrow \theta^\circ = \cos^{-1}\left(\frac{8}{4\sqrt{17}}\right) = 60\cdot98...^\circ$$

The angle between edges CA and CB is approximately 61·0° (to 1 decimal place).

3. a) For Jan 1 2000 set $t = 0$

$$p = 6\times e^{0.0138\times 0}$$
$$= 6\times e^0 = 6\times 1 = 6$$

The population was 6 billion.

b) Solve: $12 = 6\,e^{0.0138t}$

$$\Rightarrow 2 = e^{0.0138t}$$
$$\Rightarrow \log_e 2 = 0\cdot0138t$$
$$\Rightarrow t = \frac{\log_e 2}{0\cdot0138} = 50\cdot22...$$

So during 2050 the population reaches 12 billion.

At the start of 2051 the population will be more than double its level in 2000.

4. a) $\cos(a+b)^\circ$

$$= \cos a^\circ \cos b^\circ - \sin a^\circ \sin b^\circ$$
$$= \frac{3}{5}\times\frac{5}{13} - \frac{4}{5}\times\frac{12}{13}$$
$$= \frac{15}{65} - \frac{48}{65} = -\frac{33}{65}$$

b) $\sin(a+b)^\circ$

$$= \sin a^\circ \cos b^\circ + \cos a^\circ \sin b^\circ$$
$$= \frac{4}{5}\times\frac{5}{13} + \frac{3}{5}\times\frac{12}{13}$$
$$= \frac{20}{65} + \frac{36}{65} = \frac{56}{65}$$

c) $\tan(a+b)^\circ = \dfrac{\sin(a+b)^\circ}{\cos(a+b)^\circ}$

$$= -\frac{56}{33}$$

5. $f(g(x)) = f(2x+1)$

$$= \frac{2}{(2x+1)+1} = \frac{2}{2x+2}$$
$$= \frac{2}{2(x+1)} = \frac{1}{x+1}$$

Also $\dfrac{1}{2}f(x) = \dfrac{1}{2}\times\dfrac{2}{x+1}$

$$= \frac{2}{2(x+1)} = \frac{1}{x+1}$$

So $f(2x+1) = \dfrac{1}{2}f(x)$

Chapter 2 sample exam question solutions

Non-calculator

1. a) i)

$$\begin{array}{r|rrrr} -1 & 2 & 3 & 0 & -1 \\ & & -2 & -1 & 1 \\ \hline & 2 & 1 & -1 & 0 \end{array}$$

So since $h(-1) = 0$
$x + 1$ is a factor of $h(x)$

ii) $h(x) = (x+1)(2x^2 + x - 1)$
$= (x+1)(2x-1)(x+1)$
$= (x+1)^2(2x-1)$

iii) $h(x) = 0 \Rightarrow (x+1)^2(2x-1) = 0$
$\Rightarrow x+1 = 0$ or $2x - 1 = 0$
$\Rightarrow x = -1$ or $x = \frac{1}{2}$

b) The curves intersect where $f(x) = g(x)$
$\Rightarrow 2x^3 + x - 3 = -3x^2 + x - 2$
$\Rightarrow 2x^3 + x - 3 + 3x^2 - x + 2 = 0$
$\Rightarrow 2x^3 + 3x^2 - 1 = 0$
$\Rightarrow x = -1$ or $x = \frac{1}{2}$ (from (a))

For a common tangent $f'(x) = g'(x)$ at the point of intersection.

$f(x) = 2x^3 + x - 3$
$\Rightarrow f'(x) = 6x^2 + 1$
$g(x) = -3x^2 + x - 2$
$\Rightarrow g'(x) = -6x + 1$
$f'(-1) = 6 \times (-1)^2 + 1 = 7$
$g'(-1) = -6 \times (-1) + 1 = 7$
$f'\left(\frac{1}{2}\right) = 6 \times \left(\frac{1}{2}\right)^2 + 1 = \frac{5}{2}$
$g'\left(\frac{1}{2}\right) = -6 \times \frac{1}{2} + 1 = -2$

So only $x = -1$ gives a common tangent.

The y-coordinate of T is given by:
$f(-1) = 2 \times (-1)^3 + (-1) - 3$
$= -2 - 1 - 3 = -6$
So T$(-1, -6)$

2. $$\int_{-1}^{1} \frac{6x^3 - x}{3x^3}\,dx = \int_{-1}^{1} \frac{6x^3}{3x^3} - \frac{x}{3x^3}\,dx$$
$$= \int_{-1}^{1} 2 - \frac{x^{-2}}{3}\,dx$$
$$= \left[2x - \frac{x^{-1}}{3\times(-1)}\right]_{-1}^{1}$$
$$= \left[2x + \frac{1}{3x}\right]_{-1}^{1}$$
$$= \left(2\times1 + \frac{1}{3\times1}\right) - \left(2\times(-1) + \frac{1}{3\times(-1)}\right)$$
$$= 2 + \frac{1}{3} + 2 + \frac{1}{3} = 4\tfrac{2}{3}$$

3. $$y = \frac{1}{\sin^2 x} = \frac{1}{(\sin x)^2} = (\sin x)^{-2}$$
So $$\frac{dy}{dx} = -2(\sin x)^{-3} \times \cos x = -\frac{2\cos x}{\sin^3 x}$$

4. Let $$y = \frac{2x+6}{\sqrt{x}} = \frac{2x+6}{x^{\frac{1}{2}}} = \frac{2x^1}{x^{\frac{1}{2}}} + \frac{6}{x^{\frac{1}{2}}}$$
$$= 2x^{\frac{1}{2}} + 6x^{-\frac{1}{2}}$$
So $$\frac{dy}{dx} = \frac{1}{2} \times 2x^{-\frac{1}{2}} - \frac{1}{2} \times 6x^{-\frac{3}{2}}$$
$$= x^{-\frac{1}{2}} - 3x^{-\frac{3}{2}}$$

5. Differentiating a cubic expression produces a quadratic expression. So $y = f'(x)$ is a parabola. Stationary points occur when $x = 0$ and $x = b$ ($f'(0) = 0$ and $f'(b) = 0$), i.e. $y = f'(x)$ intersects the axis for these values of x:

Sketch of $y = f'(x)$

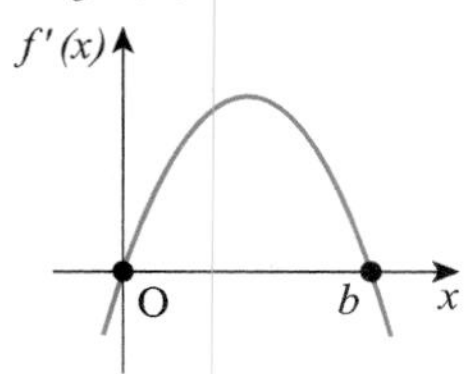

6. a) Left-hand side

$$= (\sin A + \cos B)^2 + (\cos A + \sin B)^2$$
$$= \sin^2 A + 2\sin A \cos B + \cos^2 B + \cos^2 A + 2\cos A \sin B + \sin^2 B$$
$$= (\sin^2 A + \cos^2 A) + (\sin^2 B + \cos^2 B) + 2(\sin A \cos B + \cos A \sin B)$$
$$= 1 + 1 + 2\sin(A + B)$$
$$= 2 + 2\sin(A + B)$$
$$= \text{Right-hand side}$$

b) The equation

$(\sin A + \cos B)^2 + (\cos A + \sin B)^2 = 3$ becomes $2 + 2\sin(A + B) = 3$

$$\Rightarrow 2\sin(A + B) = 1$$
$$\Rightarrow \sin(A + B) = \frac{1}{2}$$

[$(A + B)$ is in the 1st or 2nd quadrants and 1st quadrant angle is $\frac{\pi}{6}$]

So $(A + B) = \frac{\pi}{6}$ or $\pi - \frac{\pi}{6} = \frac{5\pi}{6}$

7. $2y - 3x = 6 \Rightarrow 2y = 3x + 6 \Rightarrow y = \frac{3}{2}x + 3$

The gradient of this line is $\frac{3}{2}$

tangent

gradient at this point is $\frac{3}{2}$

$$y = 2\sqrt{x+1} = 2(x+1)^{\frac{1}{2}}$$
$$\Rightarrow \frac{dy}{dx} = \frac{1}{2} \times 2(x+1)^{-\frac{1}{2}} \times 1$$
$$= (x+1)^{-\frac{1}{2}} = \frac{1}{\sqrt{(x+1)}}$$

This is the gradient formula for the curve.

So for $\frac{dy}{dx} = \frac{3}{2} \Rightarrow \frac{1}{\sqrt{(x+1)}} = \frac{3}{2}$

$$\Rightarrow 2 = 3\sqrt{(x+1)}$$
$$\Rightarrow 4 = 9 \times (x+1) \text{ (after squaring both sides)}$$
$$\Rightarrow x + 1 = \frac{4}{9}$$
$$\Rightarrow x = \frac{4}{9} - 1 = -\frac{5}{9}$$

1. a) $k\cos(x + \alpha)^\circ$

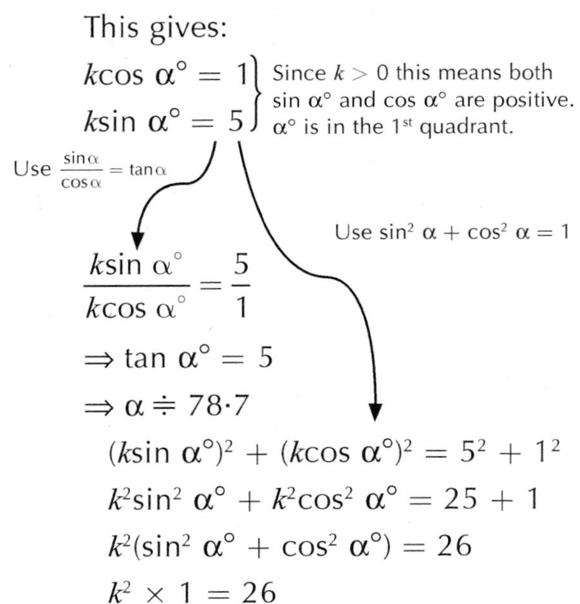

$= k\cos x^\circ \cos\alpha^\circ - k\sin x^\circ \sin\alpha^\circ$

compare $f(x) = 1\cos x^\circ - 5\sin x^\circ$

This gives:

$k\cos\alpha^\circ = 1$
$k\sin\alpha^\circ = 5$

Since $k > 0$ this means both $\sin\alpha^\circ$ and $\cos\alpha^\circ$ are positive. α° is in the 1st quadrant.

Use $\frac{\sin\alpha}{\cos\alpha} = \tan\alpha$

$\frac{k\sin\alpha^\circ}{k\cos\alpha^\circ} = \frac{5}{1}$

$\Rightarrow \tan\alpha^\circ = 5$

$\Rightarrow \alpha \doteqdot 78{\cdot}7$

Use $\sin^2\alpha + \cos^2\alpha = 1$

$(k\sin\alpha^\circ)^2 + (k\cos\alpha^\circ)^2 = 5^2 + 1^2$

$k^2\sin^2\alpha^\circ + k^2\cos^2\alpha^\circ = 25 + 1$

$k^2(\sin^2\alpha^\circ + \cos^2\alpha^\circ) = 26$

$k^2 \times 1 = 26$

$k = \sqrt{26}\ (k > 0)$

So $\cos x^\circ - 5\sin x^\circ = \sqrt{26}\cos(x + 78{\cdot}7)^\circ$

b) $f(x) = 1$ becomes

$\sqrt{26}\cos(x + 78{\cdot}7)^\circ = 1$

$\Rightarrow \cos(x + 78{\cdot}7)^\circ = \frac{1}{\sqrt{26}}$

$= 0{\cdot}1961\ldots$

The angle $(x + 78{\cdot}7)^\circ$ is in the 1st or 4th quadrants.

So $x + 78{\cdot}7 = 78{\cdot}7$

or $x + 78{\cdot}7 = 360 - 78{\cdot}7$

giving $x = 0$ or $x = 202{\cdot}6$

c) For x-axis intercept set $y = 0$ i.e. $f(x) = 0$

so $\sqrt{26}\cos(x + 78{\cdot}7)^\circ = 0$

giving $\cos(x + 78{\cdot}7)^\circ = 0$

so $x + 78{\cdot}7 = 90$ or $x + 78{\cdot}7 = 270$

If $x + 78{\cdot}7 = 90$

$\Rightarrow x = 90 - 78{\cdot}7$

$= 11.3$

This is not in the required range

If $x + 78{\cdot}7 = 270$

$\Rightarrow x = 270 - 78{\cdot}7$

$= 191.3$

Thus $a = 191{\cdot}3$

2. a) Show that $\cos^2 x^\circ - \cos 2x^\circ = 1 - \cos^2 x^\circ$

Left-hand side $= \cos^2 x^\circ - \cos 2x^\circ$

$= \cos^2 x^\circ - (2\cos^2 x^\circ - 1)$

$= \cos^2 x^\circ - 2\cos^2 x^\circ + 1$

$= 1 - \cos^2 x^\circ$

= Right-hand side

so $\cos^2 x^\circ - \cos 2x^\circ = 1 - \cos^2 x^\circ$

b) The equation $3\cos^2 x^\circ - 3\cos 2x^\circ = 8\cos x^\circ$ becomes $3(\cos^2 x^\circ - \cos 2x^\circ) = 8\cos x^\circ$

$\Rightarrow 3(1 - \cos^2 x^\circ) = 8\cos x^\circ$ (using part **a**)

$\Rightarrow 3 - 3\cos^2 x^\circ = 8\cos x^\circ$

$\Rightarrow 3\cos^2 x^\circ + 8\cos x^\circ - 3 = 0$

$\Rightarrow (3\cos x^\circ - 1)(\cos x^\circ + 3) = 0$

so $3\cos x^\circ - 1 = 0$ or $\cos x^\circ + 3 = 0$

$\cos x^\circ = \frac{1}{3}$ (x is in 1st or 4th quadrants)

or $\cos x^\circ = -3$ This equation has no solutions since $\cos x^\circ$ in never less than -1.

so $x = 70{\cdot}5$ or $360 - 70{\cdot}5$

$= 289{\cdot}5$

$289{\cdot}5$ is not in the required range. The only valid solution is $x = 70{\cdot}5$

3. $(kx + 2)(x + 3) = 8$

$kx^2 + 3kx + 2x + 6 = 8$

$kx^2 + (3k + 2)x - 2 = 0$

Discriminant $= (3k + 2)^2 - 4 \times k \times (-2)$

$= 9k^2 + 12k + 4 + 8k$

$= 9k^2 + 20k + 4$

For equal roots Discriminant $= 0$

so $9k^2 + 20k + 4 = 0$

$\Rightarrow (9k + 2)(k + 2) = 0$

$\Rightarrow 9k + 2 = 0$ or $k + 2 = 0$

$\Rightarrow k = -\frac{2}{9}$ or $k = -2$

4. a) $y = 3 + 2x^2 - x^4 \Rightarrow \frac{dy}{dx} = 4x - 4x^3$

For stationary points set $\frac{dy}{dx} = 0$

so $4x - 4x^3 = 0 \Rightarrow 4x(1 - x^2) = 0$

$\Rightarrow 4x(1 - x)(1 + x) = 0$

$\Rightarrow x = 0$ or $x = 1$ or $x = -1$

When $x = 0$, $y = 3 + 2 \times 0^2 - 0^4 = 3$ giving $(0, 3)$

When $x = 1$, $y = 3 + 2 \times 1^2 - 1^4 = 4$ giving $(1, 4)$

When $x = -1$, $y = 3 + 2 \times (-1)^2 - (-1)^4 = 4$ giving $(-1, 4)$

The stationary points are $(-1, 4)$, $(0, 3)$ and $(1, 4)$

b) $(1 + x)$

x:		−1		0		1	
$\frac{dy}{dx} = 4x(1-x)$	+		−		+		−
Shape of graph:	/	−	\	_	/	−	\

So $(-1, 4)$ and $(1, 4)$ are maximum stationary points and $(0, 3)$ is a minimum stationary point.

5. a) Let $3\cos\theta^\circ - \sin\theta^\circ$
$= k\cos(\theta + \alpha)^\circ$: so

$3\cos\theta^\circ - 1\sin\theta^\circ$

$= k\cos\theta^\circ\cos\alpha^\circ - k\sin\theta^\circ\sin\alpha^\circ$

Comparing these two expressions gives:

$$\left.\begin{aligned} -k\sin\alpha^\circ &= -1 \\ k\cos\alpha^\circ &= 3 \end{aligned}\right\} \Rightarrow \left.\begin{aligned} k\sin\alpha^\circ &= 1 \\ k\cos\alpha^\circ &= 3 \end{aligned}\right\}$$

Since both $\sin\alpha^\circ$ and $\cos\alpha^\circ$ are positive, α° is in the 1st quadrant.

$\frac{k\sin\alpha^\circ}{k\cos\alpha^\circ} = \frac{1}{3} \Rightarrow \tan\alpha^\circ = \frac{1}{3}$

so $\alpha = 18{\cdot}4$ (to 3 sig. figs)

$(k\sin\alpha^\circ)^2 + (k\cos\alpha^\circ)^2 = 1^2 + 3^2$

$k^2\sin^2\alpha^\circ + k^2\cos^2\alpha^\circ = 1 + 9$

$k^2(\sin^2\alpha^\circ + \cos^2\alpha^\circ) = 10$

$k^2 \times 1 = 10$

$k = \sqrt{10}$ $(k > 0)$

So $3\cos\theta^\circ - \sin\theta^\circ = \sqrt{10}\cos(\theta + 18{\cdot}4)^\circ$

b) $f(\theta) = \sqrt{10}\cos(\theta + 18{\cdot}4)^\circ$

Maximum value is $\sqrt{10}$ when

$\theta + 18{\cdot}4 = 0$ or 360

i.e. $\theta = -18{\cdot}4$ or $360 - 18{\cdot}4 = 341{\cdot}6$

but $0 \leq \theta < 360$

so $\theta = 341{\cdot}6$ is the only possibility

Minimum value is $-\sqrt{10}$ when

$\theta + 18{\cdot}4 = 180$, i.e. $\theta = 180 - 18{\cdot}4$

$\Rightarrow \theta = 161{\cdot}6$

c) $\sqrt{10}(3\cos\theta^\circ - \sin\theta^\circ) + 10$

$= \sqrt{10} \times \sqrt{10}\cos(\theta + 18{\cdot}4)^\circ + 10$

So minimum value is

$\sqrt{10} \times \sqrt{10} \times (-1) + 10$

$= -10 + 10 = 0$

6. a) $f(x) = x^4 + 4x^3 + 5x^2 + 14x + 24$

-2	1	4	5	14	24
		-2	-4	-2	-24
	1	2	1	12	0

So $f(x) = (x + 2)(x^3 + 2x^2 + x + 12)$

-3	1	2	1	12
		-3	3	-12
	1	-1	4	0

giving
$f(x) = (x + 2)(x + 3)(x^2 - x + 4)$

Thus $a = 2$ and $b = 3$
(or $a = 3$ and $b = 2$)

b) The equation $f(x) = 0$ becomes

$(x + 2)(x + 3)(x^2 - x + 4) = 0$

So $x + 2 = 0$ or $x + 3 = 0$ or
$x^2 - x + 4 = 0$

$\Rightarrow x = -2$ or $x = -3$

For $x^2 - x + 4 = 0$

the discriminant $= (-1)^2 - 4 \times 1 \times 4$
$= 1 - 16 = -15$

Since this is negative there are no real roots.

The only two real roots of $f(x) = 0$ are $x = -2$ and $x = -3$

7. $5 \times 3^\alpha = 2 \Rightarrow 3^\alpha = \frac{2}{5} \Rightarrow 3^\alpha = 0{\cdot}4$

So $\log_{10}3^\alpha = \log_{10}0{\cdot}4$

$\Rightarrow \alpha\log_{10}3 = \log_{10}0{\cdot}4$

$$\Rightarrow \alpha = \frac{\log_{10}0\cdot4}{\log_{10}3} = -0\cdot834\ldots$$

Thus $\cos^2 x - \sin^2 x = -0{\cdot}834\ldots$

$\Rightarrow \cos 2x = -0{\cdot}834\ldots$

($2x$ is in 2nd or 3rd quadrants)

(1st quadrant angle is $0{\cdot}584\ldots$ radians)

So $2x = \pi - 0{\cdot}584\ldots$ or $\pi + 0{\cdot}584\ldots$

So $2x = 2{\cdot}557\ldots$ only the smallest positive value is required.

So $x = 1{\cdot}278\ldots$

The required value is:

$x \doteqdot 1{\cdot}28$ (to 3 sig. figs)

Chapter 3 sample exam question solutions

Non-calculator

1. Coordinates of M:

$$M\left(\frac{-1+3}{2}, \frac{3+(-1)}{2}\right)$$

$$= M\left(\frac{2}{2}, \frac{2}{2}\right) = M(1, 1)$$

Gradient of line:

$$m_{BC} = \frac{3-0}{-1-(-2)} = \frac{3}{-1+2} = \frac{3}{1} = 3$$

So $m_{\perp} = -\frac{1}{3}$

Equation of line:

Point on line is M(1, 1) and gradient is $-\frac{1}{3}$

Required equation is $y-1=-\frac{1}{3}(x-1)$

So $3y - 3 = -(x - 1)$

So $3y - 3 = -x + 1$

giving $3y + x = 4$

2. a) The multiplier a must lie between -1 and 1, i.e. $-1 < a < 1$, for a limit to exist.

b) Since $u_{n+1} = au_n + b$ and $u_1 = 5$

Then $u_2 = au_1 + b = a \times 5 + b$
$= 5a + b$

So $5a + b = 8$

also $u_3 = au_2 + b$

$= a \times 8 + b$

$= 8a + b$

So $8a + b = 9{\cdot}5$

Now solve $\left.\begin{array}{l} 5a+b=8 \\ 8a+b=9{\cdot}5 \end{array}\right\} \Rightarrow 3a = 1{\cdot}5$

So $a = \frac{1{\cdot}5}{3} = 0{\cdot}5$

So $5a + b = 8$

gives $5 \times 0{\cdot}5 + b = 8$

So $2{\cdot}5 + b = 8 \Rightarrow b = 5{\cdot}5$

c) The recurrence relation is
$u_{n+1} = 0{\cdot}5u_n + 5{\cdot}5$

Let the limit of the sequence be L then

$L = 0{\cdot}5L + 5{\cdot}5$

$L - 0{\cdot}5L = 5{\cdot}5 \Rightarrow 0{\cdot}5L = 5{\cdot}5$

$$\Rightarrow L = \frac{5{\cdot}5}{0{\cdot}5} = 11$$

3. Consider $x^2 + y^2 - 2x - 6y + 1 = 0$.

Centre is (1, 3)

Radius $= \sqrt{1^2 + 3^2 - 1}$

$= \sqrt{9} = 3$

Now use this information on the given diagram:

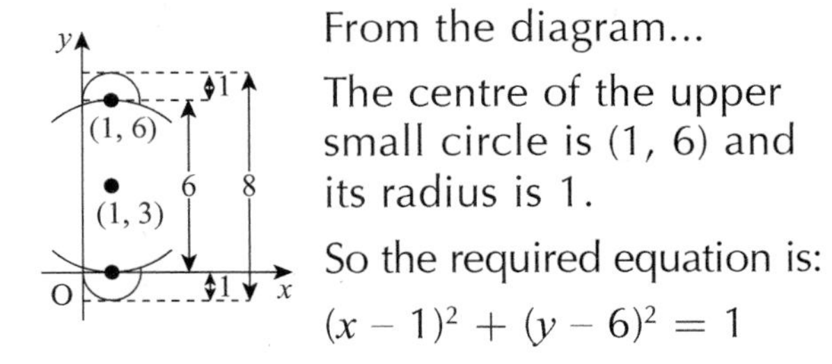

From the diagram...

The centre of the upper small circle is (1, 6) and its radius is 1.

So the required equation is:

$(x - 1)^2 + (y - 6)^2 = 1$

4. $$\int_{\frac{\pi}{6}}^{\frac{\pi}{2}} \sin x \, dx = [-\cos x]_{\frac{\pi}{6}}^{\frac{\pi}{2}}$$

$$= \left(-\cos\frac{\pi}{2}\right) - \left(-\cos\frac{\pi}{6}\right)$$

$$= -0 + \frac{\sqrt{3}}{2} = \frac{\sqrt{3}}{2}$$

Area of the shaded region is $\frac{\sqrt{3}}{2}$ unit²

5. a) For A(1, 2) and B(−3, −1) then

$$AB = \sqrt{(1-(-3))^2 + (2-(-1))^2}$$
$$= \sqrt{16+9} = \sqrt{25} = 5$$

For A(1, 2) and C(4, −2) then

$$AC = \sqrt{(1-4)^2 + (2-(-2))^2}$$
$$= \sqrt{9+16} = \sqrt{25} = 5$$

so AB = AC and triangle ABC is isosceles

b)
$$m_{AB} = \frac{2-(-1)}{1-(-3)} = \frac{3}{4}$$

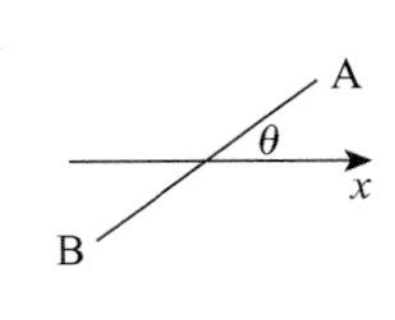

$$\tan\theta = \frac{3}{4}$$

Calculator allowed

1.

$$m_{SR} = m_{PQ} = \frac{4-(-1)}{-1-(-2)} = \frac{5}{1} = 5$$

So gradient of line is 5 and point on line is R(4, 5)

Equation is $y - 5 = 5(x - 4)$

$y - 5 = 5x - 20$

$y = 5x - 15$

2. a) Consider $y = x^3 - 2x + 1$

$$\Rightarrow \frac{dy}{dx} = 3x^2 - 2$$

when $x = 1$ (at point P) then

$$\frac{dy}{dx} = 3\times 1^2 - 2 = 3 - 2 = 1$$

The gradient of the tangent is 1, point on tangent is P(1, 0), so the equation is:

$y - 0 = 1(x - 1)$

$\Rightarrow y = x - 1$

b) To find Q, the point of intersection, solve:

$$\left.\begin{array}{l} y = x^3 - 2x + 1 \\ y = x - 1 \end{array}\right\} \Rightarrow x^3 - 2x + 1 = x - 1$$

$$\Rightarrow x^3 - 3x + 2 = 0$$

To factorise $x^3 - 3x + 2$ try dividing it by $x - 1$:

$$\begin{array}{c|cccc} 1 & 1 & 0 & -3 & 2 \\ & & 1 & 1 & -2 \\ \hline & 1 & 1 & -2 & 0 \end{array}$$

So $x^3 - 3x + 2 = (x - 1)(x^2 + x - 2)$

$= (x - 1)(x + 2)(x - 1)$

The equation becomes:

$(x - 1)(x + 2)(x - 1) = 0$

So $x = 1$ or $x = -2$

$x = 1$ gives the known point of intersection P(1, 0)

when $x = -2$ then $y = x - 1$

$= -2 - 1$

$= -3$

so Q(−2, −3)

c)

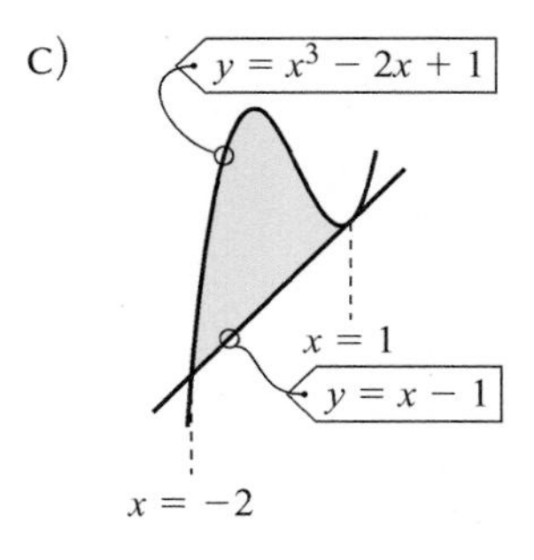

$$\int_{-2}^{1} (x^3 - 2x + 1) - (x - 1)\,dx$$

$$= \int_{-2}^{1} x^3 - 2x + 1 - x + 1\,dx$$

$$= \int_{-2}^{1} x^3 - 3x + 2\,dx$$

$$= \left[\frac{x^4}{4} - \frac{3x^2}{2} + 2x\right]_{-2}^{1}$$

$$= \left(\frac{1^4}{4} - \frac{3\times 1^2}{2} + 2\times 1\right)$$

$$- \left(\frac{(-2)^4}{4} - \frac{3\times(-2)^2}{2} + 2\times(-2)\right)$$

$$= \left(\frac{1}{4} - \frac{3}{2} + 2\right) - (4 - 6 - 4)$$

$$= \frac{1}{4} - \frac{6}{4} + 2 - 4 + 6 + 4$$

$$= -\frac{5}{4} + 8 = \frac{27}{4}$$

Required area (shaded) $= \frac{27}{4}$ unit^2

3.

$y = 4\sqrt{x}$

$y = 5 - x^2$

O 1 x y

$$\int_0^1 5 - x^2 - 4\sqrt{x}\,dx = \int_0^1 5 - x^2 - 4x^{\frac{1}{2}}\,dx$$

$$= \left[5x - \frac{x^3}{3} - \frac{4x^{\frac{3}{2}}}{\frac{3}{2}}\right]_0^1 = \left[5x - \frac{x^3}{3} - \frac{8\left(\sqrt{x}\right)^3}{3}\right]_0^1$$

$$= \left(5\times 1 - \frac{1^3}{3} - \frac{8\times\left(\sqrt{1}\right)^3}{3}\right)$$

$$- \left(5\times 0 - \frac{0^3}{3} - \frac{8\times\left(\sqrt{0}\right)^3}{3}\right)$$

$$= 5 - \frac{1}{3} - \frac{8}{3} = \frac{15}{3} - \frac{1}{3} - \frac{8}{3} = \frac{6}{3} = 2$$

The area of the end is 2 cm^2 so the volume is $2 \times 15 = 30$ cm^3.

So 30,000 require $30 \times 30{,}000$ $= 900{,}000$ cm^3.

Wastage = 11% of 900,000 $= 99{,}000$ cm^3

Total volume required $= 900{,}000 + 99{,}000$ $= 999{,}000$ cm^3

4. a) $x^2 + y^2 + 4x + 2y - 15 = 0$

Centre: $(-2, \; -1)$

So point on AB is $(-2, -1)$ and gradient is 2. The equation of AB is $y - (-1) = 2(x - (-2))$

$\Rightarrow y + 1 = 2(x + 2)$

$\Rightarrow y = 2x + 3$

The line AB, with equation $y = 2x + 3$, has y-axis intercept of $(0, 3)$.

When $x = 0$ and $y = 3$

$x^2 + y^2 + 4x + 2y - 15$

$= 0^2 + 3^2 + 4 \times 0 + 2 \times 3 - 15$

$= 0$

So $(0, 3)$ also lies on the circle. Hence, $A(0, 3)$ is the point of intersection of the line and circle and lies on the y-axis.

b)

$C(-2, -1)$

B

tangent

now $\overrightarrow{AC} = \overrightarrow{CB}$

So $\boldsymbol{c} - \boldsymbol{a} = \boldsymbol{b} - \boldsymbol{c}$

$\boldsymbol{b} = 2\boldsymbol{c} - \boldsymbol{a}$

$= 2\begin{pmatrix}-2\\-1\end{pmatrix} - \begin{pmatrix}0\\3\end{pmatrix}$

$= \begin{pmatrix}-4\\-5\end{pmatrix}$

Thus $B(-4, -5)$

So

$$m_{BC} = \frac{-1-(-5)}{-2-(-4)} = \frac{4}{2} = 2 \Rightarrow m_{\perp} = -\frac{1}{2}$$

gradient of tangent is $-\frac{1}{2}$, point on tangent is $B(-4, -5)$

so equation of tangent is

$$y - (-5) = -\frac{1}{2}(x - (-4))$$

$$\Rightarrow y + 5 = -\frac{1}{2}(x + 4)$$

$\Rightarrow 2y + 10 = -x - 4$

$\Rightarrow 2y + x = -14$

5. a) For the equation:

$$x^2 + y^2 - 2x + 2y - 7 = 0$$

Centre is $(1, \; -1)$

$$\text{Radius} = \sqrt{1^2 + (-1)^2 - (-7)}$$

$$= \sqrt{9} = 3$$

If D is the centre of the other circle then CD is parallel to the x-axis and CD $= 6$ so D$(-5, -1)$. The radius being 3 gives equation:

$$(x + 5)^2 + (y + 1)^2 = 9$$

b) For $x^2 + y^2 + 4x + 2y + c = 0$

The centre is E$(-2, -1)$

with radius $= \sqrt{(-2)^2 + (-1)^2 - c}$

$= \sqrt{5 - c}$

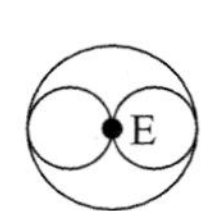

To enclose the two circles, the radius of large circle is greater than 6 units

so $\sqrt{5 - c} > 6$

$\Rightarrow 5 - c > 36$

$\Rightarrow 5 - 36 > c$

Thus $c < -31$

Index for Chapters 1-3

Index

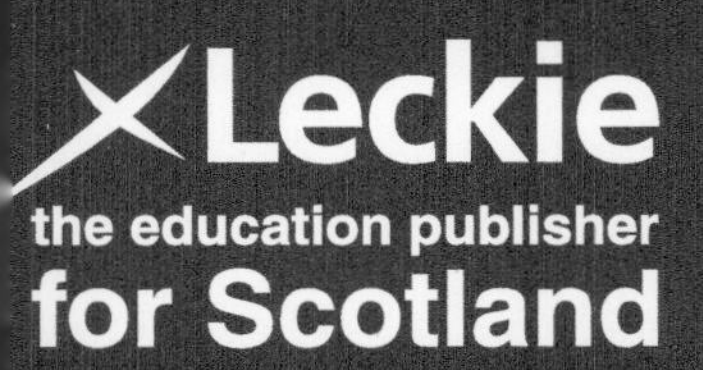

Higher MATHS

For SQA 2019 and beyond

Practice Papers

Ken Nisbet

Revision advice

Revision methods

Work out a revision timetable for each week's work in advance – remember to cover all of your subjects and to leave time for homework and breaks. For example:

Day	6pm–6.45pm	7pm–8pm	8.15pm–9pm	9.15pm–10pm
Monday	Homework	Homework	English Revision	Chemistry Revision
Tuesday	Maths Revision	Physics Revision	Homework	Free
Wednesday	Geography Revision	Modern Studies Revision	English Revision	French Revision
Thursday	Homework	Maths Revision	Chemistry Revision	Free
Friday	Geography Revision	French Revision	Free	Free
Saturday	Free	Free	Free	Free
Sunday	Modern Studies Revision	Maths Revision	Modern Studies	Homework

Make sure that you have at least one evening free a week to relax, socialise and re-charge your batteries. It also gives your brain a chance to process the information that you have been feeding it all week.

Arrange your study time into one hour or 30 minute sessions, with a break between sessions, e.g. 6pm – 7pm, 7.15pm–7.45pm, 8pm–9pm. Try to start studying as early as possible in the evening when your brain is still alert and be aware that the longer you put off starting, the harder it will be to start!

Study a different subject in each session, except for the day before an exam.

Do something different during your breaks between study sessions – have a cup of tea, or listen to some music. Don't let your 15 minutes expand into 20 or 25 minutes though!

Have your class notes and any textbooks available for your revision to hand, as well as plenty of blank paper, a pen, etc. You should take note of any topic area that you are having particular difficulty with, as and when the difficulty arises. Revisit that question later having revised that topic area, by attempting some further questions from the exercises in your textbook or referring to the revision guide.

Revising for a Maths exam is different from revising for some of your other subjects. Revision is only effective if you are trying to solve problems. You may like to make a list of 'Key Questions' with the dates of your various attempts (successful or not!). These should be questions that you have had real difficulty with.

Key Question	1st Attempt		2nd Attempt		3rd Attempt	
Textbook P56 Q3a	18/2/20	✗	21/2/20	✔	28/2/20	✔
Practice Exam A Paper 1 Q5	25/2/20	✗	28/2/20	✗	3/3/20	
2017 SQA Paper, Paper 2 Q3	27/2/20	✗	2/3/20			

The method for working this list is as follows:

1. Any attempt at a question should be dated.
2. A tick or cross should be entered to mark the success or failure of each attempt.
3. A date for your next attempt at that question should be entered. For an unsuccessful attempt: 3 days later; for a successful attempt: 1 week later

4. After two successful attempts remove that question from the list (you can assume the question has been learnt!)

Using 'The List' method for revising for your Maths Exam ensures that your revision is focused on the difficulties you have had and that you are actively trying to overcome these difficulties.

Finally, forget or ignore all or some of the advice in this section if you are happy with your present way of studying. Everyone revises differently, so find a way that works for you!

Transfer your knowledge

As well as using your class notes and textbooks to revise, these practice exam papers will also be a useful revision tool as they will help you to get used to answering exam-style questions. You may find as you work through the questions that you find an example that you haven't come across before. Don't worry! There may be several reasons for this. You may have come across a question on a topic that you have not yet covered in class. Check with your teacher to find out if this is the case. Or it may be the case that the wording or the context of the question is unfamiliar. This is often the case with reasoning questions in the Maths exam. Once you have familiarised yourself with the worked solutions, in most cases you will find that the question is using mathematical techniques with which you are familiar. In either case you should revisit that question later to check that you can successfully solve it.

Trigger words

In the practice exam papers and in the exam itself, a number of 'trigger words' will be used in the questions. These trigger words should help you identify a process or a technique that is expected in your solution to that part of the question. If you familiarise yourself with these trigger words, it will help you to structure your solutions more effectively.

Trigger word	Meaning/Explanation
Evaluate	Carry out a calculation to give an answer that is a value.
Hence	You must use the result of the previous part of the question to complete your solution. No marks will be given if you use an alternative method that does not use the previous answer (unless 'or otherwise' appears after 'hence').
Determine	Usually you should find a numerical value or values.
Expand	This means different things in different contexts: Algebraic expressions: remove brackets. Trig examples: use the addition formulae sin(A$\pm$B) or cos(A$\pm$B).
Show that	You should include every step of your working.
Simplify	This means different things in different contexts: Surds: reduce the number under the root sign to the smallest possible by removing square factors. Fractions: one fraction, cancelled down, is expected. Algebraic expressions: get rid of brackets and gather all like terms together.
Express	This usually means that you are required to rewrite what you are given in a particular form.
Algebraically	The method you use must involve algebra, e.g. you must solve an equation or simplify an algebraic equation. It is usually stated to avoid trial-and-improvement methods or reading answers from your calculator.
Justify your answer	This is a request for you to indicate clearly your reasoning. Will the examiner know how your answer was obtained?
Show all your working	Marks will be allocated for the individual steps in your working. Steps missed out may lose you marks.

In the exam

Watch your time and pace yourself carefully. Some questions you will find harder than others. Try not to get stuck on one question as you may later run out of time. Rather, return to a difficult question later. Remember also that if you have spare time towards the end of your exam, use this time to check through your solutions. Often mistakes are discovered in this checking process and can be corrected.

Become familiar with the exam instructions. The practice exam papers in this book have the exam instructions at the front of each exam. Also, remember that there is a formulae list to consult. You will find this at the front of your exam paper. However, even though these formulae are given to you, it is important that you learn them so that they are familiar to you.

Read the question thoroughly before you begin to answer it – make sure you know exactly what the question is asking you to do. If the question is in sections (e.g. 15a, 15b, 15c, etc), then it is often the case that answers obtained in the earlier sections are used in the later sections of that question.

When you have completed your solution, read it over again. Is your reasoning clear? Will the examiner understand how you arrived at your answer? If in doubt, fill in more details.

If you change your mind or think that your solution is wrong, don't score it out unless you have another solution to replace it with. Solutions that are not correct can often gain some of the marks available. Do not miss working out. Showing step-by-step working will help you gain some marks even if there is a mistake in the working.

Use these resources constructively by reworking questions later that you found difficult or impossible first time round. Remember: success in a Maths exam will only come from actively trying to solve lots of questions and only consulting notes when you are stuck. Reading notes alone is not a good way to revise for your Maths exam. Always be active, always solve problems.

Good luck!

Topic Index

Topic	A Paper 1	A Paper 2	B Paper 1	B Paper 2
Chapter 1				
Functions and graphs	4	4	2,15a	10a,b
Exponential and logarithmic functions	11	8	3,20,15a	
Trigonometric functions	1,13a		5	8a,b
Vectors	5	1c,6	4,9	5
Chapter 2				
Polynomials	2,12		11,15b	6
Trigonometric equations	13b,14	10		4,8c
Differentiation	8	1a,2,4	7,14	1,8c
Integration	3	5	1,8,12	7
Chapter 3				
The straight line	8	3	6	3
The circle	9	1b		2
Sequences and recurrence relations	6	7	13	
Applications of differentiation and integration	7	9		9,10c,d

This table shows how the various Chapter 1, Chapter 2 and Chapter 3 topics are distributed throughout the practice exam papers. You may find this information helpful if you particularly wish to focus on the revision of one particular topic.

Practice Paper A

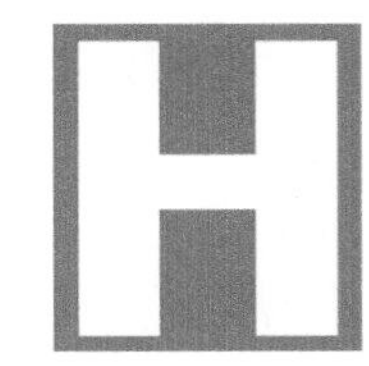

Higher Mathematics

Practice Papers for SQA Exams
Duration - 1 hour 30 minutes

Exam A
Paper 1
Non-calculator

Fill in these boxes and read what is printed below.

Full name of centre	Town

Forename(s)	Surname

Total marks — 70

Attempt ALL questions.

You may NOT use a calculator.

To earn full marks you must show your working in your answers.

State the units for your answers where appropriate.

Use blue or black ink.

FORMULAE LIST

Circle:

The equation $x^2 + y^2 + 2gx + 2fy + c = 0$ represents a circle centre $(-g, -f)$ and radius $\sqrt{g^2 + f^2 - c}$

The equation $(x - a)^2 + (y - b)^2 = r^2$ represents a circle centre (a, b) and radius r

Scalar product: $\boldsymbol{a}.\boldsymbol{b} = |\boldsymbol{a}||\boldsymbol{b}| \cos \theta$, where θ is the angle between $\boldsymbol{a}$ and $\boldsymbol{b}$

or $\boldsymbol{a}.\boldsymbol{b} = a_1b_1 + a_2b_2 + a_3b_3$ where $\boldsymbol{a} = \begin{pmatrix} a_1 \\ a_2 \\ a_3 \end{pmatrix}$ and $\boldsymbol{b} = \begin{pmatrix} b_1 \\ b_2 \\ b_3 \end{pmatrix}$

Trigonometric formulae:

$$\sin(A \pm B) = \sin A \cos B \pm \cos A \sin B$$

$$\cos(A \pm B) = \cos A \cos B \mp \sin A \sin B$$

$$\sin 2A = 2\sin A \cos A$$

$$\cos 2A = \cos^2 A - \sin^2 A$$

$$= 2\cos^2 A - 1$$

$$= 1 - 2\sin^2 A$$

Table of standard derivatives:

$f(x)$	$f'(x)$
$\sin ax$	$a \cos ax$
$\cos ax$	$-a \sin ax$

Table of standard integrals:

$f(x)$	$\int f(x)\,dx$
$\sin ax$	$-\frac{1}{a}\cos ax + C$
$\cos ax$	$\frac{1}{a}\sin ax + C$

MARKS Do not write in this margin

1. (a) Expand $\cos(x - 30)^\circ$ **1**

   (b) Hence, or otherwise, find the exact value of $\cos 15^\circ$ **3**

2. Find the range of values of k for which $(2k + 3)x^2 - kx - 1 = 0$ has real roots where $k \in \mathbb{R}$ **4**

3. Find $\int \frac{2x^5 - 3}{3x^4}\,dx$ **4**

4. Two functions are defined by $f(x) = \frac{1}{2}x + 1$ and $g(x) = 3x - \frac{1}{2}$ where $x \in \mathbb{R}$

   (a) Find a formula for $h(x)$ where $h(x) = g(f(x))$ **2**

   (b) $h^{-1}(x) = px + q$ where p and q are constants. Find the values of p and q. **3**

5. Five points are located in space.

   The following facts are known:

   - A has coordinates $(-1, 2, 3)$
   - $\overrightarrow{AB} = 4\boldsymbol{i} - 2\boldsymbol{k}$
   - $\overrightarrow{BC} = \overrightarrow{AB} + 2(\boldsymbol{i} + \boldsymbol{j} + \boldsymbol{k})$
   - $\overrightarrow{AB} = -2\overrightarrow{CD}$ and $\overrightarrow{BC} = 2\overrightarrow{DE}$

   B

   $A(-1, 2, 3)$

   C

   D

   E

   Find the coordinates of the point E. **4**

6. A sequence is defined by the recurrence relation $u_{n+1} = \frac{1}{4}u_n + 6$ with $u_5 = 12$.

   (a) Find the value of u_6. **1**

   (b) Explain why this sequence approaches a limit as $n \to \infty$ **1**

   (c) Calculate this limit. **2**

7. The diagram shows the graphs of two functions.

$$f(x) = x^3 - 6x^2 + 3x + 1$$

$$g(x) = -2x + 1$$

   (a) Show that $f(1) = g(1)$ **1**

   (b) Calculate the shaded area **5**

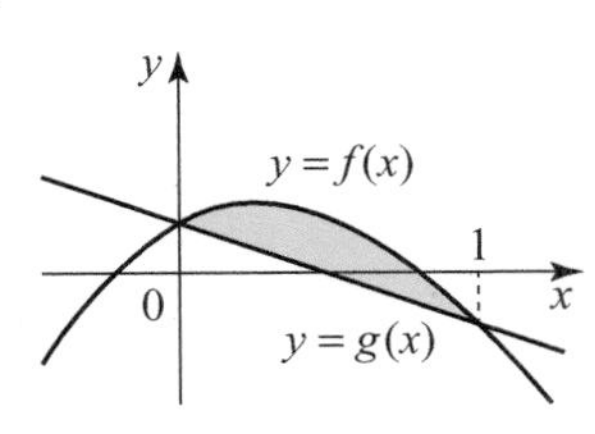

MARKS Do not write in this margin

8. The line l_1 makes an angle of $\frac{\pi}{3}$ with the positive direction of the x-axis.

Line l_2 is perpendicular to l_1

(a) Find the gradient of l_1 **1**

(b) Find the equation of l_2 given that it passes through the point $(0, -\sqrt{2})$ **3**

(c) Determine where l_2 crosses the x-axis. **2**

9. A circle touches both the x-axis and the y-axis and has its centre on the line $y = x$

The centre is at a distance of $\sqrt{2}$ units from the origin.

Find the two possible equations for the circle. **5**

10. Find the rate of change of $f(x) = \cos\left(3x - \frac{\pi}{6}\right)$ when $x = \frac{\pi}{3}$ **4**

11. (a) Express $\log_3(9\sqrt{x})$ in the form $a\log_3 x + b$ where a and b are constants. **3**

(b) Hence, or otherwise, evaluate $\log_3(9\sqrt{x})$ when $x = \sqrt{3}$ **1**

12. Two cubic graphs, $y = f(x)$ and $y = g(x)$, where $f(x) = 2x^3 + 3x + 12$ and $g(x) = 2 + 16x^2 - x^3$, are shown in the diagram.

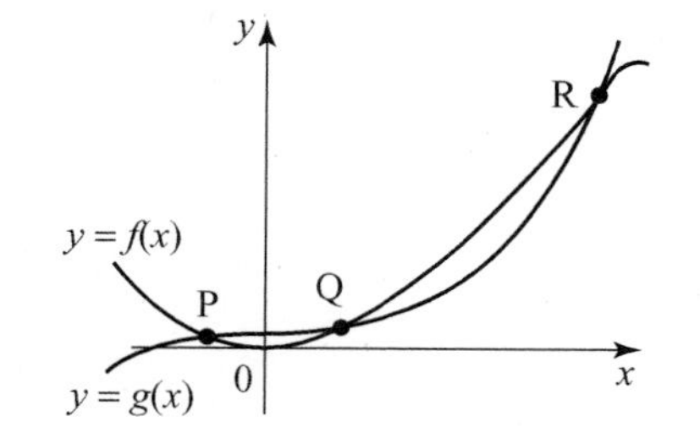

Determine the x-coordinates of each of P, Q and R, the three points of intersection of the two graphs. **8**

13. (a) Express $\sqrt{3}\cos x° - \sin x°$ in the form $k\cos(x + a)°$

where $k > 0$ and $0 \leq a \leq 90$ **4**

(b) Hence, solve the equation $\sqrt{3}\cos x° - \sin x° = 1$ for $0 < x < 360$ **3**

14. Solve the equation $\sin\theta(\sin\theta - 1) = \cos^2\theta$ for $\frac{\pi}{2} < \theta < \frac{3\pi}{2}$ **5**

[END OF QUESTION PAPER]

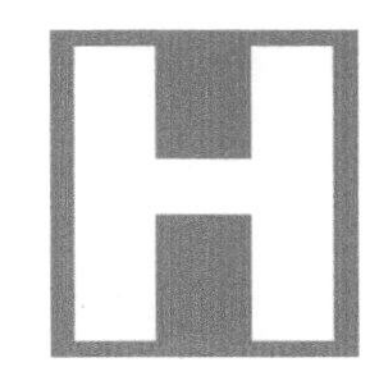

Higher Mathematics

Practice Papers for SQA Exams
Duration - 1 hour 45 minutes

Exam A
Paper 2

Fill in these boxes and read what is printed below.

Full name of centre	Town

Forename(s)	Surname

Total marks — 80

Attempt ALL questions.

You may use a calculator.

To earn full marks you must show your working in your answers.

State the units in your answer where appropriate.

You will not earn marks for answers obtained by readings from scale drawings.

Use blue or black ink.

FORMULAE LIST

Circle:

The equation $x^2 + y^2 + 2gx + 2fy + c = 0$ represents a circle centre $(-g, -f)$ and radius $\sqrt{g^2 + f^2 - c}$

The equation $(x - a)^2 + (y - b)^2 = r^2$ represents a circle centre (a, b) and radius r

Scalar Product: $\boldsymbol{a}.\boldsymbol{b} = |\boldsymbol{a}||\boldsymbol{b}| \cos \theta$, where θ is the angle between $\boldsymbol{a}$ and $\boldsymbol{b}$

or $\boldsymbol{a}.\boldsymbol{b} = a_1b_1 + a_2b_2 + a_3b_3$ where $\boldsymbol{a} = \begin{pmatrix} a_1 \\ a_2 \\ a_3 \end{pmatrix}$ and $\boldsymbol{b} = \begin{pmatrix} b_1 \\ b_2 \\ b_3 \end{pmatrix}$

Trigonometric formulae:

$$\sin (A \pm B) = \sin A \cos B \pm \cos A \sin B$$

$$\cos (A \pm B) = \cos A \cos B \mp \sin A \sin B$$

$$\sin 2A = 2\sin A \cos A$$

$$\cos 2A = \cos^2 A - \sin^2 A$$

$$= 2\cos^2 A - 1$$

$$= 1 - 2\sin^2 A$$

Table of standard derivatives:

$f(x)$	$f'(x)$
$\sin ax$ $\cos ax$	$a \cos ax$ $-a \sin ax$

Table of standard integrals:

$f(x)$	$\int f(x)\,dx$
$\sin ax$	$-\frac{1}{a} \cos ax + C$
$\cos ax$	$\frac{1}{a} \sin ax + C$

MARKS
Do not write in this margin

1. The diagram shows a cubic curve with equation

$y = x^2 - \frac{1}{3}x^3$

A tangent PQ to the curve has point of contact $M\,(3, 0)$.

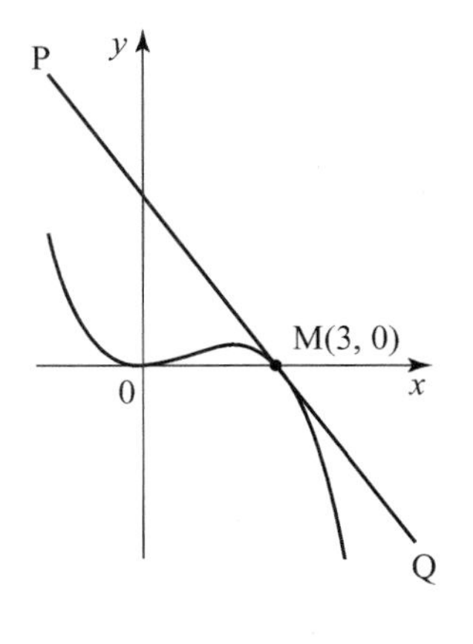

(a) Find the equation of PQ. **4**

A circle has equation $x^2 + y^2 - 4x - 26y + 163 = 0$

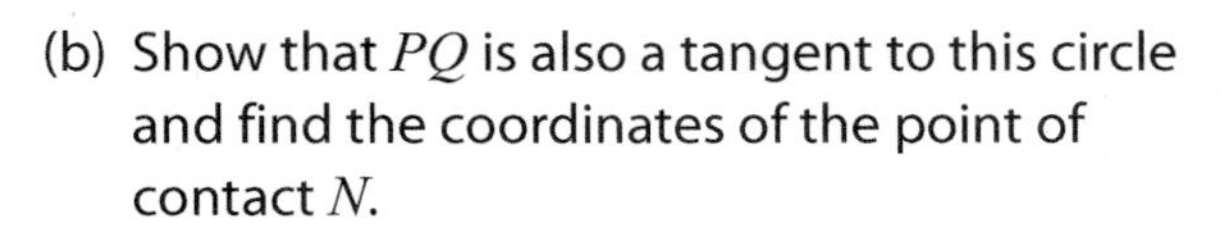

(b) Show that PQ is also a tangent to this circle and find the coordinates of the point of contact N. **6**

(c) Find the ratio in which the y-axis cuts the line MN. **3**

2. (a) Find the stationary points on the curve with equation $y = x^3 - 3x^2 + 4$ and justify their nature. **7**

(b) (i) Show that $(x + 1)(x - 2)^2 = x^3 - 3x^2 + 4$

(ii) Hence, sketch the graph of $y = x^3 - 3x^2 + 4$ **4**

3. Triangle PQR has coordinates $P(-3, -4)$, $Q(-3, 4)$ and $R(5, 12)$.

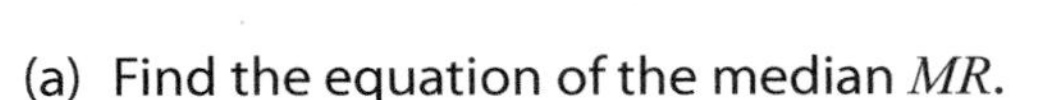

(a) Find the equation of the median MR. **3**

(b) Find the equation of the altitude NQ. **3**

(c) Median MR and altitude NQ intersect at point S. Find the coordinates of S. **3**

(d) The point $T\,(2, 9)$ lies on QR. Show that ST is parallel to PR. **2**

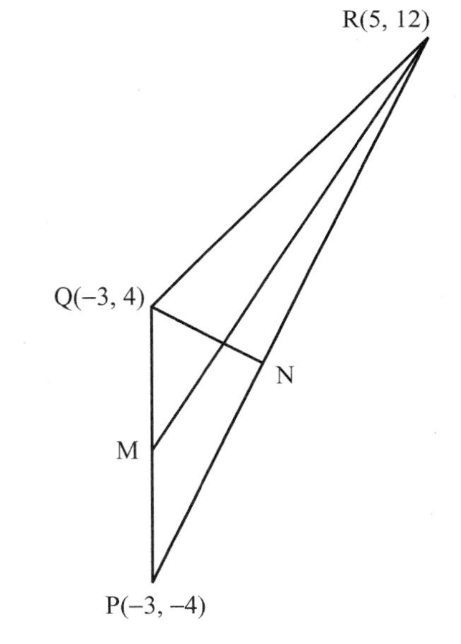

4. (a) Express $2x^2 - 8x + 9$ in the form $a(x - b)^2 + c$. **3**

(b) A function g is defined by $g(x) = \frac{2}{3}x^3 - 4x^2 + 9x + 1$. Find $g'(x)$ **2**

(c) Hence, or otherwise, explain why the curve with equation $y = g(x)$ is strictly increasing for all values of x. **2**

5. (a) Show that $\cos 2x + 1 = 2\cos^2 x$ **1**

(b) Hence, or otherwise, find $\int 2\cos^2 x\, dx$ **2**

MARKS Do not write in this margin

6. 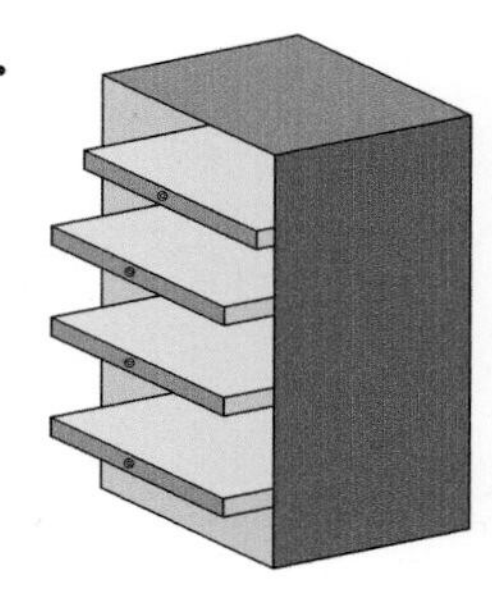

This set of drawers is being 'modelled' on a computer software design package as a cuboid as shown. The edges of the cuboid are parallel to the x, y and z-axes. Three of the vertices are $P(-1, -1, -1)$, $S(-1, 4, -1)$ and $V(3, 4, 5)$.

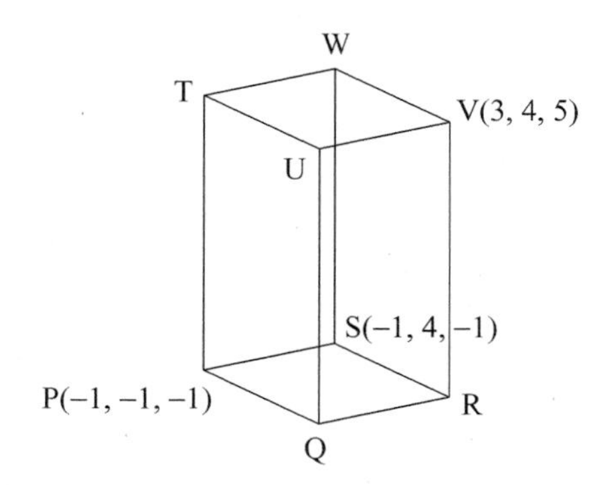

(a) Write down the lengths of PQ, QR and RV. **1**

(b) Write down the components of $\overrightarrow{VS}$ and $\overrightarrow{VP}$ and hence calculate the size of angle PVS. **7**

7. The islanders living in Tarbert on the island of Harris are planning to build a new sewage processing plant. Central to the plant is the seepage pit that allows most of the week's sewage to seep harmlessly through the soil and drain away. Sewage is pumped into the pit at the start of each week. These are two possible sites with the following specifications:

	Seepage rate	**Pumping capacity**
Upland site:	65% of 1 week's sewage	2000 litres at start of week
Lowland site:	75% of 1 week's sewage	2500 litres at start of week

(a) Write down a recurrence relation for each site. Use u_n to represent the amount of litres of sewage stored at the Upland site immediately after pumping at the start of the nth week, and let v_n be the equivalent volume at the Lowland site. Clearly label each relation with the site name. **2**

(b) The size of the storage tank at each site is determined by the maximum volume of sewage that will remain at the site in the long term. Which site requires the smaller tank in the long term? **4**

MARKS
Do not write in this margin

8. Atmospheric pressure decreases exponentially as you rise above sea-level. It is known that the atmospheric pressure, P(h), at a height h kilometres above sea-level is given by $P(h) = P_0e^{-kh}$ where P_0 is the pressure at sea-level ($h = 0$).

(a) Given that at a height of 4·95 km the atmospheric pressure is half that at sea-level, calculate the value of k correct to 4 decimal places. **3**

(b) Mount Everest is 8850 metres high. What is the percentage decrease in air pressure at the top of Mount Everest, compared to the pressure at sea-level? **2**

9. The diagram shows the curve with equation $y = 6 + 4x - x^2$ and the straight line with equation $y = x + 2$. The line intersects the curve at points S and T as shown.

(a) Calculate the exact value of the area enclosed by the curve and the line. **7**

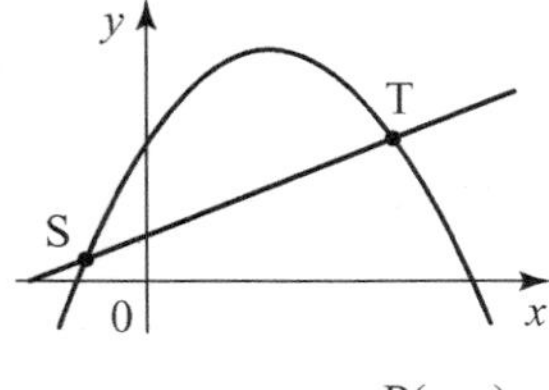

(b) A point $P(x, y)$ lies on the curve between S and T. It is known that the area of triangle PST (shaded in the diagram) is given by:

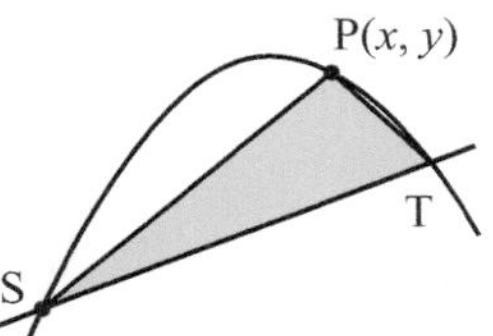

$$A(x) = -\frac{5}{2}x^2 + \frac{15}{2}x + 10$$

Calculate the maximum value of this area and hence determine what fraction this maximum value is of the area enclosed by the curve and the line from part (a). **4**

10. In right-angled triangle PQR, RS is the bisector of angle PRQ.

$PR = 5$ units and $PQ = 12$ units. Show that the exact value of $\cos\theta$ is $\frac{3\sqrt{13}}{13}$ **5**

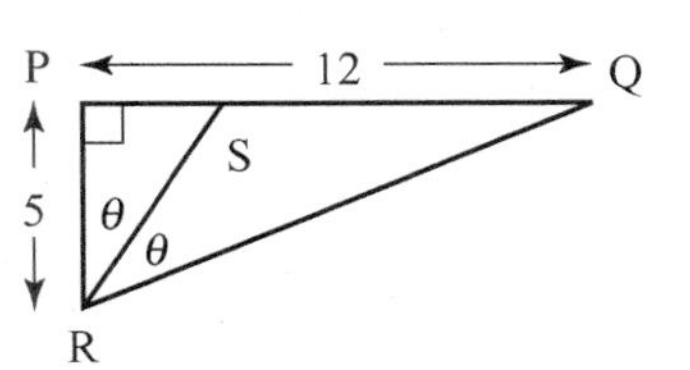

[END OF QUESTION PAPER]

Practice Paper B

Higher Mathematics

Practice Papers for SQA Exams

Duration - 1 hour 30 minutes

Exam B

Paper 1

Non-calculator

Fill in these boxes and read what is printed below.

Full name of centre

Town

Forename(s)

Surname

Total marks — 70

Attempt ALL questions.

You may NOT use a calculator.

To earn full marks you must show your working in your answers.

State the units for your answers where appropriate.

You will not earn marks for answers obtained by readings from scale drawings.

Use blue or black ink.

FORMULAE LIST

Circle:

The equation $x^2 + y^2 + 2gx + 2fy + c = 0$ represents a circle centre $(-g, -f)$ and radius $\sqrt{g^2 + f^2 - c}$

The equation $(x - a)^2 + (y - b)^2 = r^2$ represents a circle centre (a, b) and radius r

Scalar product: $\boldsymbol{a}.\boldsymbol{b} = |\boldsymbol{a}||\boldsymbol{b}| \cos \theta$, where θ is the angle between $\boldsymbol{a}$ and $\boldsymbol{b}$

or $\boldsymbol{a}.\boldsymbol{b} = a_1b_1 + a_2b_2 + a_3b_3$ where $\boldsymbol{a} = \begin{pmatrix} a_1 \\ a_2 \\ a_3 \end{pmatrix}$ and $\boldsymbol{b} = \begin{pmatrix} b_1 \\ b_2 \\ b_3 \end{pmatrix}$

Trigonometric formulae:

$$\sin (A \pm B) = \sin A \cos B \pm \cos A \sin B$$

$$\cos (A \pm B) = \cos A \cos B \mp \sin A \sin B$$

$$\sin 2A = 2\sin A \cos A$$

$$\cos 2A = \cos^2 A - \sin^2 A$$

$$= 2\cos^2 A - 1$$

$$= 1 - 2\sin^2 A$$

Table of standard derivatives:

$f(x)$	$f'(x)$
$\sin ax$	$a \cos ax$
$\cos ax$	$-a \sin ax$

Table of standard integrals:

$f(x)$	$\int f(x)\,dx$
$\sin ax$	$-\frac{1}{a} \cos ax + C$
$\cos ax$	$\frac{1}{a} \sin ax + C$

MARKS
Do not write in this margin

1. Find $\int (2x^2 + 3)dx$ — **3**

2. A function is defined on the set of real numbers by:

$$f(x) = \frac{1}{x+1} \quad (x \neq -1)$$

(a) Find an expression for $f^{-1}(x)$ — **2**

(b) State a suitable domain for f^{-1} — **1**

3. The diagram shows a sketch of part of the graph $y = \log_3 x$

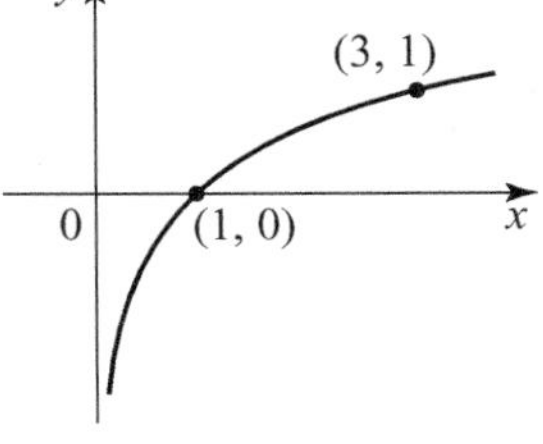
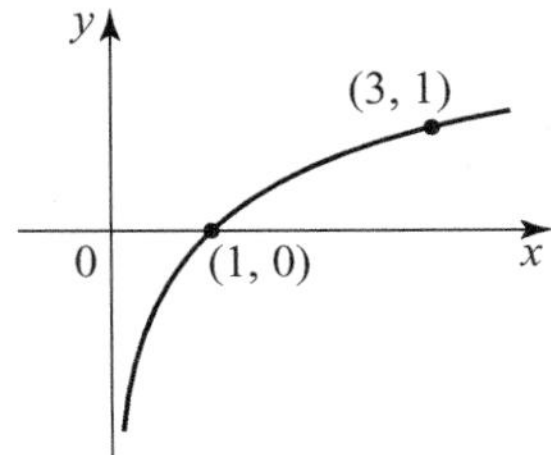

(a) Draw a sketch of the related graph
$y = 2\log_3(x + 3)$
showing clearly where it crosses the x-axis and y-axis. — **2**

(b) Hence, sketch the graph $y = \log_3 \frac{1}{(x+3)^2}$ — **2**

4. The diagram shows a cube with edge length 1 unit. Vectors $\boldsymbol{p}$ and $\boldsymbol{q}$ are represented by the line segments as shown.

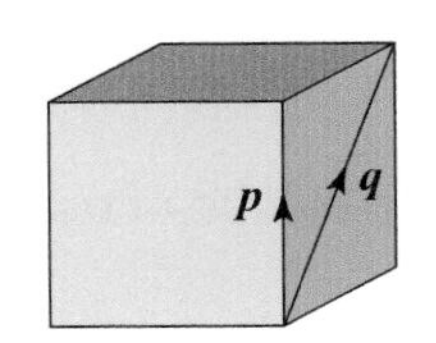

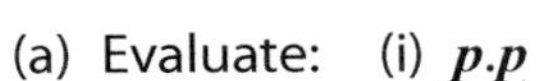

(a) Evaluate: (i) $\boldsymbol{p}.\boldsymbol{p}$ (ii) $\boldsymbol{q}.\boldsymbol{q}$ (iii) $\boldsymbol{p}.\boldsymbol{q}$ — **3**

(b) If $\boldsymbol{r}$ is the resultant of $2\boldsymbol{p} + \boldsymbol{q}$ evaluate $\boldsymbol{r}.\boldsymbol{r}$ — **3**

(c) Hence, write down the value of $|\boldsymbol{r}|$ — **1**

5. (a) Find the exact value of $\cos\left(A + \frac{5\pi}{6}\right) - \sin\left(A + \frac{4\pi}{3}\right)$ by expanding and simplifying. — **3**

(b) Hence, show that $\cos\left(\frac{11\pi}{12}\right) = \sin\left(\frac{17\pi}{12}\right)$ — **1**

6. (a) $A(-2, 1)$, $B(3, 2)$, and $C(3, -1)$ are vertices of a triangle.

Find the equation of the line through C parallel to AB. — **3**

(b) Find the coordinates of the point where this line meets the x-axis — **1**

7. Calculate the rate of change of $f(t) = \frac{2}{3t}$, $t \neq 0$, when $t = 2$. — **3**

8. Find $\int (7\cos(3x - 1) - 3\sin x)\, dx$. — **3**

9. If $\boldsymbol{p} = \begin{pmatrix} -1 \\ 2 \\ 3 \end{pmatrix}$ and $\boldsymbol{q} = \begin{pmatrix} 1 \\ 3 \\ 2 \end{pmatrix}$

(a) Write down the components of $\boldsymbol{p} + \boldsymbol{q}$ and $\boldsymbol{p} - \boldsymbol{q}$. — **1**

(b) Hence, show that $\boldsymbol{p} + \boldsymbol{q}$ and $\boldsymbol{p} - \boldsymbol{q}$ are perpendicular. — **2**

MARKS Do not write in this margin

10. Molten metal in a cooling chamber cools according to the law $T_t = T_0 \times 10^{-kt}$ where T_0 is the initial temperature and T_t is the temperature after t minutes. All temperatures are measured in °C.

(a) A metal at 3000°C takes 20 minutes to cool to 300°C. Calculate the value of k. **2**

(b) What will the temperature of the metal be after a further 20 minutes? (Show clearly your reasoning.) **2**

11. (a) Show that $(x + 4)$ is a factor of $x^3 + 6x^2 - 32$. **3**

(b) Hence, or otherwise, factorise $x^3 - 6x^2 - 32$ fully. **2**

12. (a) Find $\int \sqrt{1+3x}\,dx$ **3**

(b) Hence, show that $\int_0^1 \sqrt{1+3x}\,dx = \frac{14}{9}$ **2**

13. The terms of a sequence are generated by the recurrence relation

$$u_{n+1} = pu_n - 1 \text{ with } u_0 = 3$$

(a) Express u_1 and u_2 in terms of p. **2**

(b) Given that $u_2 = 1$ find the value of p that generates a sequence with a limit. **3**

(c) Calculate the value of this limit. **2**

14. The diagram shows a sketch of the curve with equation $y = \frac{1}{16}x^4 - \frac{1}{8}x^2 + x$. The line $y = x + c$ is a tangent to this curve.

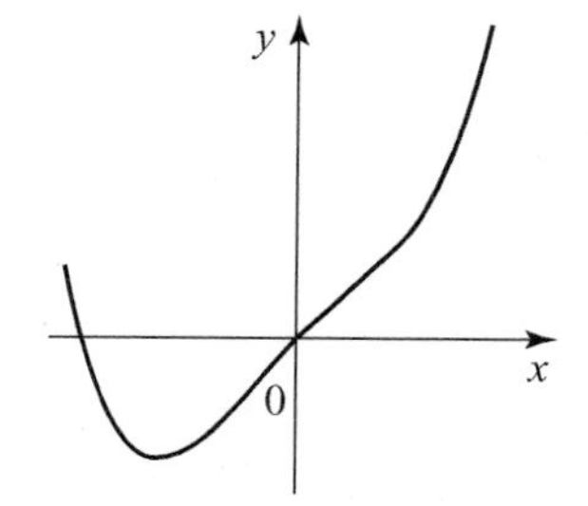

Find the possible values for c. For each value, find the coordinates of the point of contact of the tangent. **7**

15. Functions f and g are defined by $f(x) = 2x - 1$ and $g(x) = \log_{12} x$ on suitable domains.

(a) Show that the equation $f(g(x)) + g(f(x)) = 0$ has a solution $x = 2$. **6**

(b) Show that the equation has no other real solutions. **2**

[END OF QUESTION PAPER]

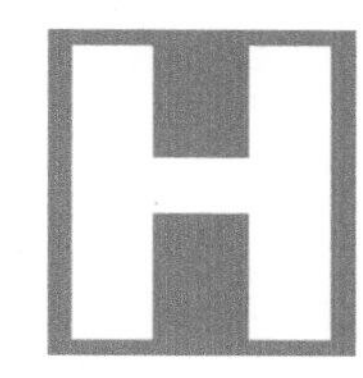

Higher Mathematics

Practice Papers for SQA Exams
Duration - 1 hour 45 minutes

Exam B
Paper 2

Fill in these boxes and read what is printed below.

Full name of centre	Town

Forename(s)	Surname

Total marks — 80

Attempt ALL questions.

You may NOT use a calculator.

To earn full marks you must show your working in your answers.

State the units in your answer where appropriate.

You will not earn marks for answers obtained by readings from scale drawings.

Use blue or black ink.

FORMULAE LIST

Circle:

The equation $x^2 + y^2 + 2gx + 2fy + c = 0$ represents a circle centre $(-g, -f)$ and radius $\sqrt{g^2 + f^2 - c}$

The equation $(x - a)^2 + (y - b)^2 = r^2$ represents a circle centre (a, b) and radius r

Scalar Product: $\boldsymbol{a}.\boldsymbol{b} = |\boldsymbol{a}||\boldsymbol{b}| \cos \theta$, where θ is the angle between $\boldsymbol{a}$ and $\boldsymbol{b}$

or $\boldsymbol{a}.\boldsymbol{b} = a_1b_1 + a_2b_2 + a_3b_3$ where $\boldsymbol{a} = \begin{pmatrix} a_1 \\ a_2 \\ a_3 \end{pmatrix}$ and $\boldsymbol{b} = \begin{pmatrix} b_1 \\ b_2 \\ b_3 \end{pmatrix}$

Trigonometric formulae:

$$\sin (A \pm B) = \sin A \cos B \pm \cos A \sin B$$

$$\cos (A \pm B) = \cos A \cos B \mp \sin A \sin B$$

$$\sin 2A = 2\sin A \cos A$$

$$\cos 2A = \cos^2 A - \sin^2 A$$

$$= 2\cos^2 A - 1$$

$$= 1 - 2\sin^2 A$$

Table of standard derivatives:

$f(x)$	$f'(x)$
$\sin ax$	$a \cos ax$
$\cos ax$	$-a \sin ax$

Table of standard integrals:

$f(x)$	$\int f(x)\,dx$
$\sin ax$	$-\frac{1}{a} \cos ax + C$
$\cos ax$	$\frac{1}{a} \sin ax + C$

MARKS
Do not write in this margin

1. A function f is defined by the formula $f(x) = x^3 + 3x^2 - 4$

(a) Find the coordinates of the stationary points on the graph with equation $y = f(x)$ and determine their nature. **6**

(b) Given that $x^3 + 3x^2 - 4 = (x + 2)^2 (x - 1)$ find the coordinates of the points where the curve $y = f(x)$ crosses the x and y-axes and hence sketch the curve. **4**

2. Three circles have equations as follows:

Circle A: $x^2 + y^2 + 4x - 6y + 5 = 0$

Circle B: $(x - 2)^2 + (y + 1)^2 = 2$

Circle C: $(x - 2)^2 + (y + 1)^2 = 40$

(a) (i) State the centre of circle A. **1**

(ii) Show that the radius of circle A is $2\sqrt{2}$. **1**

(b) (i) Calculate the distance between the centres of circles A and B writing your answer as a surd in its simplest form. **2**

(ii) Hence, show that circles A and B do not intersect. **2**

(c) Circles A and C intersect at points P and Q. Chord PQ has equation $y = x + 5$. Find the coordinates of points P and Q if P lies to the left of Q. **5**

3. The diagram shows kite $ABCD$ with diagonal AC drawn. The vertices of the kite are $A(-5, 2)$, $B(-3, 8)$, $C(3, 6)$ and $D(5, -8)$.

The dotted line shows the perpendicular bisector of AB.

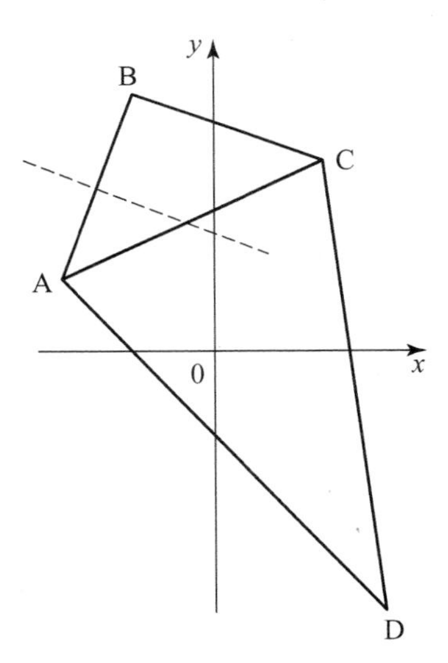

(a) Show that the perpendicular bisector of AB has equation $3y + x = 11$. **4**

(b) Find the equation of the median from C in triangle ACD. **3**

(c) The perpendicular bisector of AB and the median from C in triangle ACD meet at the point S. Find the coordinates of S. **3**

MARKS
Do not write in this margin

4. Solve the equation

$$3 \cos 2x° + 9 \cos x° = \cos^2 x° - 7 \text{ for } 0 \leq x < 360$$ **5**

5. $OABC, DEFG$ is cuboid.

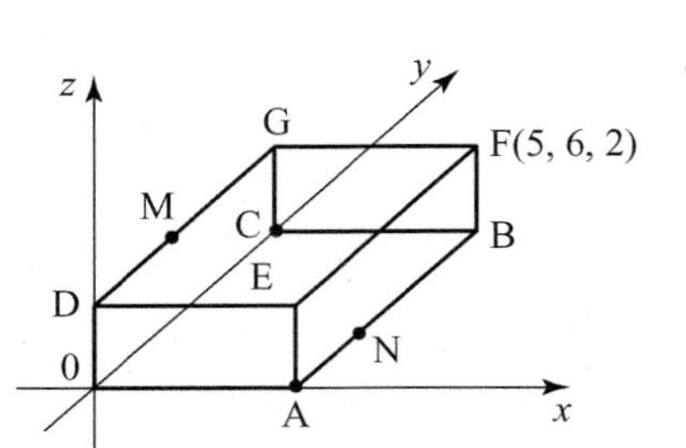

The vertex F is the point (5, 6, 2).

M is the midpoint of DG.

N divides AB in the ratio 1 : 2.

(a) Find the coordinates of M and N. **2**

(b) Write down the components of $\overrightarrow{MB}$ and $\overrightarrow{MN}$. **2**

(c) Find the size of angle BMN. **5**

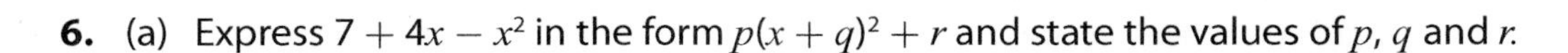

6. (a) Express $7 + 4x - x^2$ in the form $p(x + q)^2 + r$ and state the values of p, q and r. **3**

(b) Find the range of values for k such that $2x^2 - 3x + 1 - k = 0$ has no real roots. **3**

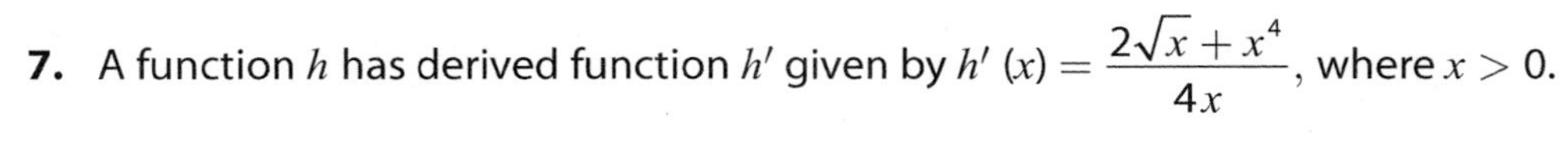

7. A function h has derived function h' given by $h'(x) = \dfrac{2\sqrt{x} + x^4}{4x}$, where $x > 0$.

If $h(4) = 17$ express $h(x)$ in terms of x. **4**

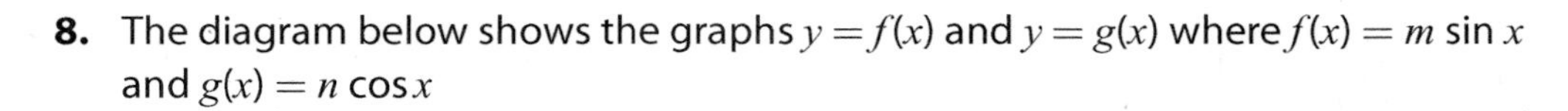

8. The diagram below shows the graphs $y = f(x)$ and $y = g(x)$ where $f(x) = m \sin x$ and $g(x) = n \cos x$

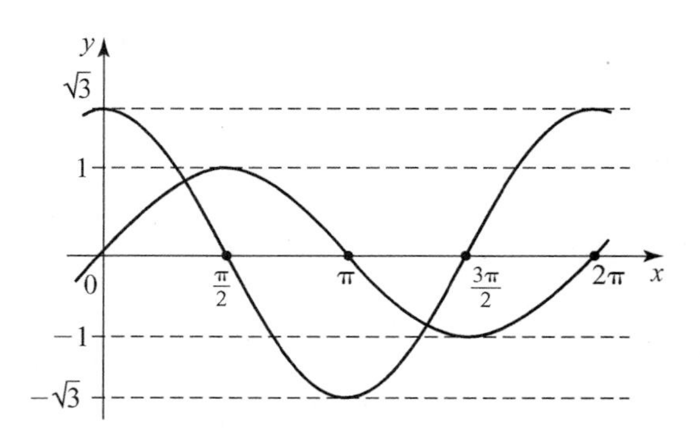

(a) Write down the values of m and n. **2**

(b) Write $f(x) - g(x)$ in the form $k \sin(x - a)$ where $k > 0$ and $0 < a < \dfrac{\pi}{2}$ **4**

(c) Hence find, in the interval $0 \leq x \leq \pi$ the x-coordinate of the point on the curve $y = f(x) - g(x)$ where the gradient is 2. **2**

MARKS Do not write in this margin

9. The diagram shows the graph with equation $y = x^4 - 1$

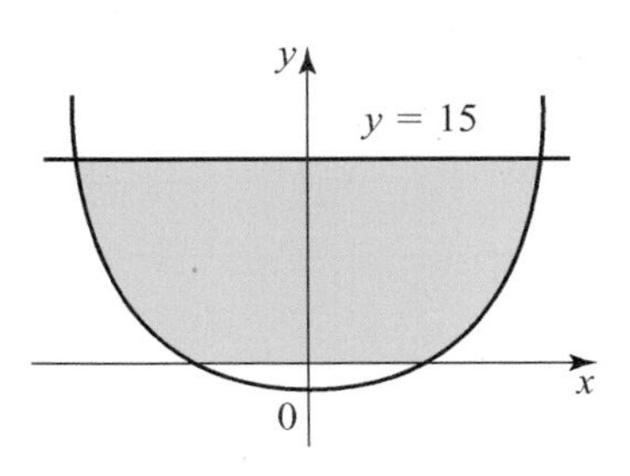

The graph has the y-axis as an axis of symmetry.

The shaded area lies between the curve, the x-axis and the line $y = 15$.

Calculate the exact value of the shaded area. **8**

10. The diagram shows a parabola with equation $y = f(x)$ passing through the points $(-3, 0)$, $(0, 9)$ and $(3, 0)$.

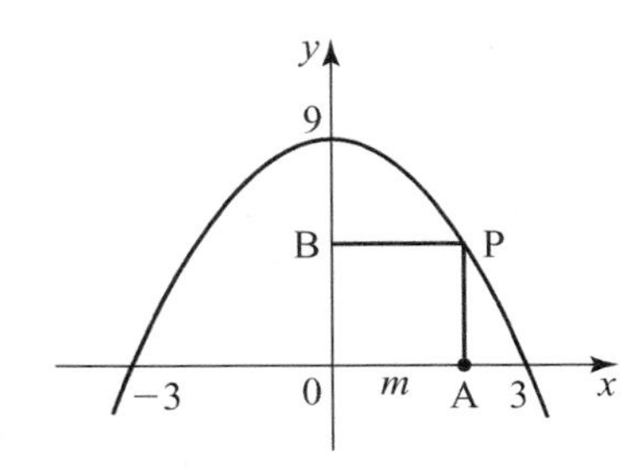

$OAPB$ is a rectangle with A and B lying on the axes and P lying on the parabola as shown. $OA = m$, $0 < m < 3$

(a) If $f(x)$ is of the form $-x^2 + a$ where a is a constant, determine the value of a. **1**

(b) Show that $AP = 9 - m^2$ **1**

(c) Find the value of m for which the area of the rectangle has a maximum. **6**

(d) Find the exact value of this maximum area. **1**

[END OF QUESTION PAPER]

Answers

Answers to Exam A

Exam A: Paper 1

Q1. (a)
$\cos(x-30)^\circ = \cos x^\circ \cos 30^\circ + \sin x^\circ \sin 30^\circ$ ✓

1 mark

Expansion
- You are given this formula during your exam: $\cos(A \pm B) = \cos A \cos B \mp \sin A \sin B$

Q1. (b)
Let $x = 45$ giving ✓

$\cos 15^\circ = \cos(45-30)^\circ$

$= \cos 45^\circ \cos 30^\circ + \sin 45^\circ \sin 30^\circ$ ✓

$= \frac{1}{\sqrt{2}} \times \frac{\sqrt{3}}{2} + \frac{1}{\sqrt{2}} \times \frac{1}{2}$

$= \frac{\sqrt{3}}{2\sqrt{2}} + \frac{1}{2\sqrt{2}} = \frac{\sqrt{3}+1}{2\sqrt{2}}$ ✓

3 marks

Use of $x = 45$
- 1st mark is for evidence of $45 - 30$

Exact values
- $\sin 45^\circ = \cos 45^\circ = \frac{1}{\sqrt{2}}$
  $\sin 30^\circ = \frac{1}{2}$ $\quad \cos 30^\circ = \frac{\sqrt{3}}{2}$
- You should be able to work out these exact values by drawing the appropriate triangles.

Simplification
- A single fraction is required for this mark.
- Addition and multiplication of fractions:
  $\frac{a}{b} \times \frac{c}{d} = \frac{ac}{bd}$ $\quad \frac{a}{c} + \frac{b}{c} = \frac{a+b}{c}$

see p 26, p 32

Q2.
Compare $(2k+3)x^2 - kx - 1 = 0$
with: $\quad ax^2 + bx + c = 0$
This gives: $a = 2k+3$, $b = -k$ and $c = -1$
Discriminant $= b^2 - 4ac$

$= (-k)^2 - 4(2k+3) \times (-1)$ ✓

$= k^2 + 4(2k+3)$

$= k^2 + 8k + 12$ ✓

For real roots set Discriminant ≥ 0 ✓

$\Rightarrow k^2 + 8k + 12 \geq 0 \Rightarrow (k+6)(k+2) \geq 0$

$\Rightarrow k \leq -6$ or $k \geq -2$ $\quad$ Graph:

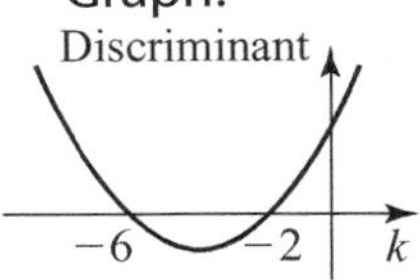

So the roots are real for $k \leq -6$ or $k \geq -2$ ✓

4 marks

Substitution
- Correct identification of a, b and c and substitution into $b^2 - 4ac$
- Be careful to include the negative signs

Simplification
- Squaring a negative gives a positive answer
- Note that you have $-4 \times (-1) = 4$

Condition
- You must clearly state the condition for real roots: Discriminant ≥ 0

Solution
- Notice $(k+6)(k+2) = 0 \Rightarrow k = -6$ or $k = -2$ and for $(k+6)(k+2) \geq 0$ you are looking for where the graph (showing the values of the discriminant) lies above or on the k-axis.

see pp 52–54

Q3.

$\int \frac{2x^5 - 3}{3x^4} dx = \int \frac{2x^5}{3x^4} - \frac{3}{3x^4} dx$ ✓

$= \int \frac{2x}{3} - \frac{1}{x^4} dx = \int \frac{2x}{3} - x^{-4} dx$ ✓

$= \frac{2x^2}{3 \times 2} - \frac{x^{-3}}{-3} + C = \frac{x^2}{3} + \frac{x^{-3}}{3} + C$ ✓

$= \frac{1}{3}x^2 + \frac{1}{3x^3} + C$ ✓

4 marks

Preparation
- This mark is for correctly preparing for differentiation: terms $\frac{2}{3}x$ and x^{-4} appear after splitting the fraction into two terms.

Integration of 1st term
- Correct integration of $\frac{2}{3}x$

Integration of 2nd term
- Correct integration of x^{-4}
- Increasing -4 by 1 gives -3 not -5

Simplification & constant
- Forget the constant C and you lose the mark!

see p 77

Q4. (a)

$h(x) = g(f(x))$

$= g(\frac{1}{2}x + 1)$ ✓

$= 3\,(\frac{1}{2}x + 1) - \frac{1}{2}$

$= \frac{3}{2}x + 3 - \frac{1}{2} = \frac{3}{2}x + \frac{5}{2}$ ✓

2 marks

Substitution
- Replacement of $f(x)$ by $\frac{1}{2}x + 1$ for 1st mark.

Substitution & simplification
- Notice that g acts on $\frac{1}{2}x + 1$ by multiplying it by 3 then subtracting $\frac{1}{2}$
- The expression must be simplified to gain this mark.

Q4. (b)

$h(x) = \frac{3}{2}x + \frac{5}{2}$

Suppose $b = \frac{3}{2}a + \frac{5}{2}$

$\Rightarrow 2b = 3a + 5 \Rightarrow 2b - 5 = 3a$ ✓

$\Rightarrow \frac{2b-5}{3} = a \Rightarrow a = \frac{2}{3}b - \frac{5}{3}$

So $h^{-1}(x) = \frac{2}{3}x - \frac{5}{3}$ ✓

$\Rightarrow p = \frac{2}{3}$ and $q = -\frac{5}{3}$ ✓

3 marks

Change the subject
- Evidence that you have used a correct strategy is required to gain this mark.

Inverse
- When $x = a$ goes in to $h(x)$ then b comes out. When b goes in to $h^{-1}(x)$ then a comes out. Start with b in terms of a, finish with a in terms of b. Replace b by x for the inverse formula.

Values
- The question asks for the values of p and q.

see p 12

Q5. $\overrightarrow{AB} = \begin{pmatrix} 4 \\ 0 \\ -2 \end{pmatrix}$ with $A\,(-1, 2, 3)$

so $B\,(-1 + 4, 2 + 0, 3 - 2) = B\,(3, 2, 1)$ ✓

$\overrightarrow{BC} = \begin{pmatrix} 4 \\ 0 \\ -2 \end{pmatrix} + \begin{pmatrix} 2 \\ 2 \\ 2 \end{pmatrix} = \begin{pmatrix} 6 \\ 2 \\ 0 \end{pmatrix}$

so $C\,(3 + 6, 2 + 2, 1 + 0) = C\,(9, 4, 1)$ ✓

$\overrightarrow{AB} = -2\overrightarrow{CD} \Rightarrow \overrightarrow{CD} = -\frac{1}{2}\overrightarrow{AB} = -\frac{1}{2}\begin{pmatrix} 4 \\ 0 \\ -2 \end{pmatrix} = \begin{pmatrix} -2 \\ 0 \\ 1 \end{pmatrix}$

so $D(9 - 2, 4 + 0, 1 + 1) = D(7, 4, 2)$ ✓

$\overrightarrow{DE} = \frac{1}{2}\overrightarrow{BC} = \frac{1}{2}\begin{pmatrix} 6 \\ 2 \\ 0 \end{pmatrix} = \begin{pmatrix} 3 \\ 1 \\ 0 \end{pmatrix}$

so $E(7 + 3, 4 + 1, 2 + 0) = E(10, 5, 2)$ ✓

4 marks

1st point
- This solution uses the following result: if $P\,(x, y, z)$ and $\overrightarrow{PQ} = \begin{pmatrix} a \\ b \\ c \end{pmatrix}$ then you add the components of $\overrightarrow{PQ}$ to the corresponding coordinates of P to get $Q\,(x + a, y + b, z + c)$

2nd point
- It is usually easier to deal with $\begin{pmatrix} a \\ b \\ c \end{pmatrix}$ rather than $a\boldsymbol{i} + b\boldsymbol{j} + c\boldsymbol{k}$ as is shown in this solution.

3rd point
- Using the result: $k\begin{pmatrix} a \\ b \\ c \end{pmatrix} = \begin{pmatrix} ka \\ kb \\ kc \end{pmatrix}$

Answer
- The coordinates of point E are required.

see pp 36–39

Q6. (a)

$u_6 = \frac{1}{4}u_5 + 6 = \frac{1}{4} \times 12 + 6 = 3 + 6 = 9$ ✓

1 mark

Evaluation
- Use $u_6 = \frac{1}{4}u_5 + 6$ and substitute $u_5 = 12$.

Q6. (b)

A limit exists for this linear recurrence relation since the multiplier $\frac{1}{4}$ lies between -1 and 1. ✓

1 mark

Existence of limit
- Mention 'linear' (the type of recurrence relation).

Q6. (c)

Let the limit be L

So $L = \frac{1}{4}L + 6 \Rightarrow 4L = L + 24$ ✓

$\Rightarrow 3L = 24 \Rightarrow L = 8$ ✓

2 marks

Strategy
- Or you can use $\frac{6}{1-\frac{1}{4}}$ (from the formula!)

Calculation
- Multiply both sides of the equation by 4.

see pp 102–104

Q7. (a)

$f(1) \quad = 1^3 - 6 \times 1^2 + 3 \times 1 + 1$

$\quad = 1 - 6 + 3 + 1 = -1$

$g(1) \quad = -2 \times 1 + 1 = -1$

so $f(1) = g(1)$ ✓

1 mark

Proof
- Be careful with your setting out here.
  Avoid $1 - 6 + 3 + 1 = -2 \times 1 + 1$ so $-1 = -1$
  Better to evaluate $f(1)$ and $g(1)$ separately then state $f(1) = g(1)$.

Q7. (b)

Shaded area $= \int_0^1 f(x) - g(x)\,dx$

$= \int_0^1 x^3 - 6x^2 + 3x + 1 - (-2x + 1)\ dx$

$= \int_0^1 x^3 - 6x^2 + 3x + 1 + \ 2x - 1\ dx$

$= \int_0^1 x^3 - 6x^2 + 5x\ dx$

$= \left[\frac{x^4}{4} - \frac{6x^3}{3} + \frac{5x^2}{2}\right]_0^1$ ✓

$= \frac{1^4}{4} - \frac{6 \times 1^3}{3} + \frac{(5 \times 1^2)}{2} - 0$ ✓

$= \frac{1}{4} - 2 + \frac{5}{2} = \frac{3}{4}$ unit^2 ✓

5 marks

Limits
- 0 at the bottom, 1 at the top.

Strategy
- 'upper curve' minus 'lower curve'.

Integration
- Correct integration gains this mark.

Substitution
- Substitute $x = 1$ and $x = 0$ correctly for this mark.

Evaluation
- This mark is for correct calculation of the area.

see pp 110–111

Q8. (a) ✓

$m_{L_1} = \tan\ \frac{\pi}{3} = \sqrt{3}$

1 mark

Gradient
- Using the result $m = \tan\theta$

Q8. (b)

$m_{L_1} = \sqrt{3} \Rightarrow m_{\perp} = -\frac{1}{\sqrt{3}} \Rightarrow m_{L_2} = -\frac{1}{\sqrt{3}}$ ✓

Gradient of $l_2 = -\frac{1}{\sqrt{3}}$ and point on l_2 is $(0, -\sqrt{2})$

$\Rightarrow$ equation of l_2 is $y - (-\sqrt{2}) = -\frac{1}{\sqrt{3}}(x - 0)$ ✓

$\Rightarrow y + \sqrt{2} = -\frac{1}{\sqrt{3}}x \Rightarrow \sqrt{3}y + \sqrt{6} = -x$

$\Rightarrow \sqrt{3}y + x = -\sqrt{6}$ ✓

3 marks

Perpendicular gradient
- Using the result $m = \frac{a}{b} \Rightarrow m_{\perp} = -\frac{b}{a}$

Equation
- If the gradient $= m$ and point on line is (a, b) then the equation is $y - b = m(x - a)$

Simplification
- The final mark is for 'tidying up the equation'.

Q8. (c)

For the x-axis intercept set $y = 0$ ✓

so $\sqrt{3} \times 0 + x = -\sqrt{6} \Rightarrow x = -\sqrt{6}$

l_2 crosses the x-axis at the point $(-\sqrt{6}, 0)$ ✓

2 marks

Strategy

- Evidence of $y = 0$ for this mark.

Intercept

- Calculation of $-\sqrt{6}$ will gain this mark.

see pp 88–89

Q9. This diagram shows the two possible positions for the circles:

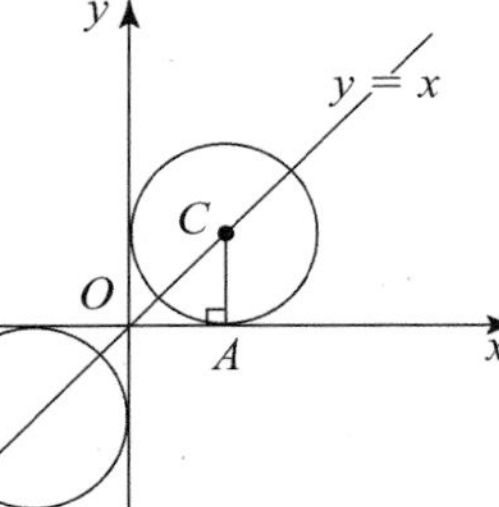

✓

$x^2 + x^2 = (\sqrt{2})^2 \Rightarrow 2x^2 = 2$ ✓

$\Rightarrow x^2 = 1 \Rightarrow x = 1$ (positive)

so $C(1, 1)$ and radius $= 1$ ✓

Equation is: $(x - 1)^2 + (y - 1)^2 = 1^2 = 1$ ✓

By symmetry the other circle has centre $(-1, -1)$ and radius $= 1$

Equation is: $(x + 1)^2 + (y + 1)^2 = 1$ ✓

5 marks

Strategy

- Evidence of use of Pythagoras' Theorem to calculate the radius and centre.
- It is important in questions like this one to draw a diagram.

Calculation

- Correct calculation of $x = 1$
- x measures a length so is a positive quantity.

Centre & radius

- A clear statement showing the coordinates of the centre and the value of the radius.

Equation

- The equation of a circle with centre (a, b) and radius r is $(x - a)^2 + (y - b)^2 = r^2$
  In this case $a = 1$, $b = 1$ and $r = 1$
- You must not leave 1^2 but indicate that $1^2 = 1$

2nd equation

- In this case $a = -1$, $b = -1$ and $r = 1$ are the values used in $(x - a)^2 + (y - b)^2 = r^2$

see p 94

Q10. $f'(x) = -\sin(3x - \frac{\pi}{6}) \times 3$ ✓

$= -3\sin(3x - \frac{\pi}{6})$ ✓

so $f'(\frac{\pi}{3}) = -3\sin(3 \times \frac{\pi}{3} - \frac{\pi}{6})$ ✓

$= -3\sin(\pi - \frac{\pi}{6}) = -3\sin\frac{5\pi}{6}$

$= -3 \times \frac{1}{2} = -\frac{3}{2}$ ✓

4 marks

Strategy

- You should know that 'rate of change' means you should differentiate, i.e. find $f'(x)$

Differentiate cos( )

- On your formula sheet you are told:
  $f(x)$: $\cos ax$ $\quad f'(x)$: $-a\sin ax$

Chain rule

- You are using the 'chain rule' with the factor 3 appearing when you differentiate $3x - \frac{\pi}{6}$

Evaluation

- $\frac{5\pi}{6}$ is a 2nd quadrant angle where $\sin\frac{5\pi}{6}$ is positive and the same as $\sin\frac{\pi}{6} = \frac{1}{2}$

see pp 72–73

Q11. (a)

$\log_3(9\sqrt{x}) = \log_3(9 \times x^{\frac{1}{2}})$

$= \log_3 9 + \log_3 x^{\frac{1}{2}}$ ✓

$= 2 + \frac{1}{2}\log_3 x$ ✓

$= \frac{1}{2}\log_3 x + 2$ ✓

3 marks

Addition rule

- Correct use of the rule:
  $\log(mn) = \log m + \log n$

Simplification

- $\log_3 9 = 2$ is equivalent to $3^2 = 9$
- It is useful to read '$\log_3 9$' as 'what power of 3 gives 9?' with the answer: 2

Power rule

- Use of the log rule: $\log a^n = n\log a$ to change $\log_3 x^{\frac{1}{2}}$ to $\frac{1}{2}\log_3 x$

Q11. (b)

$x = \sqrt{3}$ gives $\frac{1}{2}\log_3\sqrt{3} + 2$

$= \frac{1}{2} \times \frac{1}{2} + 2 = \frac{1}{4} + 2 = \frac{9}{4}$ ✓

1 mark

Evaluation

- Read $\log_3\sqrt{3}$ as 'what power of 3 gives $\sqrt{3}$?' the answer being $\frac{1}{2}$

see p 22

Q12.

For points of intersection:

Solve $f(x) = g(x)$

$\Rightarrow 2x^3 + 3x + 12 = 2 + 16x^2 - x^3$ ✓

$\Rightarrow 3x^3 - 16x^2 + 3x + 10 = 0$ ✓

$$\begin{array}{c|cccc} 1 & 3 & -16 & 3 & 10 \\ & & 3 & -13 & -10 \\ \hline & 3 & -13 & -10 & 0 \end{array}$$ ✓

0 remainder $\Rightarrow$ 1 is a root

So $x - 1$ is a factor ✓

The equation becomes:

$(x - 1)(3x^2 - 13x - 10) = 0$ ✓

$\Rightarrow (x - 1)(3x + 2)(x - 5) = 0$ ✓

$\Rightarrow x - 1 = 0$ or $3x + 2 = 0$ ✓

or $x - 5 = 0$

$x = 1$ or $x = -\frac{2}{3}$ or $x = 5$

So the x-coordinates are:

For point P: $-\frac{2}{3}$

For point Q: 1

For point R: 5 ✓

8 marks

Strategy
- Setting the two formulae equal to each other and solving is the strategy in this question.

Cubic equation
- Recognising a cubic and rearranging terms in order gains a mark.

Strategy
- Trying particular values of x. In this case $x = 1$ was successful.

Interpretation
- $x = 1$ is a solution of the equation so $x - 1$ is a factor

Factorising
- Three marks are allocated to this:

1 mark: starting to factorise $(x - 1)(3x^2 \ldots)$
1 mark: quadratic factor $3x^2 - 13x - 10$
1 mark: completing the factorisation $(x - 1)(3x + 2)(x - 5)$

Interpretation
- From the diagram you can assign the three values to the points P, Q and R.

see pp 58–61

Q13. (a)

$\sqrt{3}\cos x^\circ - \sin x^\circ = k\cos(x + a)^\circ$ ✓

$\sqrt{3}\cos x^\circ - \sin x^\circ$
$= k\cos x^\circ \cos a^\circ - k\sin x^\circ \sin a^\circ$

Comparing coefficients of $\cos x^\circ$ and $\sin x^\circ$:

$$\left.\begin{array}{l} k\cos a^\circ = \sqrt{3} \\ k\sin a^\circ = 1 \end{array}\right\}$$ ✓

Dividing gives:

$\frac{k\sin a^\circ}{k\cos a^\circ} = \frac{1}{\sqrt{3}} \Rightarrow \tan a^\circ = \frac{1}{\sqrt{3}} \Rightarrow a = 30$ ✓

Squaring and adding gives:

$(k\sin a^\circ)^2 + (k\cos a^\circ)^2 = 1^2 + (\sqrt{3})^2$
$= 1 + 3 = 4$

$\Rightarrow k^2(\sin^2 a^\circ + \cos^2 a^\circ) = 4$

$\Rightarrow k^2 \times 1 = 4 \Rightarrow k = 2\ (k > 0)$ ✓

so $\sqrt{3}\cos x^\circ - \sin x^\circ = 2\cos(x + 30)^\circ$

4 marks

Addition formula
- Expanding $\cos(x + a)^\circ$ is the essential first step. The formula: $\cos(A \pm B) = \cos A \cos B \mp \sin A \sin B$ is on your formulae sheet in the exam.

Coefficients

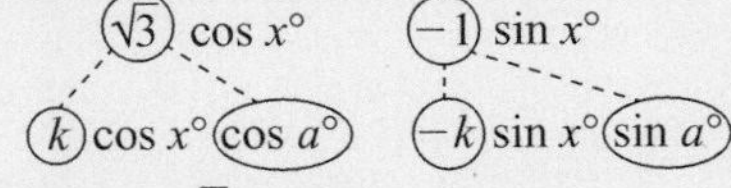

- This gives $k\cos a^\circ = \sqrt{3}$ and $k\sin a^\circ = 1$

Calculation
- Here you are using $\frac{\sin a^\circ}{\cos a^\circ} = \tan a^\circ$ along with cancelling the factor k to find angle a°.

Calculation
- The result $\sin^2 a^\circ + \cos^2 a^\circ = 1$ leads to the disappearance of a° allowing k to be found.
- k is always positive in this context.

Q13. (b)

$\sqrt{3}\cos x° - \sin x° = 1 \Rightarrow 2\cos(x+30)° = 1$

$\Rightarrow \cos(x+30)° = \frac{1}{2}$ ✓

$(x+30)°$ is a 1st or 4th quadrant angle since $\cos(x+30)°$ is positive.

1st quadrant: $\cos^{-1}\frac{1}{2} = 60°$

$\Rightarrow x + 30 = 60 \Rightarrow x = 30$ ✓

4th quadrant:

$x + 30 = 360 - 60 = 300$

$\Rightarrow x = 300 - 30 = 270$ ✓

So for $0 < x < 360$ the solutions are $x = 30, 270$

3 marks

Set up equation

- Rewrite the equation using the result from (**a**)
- The aim is to rearrange the equation to get: cos(an angle) = a number

1st solution

- The 1st quadrant angle is 60° but in this case it's not $x°$ but $(x+30)°$ that equals 60°

2nd solution

- Cosine is positive in the 1st and 4th quadrants. Here is the diagram:

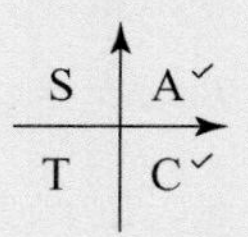

see pp34–35, p 64

Q14.

$\sin\theta(\sin\theta - 1) = \cos^2\theta$

$\Rightarrow \sin^2\theta - \sin\theta = \cos^2\theta$

$\Rightarrow \sin^2\theta - \sin\theta = 1 - \sin^2\theta$ ✓

$\Rightarrow 2\sin^2\theta - \sin\theta - 1 = 0$ ✓

$\Rightarrow (2\sin\theta + 1)(\sin\theta - 1) = 0$ ✓

$\Rightarrow 2\sin\theta + 1 = 0$ or $\sin\theta = 1$

$\Rightarrow \sin\theta = -\frac{1}{2}$ or $\sin\theta = 1$ ✓

For $\sin\theta = -\frac{1}{2}$

θ is in 3rd or 4th quadrants (−ve)

1st quadrant angle is $\frac{\pi}{6}$

However, since $\frac{\pi}{2} < \theta < \frac{3\pi}{2}$ the 4th quadrant solution is not valid.

So $\theta = \pi + \frac{\pi}{6} = \frac{6\pi}{6} + \frac{\pi}{6} = \frac{7\pi}{6}$

For $\sin\theta = 1$

Using the graph gives $\theta = \frac{\pi}{2}$

but since $\frac{\pi}{2} < \theta < \frac{3\pi}{2}$

this solution is invalid.

So $\theta = \frac{7\pi}{6}$ is the only solution. ✓

5 marks

Strategy

- Looking at the 'make-up' of the equation will lead you to produce an equation in $\sin\theta$ – so use $\sin^2\theta + \cos^2\theta = 1$ to replace $\cos^2\theta$ by $1 - \sin^2\theta$.

'Standard form'

- The equation is a 'quadratic' in $\sin\theta$ and should be arranged in the standard quadratic form: $ax^2 + bx + c = 0$ in this case $a\sin^2\theta + b\sin\theta + c = 0$.

Factorisation

- $2x^2 - x - 1 = (2x + 1)(x - 1)$ so likewise: $2\sin^2\theta - \sin\theta - 1 = (2\sin\theta + 1)(\sin\theta - 1)$

Solve for sin θ

- In this case the 'roots' of the quadratic equation are $-\frac{1}{2}$ and 1. These are the possible values for $\sin\theta$.

Solve for θ

- In general for values of $\sin\theta$ or $\cos\theta$ of −1, 0 or 1 you should use the sine or cosine graph to determine the angles. In this case the value $(\frac{\pi}{2})$ is not in the allowed interval $(\frac{\pi}{2} < \theta < \frac{3\pi}{2})$.
- For $\sin\theta = -\frac{1}{2}$ the 4th quadrant solution is not in the allowed interval $(\frac{\pi}{2} < \theta < \frac{3\pi}{2})$.

see pp 62–63

Exam A: Paper 2

Q1. (a)

$y = x^2 - \frac{1}{3}x^3 \Rightarrow \frac{dy}{dx} = 2x - x^2$ ✓✓

At M (3, 0) $x = 3$

So $\frac{dy}{dx} = 2 \times 3 - 3^2 = -3$ ✓

gradient of tangent is –3.

Point on tangent is (3, 0).

Equation of tangent is:

$y - 0 = -3(x - 3)$

$\Rightarrow y = -3x + 9.$ ✓

4 marks

Strategy
- Knowing to differentiate gains you the 1st mark.

Differentiate
- The correct result gains the 2nd mark.

Calculation
- Where are you on the curve? At the place where $x = 3$, so use $x = 3$ and the gradient formula.

Equation
- Using $y - b = m(x - a)$. In this case $m = -3$ and (a, b) is (3, 0).

see p 68

Q1. (b)

To find the points of intersection of line and circle

Solve: $\left.\begin{array}{l} y = -3x + 9 \\ x^2 + y^2 - 4x - 26y + 163 = 0 \end{array}\right\}$ ✓

Substitute $y = -3x + 9$ in the circle equation: ✓

$x^2 + (-3x + 9)^2 - 4x - 26(-3x + 9) + 163 = 0$

$\Rightarrow x^2 + 9x^2 - 54x + 81 - 4x + 78x - 234 + 163 = 0$

$\Rightarrow 10x^2 + 20x + 10 = 0$ ✓

$\Rightarrow 10(x^2 + 2x + 1) = 0$

$\Rightarrow 10(x + 1)(x + 1) = 0$

$\Rightarrow x = -1$ ✓

Since there is only one solution, the line is a tangent to the circle. ✓

when $x = -1$ $y = -3 \times (-1) + 9 = 12$

So N(−1, 12) is the point of contact. ✓

6 marks

Rearrangement
- The form $y = -3x + 9$ is necessary for the subsequent substitution. Do not use $x = \frac{1}{3}y + 3$ as this involves fraction work leading to errors.

Strategy
- Evidence of substitution gains you the strategy mark.

'Standard form'
- This is a quadratic equation and should be written in the 'standard' way, i.e. $ax^2 + bx + c = 0$.

Solution
- Notice that removing a common factor reduces the magnitude of the coefficients and makes the rest of the factorisation easier.

Proof
- A clear statement is required — there was one point of intersection so you have a tangent.

Calculation
- A point was asked for, so you must give the coordinates, $x = -1$ and $y = 12$ is not enough.

see p 96

Q1. (c)

when $x = 0$ $y = -3 \times 0 + 9 = 9$ ✓

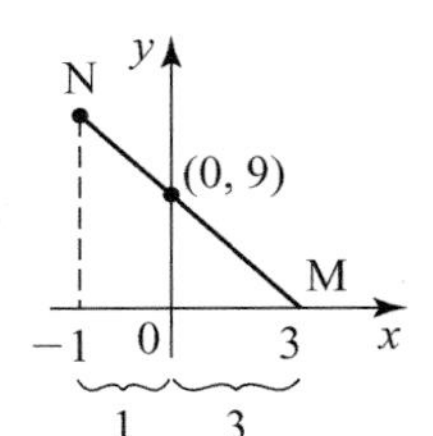

The required ratio is 3 : 1. ✓ ✓

3 marks

Strategy
- Knowing where the y-intercept is gains you this strategy mark.

Strategy
- An alternative approach uses vectors. If y-intercept is P: $\overrightarrow{MP} = \begin{pmatrix} -3 \\ 9 \end{pmatrix}$ and $\overrightarrow{PN} = \begin{pmatrix} -1 \\ 3 \end{pmatrix}$ so $\overrightarrow{MP} = 3\overrightarrow{PN}$ etc.

Ratio
- Deducing the correct ratio gains you this final mark. Note 1 : 3 will not gain this mark.

see p 40

Q2. (a)

$y = x^3 - 3x^2 + 4$

$\Rightarrow \dfrac{dy}{dx} = 3x^2 - 6x$ ✓✓

For stationary points set $\dfrac{dy}{dx} = 0$ ✓

$\Rightarrow 3x^2 - 6x = 0$

$\Rightarrow 3x(x - 2) = 0 \Rightarrow x = 0$ or 2 ✓

x:	0		2	
$\dfrac{dy}{dx} = 3x(x-2)$	+	−		+
Shape of graph:	/ —	\	—	/
Nature:	max		min	

✓

when $x = 0$ $y = 0^3 - 3 \times 0^2 + 4 = 4$
So (0, 4) is a maximum stationary point. ✓

When $x = 2$ $y = 2^3 - 3 \times 2^2 + 4 = 0$
So (2, 0) is a minimum stationary point. ✓

7 marks

Differentiate
- 1 mark for knowing to differentiate.
- 1 mark for correctly differentiating.

Strategy
- Setting $\frac{dy}{dx} = 0$ allows you to home in on the x-coordinates of the stationary points.

Solving Equation
- The common factor here is $3x$.

Justification
- 'justify their nature' means that you need a 'nature table' as shown in the solution.

y-coordinates
- 1 mark is allocated to the calculation of the two corresponding y-coordinates (4 and 0).

Interpretation
- Clear statements should be made about the type (max or min) of each stationary point.

see pp 68–69

Q2. (b)(i)

$(x+1)(x-2)^2$

$= (x+1)(x-2)(x-2)$

$= (x+1)(x^2-4x+4)$

$= x^3 - 4x^2 + 4x + x^2 - 4x + 4$ ✓

$= x^3 - 3x^2 + 4$

1 mark

Expand

- Since the answer $x^3 - 3x^2 + 4$ is given it is important that you show clearly all your steps in the expansion of the brackets.
- Notice for $(x+1)(x^2-4x+4)$ you have: $x(x^2-4x+4)$ and $1(x^2-4x+4)$.

Q2. (b)(ii)

For x-intercepts set

$y = 0$

So $(x+1)(x-2)^2 = 0$

$\Rightarrow x+1 = 0$ or $x-2 = 0$

$\Rightarrow x = -1$ or $x = 2$.

Intercepts are $(-1, 0)$, $(2, 0)$

For y-intercepts set $x = 0$ ✓

So $y = 0^3 - 3 \times 0^2 + 4 = 4$ Intercept is $(0, 4)$ ✓

Sketch: ✓

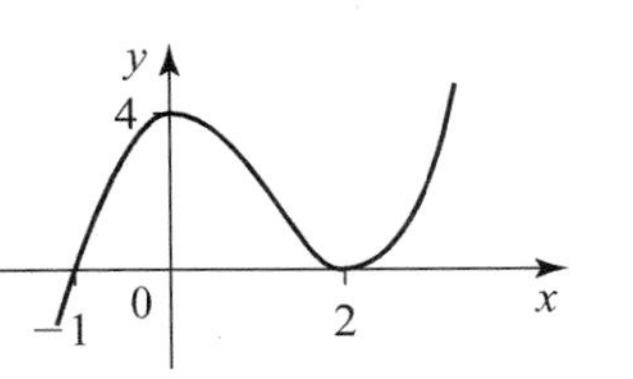

3 marks

***x*-intercepts**

- You should write down the coordinates of both x-axis intercepts, i.e. $(-1, 0)$ and $(2, 0)$.

***y*-intercept**

- The y-axis intercept is expected to be clearly indicated when sketching. It is good practice, as with the x-intercepts, to write down the coordinates, i.e. $(0,4)$.

Sketch

- Your sketch should clearly show the main features: shape and intercepts and stationary points.

see p 70

Q3. (a)

$M\left(\frac{-3+(-3)}{2}, \frac{4+(-4)}{2}\right) = M(-3,0)$ ✓

For $M(-3,0)$ and $R(5,12)$

$m_{MR} = \frac{12-0}{5-(-3)} = \frac{12}{8} = \frac{3}{2}$ ✓

gradient of median is $\frac{3}{2}$

point on median is $(-3, 0)$

so equation is $y - 0 = \frac{3}{2}(x-(-3))$

$\Rightarrow 2y = 3(x+3) \Rightarrow 2y = 3x + 9$

$\Rightarrow 2y - 3x = 9$ ✓

3 marks

Interpretation

- Did you know what a median is? The line from a vertex to the midpoint of the opposite side.

Gradient

- The formula used here is: $A(x_1, y_1)$, $B(x_2, y_2)$ $m_{AB} = \frac{y_2 - y_1}{x_2 - x_1}$
- Perpendicular gradients are not required for medians, only for altitudes.

Equation

- Using $y - b = m(x - a)$ with $m = \frac{3}{2}$ and (a, b) being $(-3, 0)$

see pp 92–93

Q3. (b)

For $P(-3, -4)$ and $R(5, 12)$ ✓

$$m_{PR} = \frac{12-(-4)}{5-(-3)} = \frac{16}{8} = 2$$

$$\Rightarrow m_{\perp} = -\frac{1}{2}$$ ✓

gradient of altitude is $-\frac{1}{2}$

point on altitude is $Q(-3,4)$

equation is $y - 4 = -\frac{1}{2}(x-(-3))$

$\Rightarrow 2y - 8 = -(x + 3)$

$\Rightarrow 2y - 8 = -x - 3$

$\Rightarrow 2y + x = 5$ ✓

3 marks

Strategy
- Finding the gradient of the 'base' PR is the essential 1st step here.

Perpendicular gradient
- Did you know what an altitude is? A line from a vertex perpendicular to the opposite side.
- The result used is $m = \frac{a}{b} \Rightarrow m_{\perp} = -\frac{b}{a}$ invert and change the sign.

Equation
- Using $y - b = m(x - a)$ with $m = -\frac{1}{2}$ and (a, b) being the point $(-3,4)$.

see pp 89–90, p92

Q3. (c)

To find the point of intersection:

Solve $\left.\begin{array}{l} 2y + x = 5 \\ 2y - 3x = 9 \end{array}\right\}$ ✓

Subtract: $4x = -4 \Rightarrow x = -1$ ✓

Substitute $x = -1$ in $2y + x = 5$

$\Rightarrow 2y - 1 = 5 \Rightarrow 2y = 6$

$\Rightarrow y = 3$

The point of intersection is

S$(-1, 3)$ ✓

3 marks

Strategy
- Simultaneous equations: if your method is clear you will gain this strategy mark.

Calculation
- Correct calculation of either x or y gains the 2nd mark.

Calculation
- Correct calculation of the other variable gains you this last mark.

see p 90

Q3. (d)

From (b) $m_{PR} = 2$

For $S(-1, 3)$ and $T(2, 9)$

$$m_{ST} = \frac{9-3}{2-(-1)} = \frac{6}{3} = 2$$ ✓

So $m_{ST} = m_{PR} = 2$

and so ST is parallel to PR. ✓

2 marks

Gradient
- You must find the gradient of this new line ST if you are to compare its slope with that of line PR.

Parallel lines
- If the gradients of two lines are equal then the lines are parallel.
- Clear statements are needed here so that it is obvious you understand the result: equal gradients $\Leftrightarrow$ parallel lines.

see p 88

Q4. (a)

$2x^2 - 8x + 9 = 2(x^2 - 4x) + 9$ ✓

$= 2[(x-2)(x-2) - 4] + 9$ ✓

$= 2(x-2)^2 - 8 + 9$

$= 2(x-2)^2 + 1$ ✓

3 marks

Common factor
- It is easier to deal with $x^2 - 4x$ where the coefficient of x^2 is 1. This is why a common factor of 2 is taken out of the first two terms.

Complete the square
- $x^2 - 4x$ ...*Half* -4 to get -2 and now write $(x-2)(x-2)$. Multiplying out these brackets gives $x^2 - 4x + 4$. Remove $+4$ by subtracting 4 and so you are back to $x^2 - 4x$ as required. You can now remove the square brackets by multiplying by the 2.

Answer
- The form $a(x-b)^2 + c$ was asked for. Make sure your answer is in this given form.

see p 14

Q4. (b)

$g(x) = \frac{2}{3}x^3 - 4x^2 + 9x + 1$

$\Rightarrow g'(x) = 3 \times \frac{2}{3}x^2 - 2 \times 4x + 9$ ✓

$= 2x^2 - 8x + 9$ ✓

2 marks

Start differentiation
- First two terms correct.

Finish differentiation
- Correct constant at the end.

see p 66

Q4. (c)

The gradient of $y = g(x)$ is given by

$g'(x) = 2(x-2)^2 + 1$

now $(x-2)^2 \geq 0$ ✓

$\Rightarrow 2(x-2)^2 \geq 0$

$\Rightarrow 2(x-2)^2 + 1 > 0$

$\Rightarrow g'(x) > 0$

A positive gradient means that $y = g(x)$ is strictly increasing for all values of x. ✓

2 marks

Strategy
- You are linking this part back to part (a).
- Note that a squared term is always positive or zero.

Explanation
- Here you must explain clearly that the expression for the gradient is always greater than zero, i.e. it is always positive.

Q5. (a)

$\cos 2x + 1 = (2\cos^2 x - 1) + 1$ ✓

$= 2\cos^2 x - 1 + 1$

$= 2\cos^2 x$

1 mark

Use of formula
- It must be very clear that you are replacing $\cos 2x$ by $2\cos^2 x - 1$. This is one of the Double Angle formulae.

Q5. (b)

$\int 2\cos^2 x \, dx = \int (\cos 2x + 1)$

$= \frac{\sin 2x}{2} + x + c$ ✓ ✓

$= \frac{1}{2}\sin 2x + x + c$

2 marks

Integration
- This mark is for correctly integrating the terms $\cos 2x$
- The formula for integrating $\cos ax$ is given on the formula sheet at the front of your exam paper.

Completing the integration
- Integrating the constant 1 gives x
- If you forget the constant of integration c then you will not be awarded this final mark.

see p 81

Q6. (a)

$PQ = 4$ units

$QR = 5$ units

$RV = 6$ units ✓

1 mark

Interpretation

- The difficulty is that the axes are not shown. Look for a single change in coordinates: $P(-1, -1, -1)$ to $S(-1, 4, -1)$: This is 5 units parallel to the y-axis, since only the y-coordinate has changed.

see p 36

Q6. (b)

V(3, 4, 5)
P(−1, −1, −1)
S(−1, 4, −1)

$\overrightarrow{VS} = \boldsymbol{s} - \boldsymbol{v}$

$= \begin{pmatrix} -1 \\ 4 \\ -1 \end{pmatrix} - \begin{pmatrix} 3 \\ 4 \\ 5 \end{pmatrix} = \begin{pmatrix} -4 \\ 0 \\ -6 \end{pmatrix}$ ✓

$\overrightarrow{VP} = \boldsymbol{p} - \boldsymbol{v} = \begin{pmatrix} -1 \\ -1 \\ -1 \end{pmatrix} - \begin{pmatrix} 3 \\ 4 \\ 5 \end{pmatrix} = \begin{pmatrix} -4 \\ -5 \\ -6 \end{pmatrix}$ ✓

$\overrightarrow{VS}.\overrightarrow{VP} = \begin{pmatrix} -4 \\ 0 \\ -6 \end{pmatrix} . \begin{pmatrix} -4 \\ -5 \\ -6 \end{pmatrix}$

$= -4 \times (-4) + 0 \times (-5) + (-6) \times (-6)$

$= 52$ ✓

$\left|\overrightarrow{VS}\right| = \left|\begin{pmatrix} -4 \\ 0 \\ -6 \end{pmatrix}\right| = \sqrt{(-4)^2 + 0^2 + (-6)^2}$

$= \sqrt{16 + 0 + 36} = \sqrt{52}$ ✓

$\left|\overrightarrow{VP}\right| = \left|\begin{pmatrix} -4 \\ -5 \\ -6 \end{pmatrix}\right| = \sqrt{(-4)^2 + (-5)^2 + (-6)^2}$

$= \sqrt{16 + 25 + 36} = \sqrt{77}$ ✓

So $\cos P\hat{V}S = \dfrac{\overrightarrow{VS}.\overrightarrow{VP}}{\left|\overrightarrow{VS}\right|\left|\overrightarrow{VP}\right|} = \dfrac{52}{\sqrt{52}\sqrt{77}}$ ✓

$\Rightarrow P\hat{V}S = \cos^{-1}\left(\dfrac{52}{\sqrt{52}\sqrt{77}}\right) = 34{\cdot}73\ldots^\circ$

Angle $PVS = 34{\cdot}7^\circ$ (to 1 decimal place) ✓

7 marks

Components

- The result $\overrightarrow{AB} = \boldsymbol{b} - \boldsymbol{a}$, where $\boldsymbol{a}$ is the position vector of ***A*** and $\boldsymbol{b}$ is the position vector of ***B***, is used frequently to find components.
- Notice that the vector arrows point outwards from the vertex of the angle. You must always do this when calculating the angle between vectors.

Dot product

- The result is $\begin{pmatrix} x_1 \\ y_1 \\ z_1 \end{pmatrix} . \begin{pmatrix} x_2 \\ y_2 \\ z_2 \end{pmatrix} = x_1x_2 + y_1y_2 + z_1z_2$

Lengths

- The result used is: $\left|\begin{pmatrix} a \\ b \\ c \end{pmatrix}\right| = \sqrt{a^2 + b^2 + c^2}$
- Careful with negatives. Squaring always produces a positive or zero quantity – never a negative quantity.

Angle formula

- The result used is: $\cos\theta^\circ = \dfrac{\boldsymbol{a} . \boldsymbol{b}}{|\boldsymbol{a}||\boldsymbol{b}|}$

Calculation

- On the calculator:

$\cos^{-1}\left(52 \div \left(\sqrt{52} \times \sqrt{77}\right)\right)$

- Radians acceptable: 0·606

see pp 42–43

Q7. (a)

Upland: $u_{n+1} = 0{\cdot}35u_n + 2000$ ✓

Lowland:

$v_{n+1} = 0{\cdot}25\, v_n + 2500$ ✓

2 marks

Recurrence relations

- Remember to use the 'percentage left' in the recurrence relation, e.g. 65% is lost so 35% remains – so use 0·35 not 0·65.

Q7. (b)

Both recurrence relations have a limit since their multipliers (0·35 and 0·25) lie between −1 and 1. ✓

For Upland let the limit be L. ✓

Then $L = 0{\cdot}35L + 2000$

$\Rightarrow L - 0{\cdot}35L = 2000$

$\Rightarrow 0{\cdot}65L = 2000$

$\Rightarrow \; L = \dfrac{2000}{0.65} = 3076.9\ldots$ ✓

For Lowland let the limit be M.

Then $M = 0{\cdot}25M + 2500$

$\Rightarrow M - 0{\cdot}25M = 2500$

$\Rightarrow 0{\cdot}75M = 2500$

$\Rightarrow \; M = \dfrac{2500}{0.75} = 3333.3\ldots$

In the long run the Upland site requires the smaller tank (3077 litres) compared to the Lowland site (3334 litres). ✓

4 marks

Limit condition

- 1 mark is allocated to a clear statement justifying the use of a limit. There is a limit only if the multiplier lies between −1 and 1.

Strategy

- To find the limit: if you apply the recurrence relation you produce the same result, e.g. $L = 0{\cdot}35L + 2000$.

Limit

- Calculation of a correct limit gains 1 mark.

Decision

- For the final mark, you must correctly calculate the other limit AND clearly make your decision based on a comparison.

see pp 104–106

Q8. (a)

$P(h) = P_0 e^{-kh}$

In this case $P(4{\cdot}95) = \frac{1}{2}P_0$ ✓

$\Rightarrow P_0 e^{-k \times 4{\cdot}95} = \frac{1}{2}P_0 \Rightarrow e^{-4{\cdot}95k} = \frac{1}{2}$

$\Rightarrow \log_e \frac{1}{2} = -4{\cdot}95k \Rightarrow k = \dfrac{\log_e \frac{1}{2}}{-4.95}$ ✓

$\Rightarrow k = 0{\cdot}1400$ (to 4 decimal places) ✓

3 marks

Interpretation

- You are finding k given that you know $P(h)$ and you know h

$P(h) = P_0 e^{-kh}$

This is $\frac{1}{2}P_0$ — This is 4·95

Log statement

- Using the result: $b^c = a \leftrightarrow \log_b a = c$

$e^{-4{\cdot}95k} = \frac{1}{2}$ becomes $\log_e \frac{1}{2} = -4{\cdot}95k$

Calculation

- Remember the [ln] button for '$\log_e$'.

see p 24

Q8. (b)

The formula is $P(h) = P_0 e^{-0{\cdot}14h}$

For top of Everest $h = 8{\cdot}85$

$P(8{\cdot}85) = P_0 e^{-0{\cdot}14 \times 8{\cdot}85}$

$= P_0 \times 0{\cdot}2896\ldots$ ✓

This is a reduction of

$100\% - 28{\cdot}96\ldots\%$

$= 71{\cdot}03\ldots\% \doteqdot 71\%$ ✓

2 marks

Substitution

- Careful with the units. In the original formula h is in kilometres so 8·85 is used, not 8850.

Calculation

- $\boxed{e^x}$ button is used for $e^{-0{\cdot}14 \times 8{\cdot}85}$ so:

  $\boxed{e^x}\boxed{(}\boxed{(-)}\boxed{0}\boxed{\cdot}\boxed{1}\boxed{4}\boxed{\times}\boxed{8}\boxed{\cdot}\boxed{8}\boxed{5}\boxed{)}\boxed{\text{EXE}}$

  or the equivalent on your calculator!

see p 24

Q9. (a) To find the points of intersection

Solve: $\left.\begin{array}{l} y = x + 2 \\ y = 6 + 4x - x^2 \end{array}\right\}$

so $x + 2 = 6 + 4x - x^2$ ✓

$\Rightarrow x^2 - 3x - 4 = 0$

$\Rightarrow (x + 1)(x - 4) = 0$

$\Rightarrow x = -1$ or $x = 4$ ✓

Area =

$\int_{-1}^{4} (6 + 4x - x^2) - (x + 2)\, dx$ ✓

$= \int_{-1}^{4} 6 + 4x - x^2 - x - 2\, dx$

$= \int_{-1}^{4} 4 + 3x - x^2 dx = \left[4x + \frac{3x^2}{2} - \frac{x^3}{3}\right]_{-1}^{4}$ ✓ ✓

$= \left(4 \times 4 + \frac{3 \times 4^2}{2} - \frac{4^3}{3}\right)$

$- \left(4 \times (-1) + \frac{3 \times (-1)^2}{2} - \frac{(-1)^3}{3}\right)$

$= 16 + 24 - \frac{64}{3} + 4 - \frac{3}{2} - \frac{1}{3}$ ✓

$= 44 - \frac{65}{3} - \frac{3}{2} = \frac{264}{6} - \frac{130}{6} - \frac{9}{6}$

$= \frac{264 - 130 - 9}{6} = \frac{125}{6}$

Enclosed area $= \frac{125}{6}$ unit² ✓

7 marks

Strategy

- To find the area you will be integrating, you need to know the limits. Hence, the need to set the equations equal to each other to find the points of intersection. The x values for these points are the limits used in the integration.

Solve Equation

- Never attempt to arrange a quadratic equation with a negative coefficient for the x^2 term – the factorising is then more difficult.

Strategy

- This mark is for knowing how to find the enclosed area:

  $\int$(top curve) minus (bottom curve) dx

Limits

- Work left to right on diagram (−1 to 4) and

  bottom to top on integral: $\int_{-1}^{4}$

Integration

- Here you are using $\int ax^n\, dx = \frac{ax^{n+1}}{n+1}$.

  No constant is needed when there are limits on the integral sign.

Substitution

- Careful with the order:

  $\left(\substack{x=4 \\ \text{Substitution}}\right) - \left(\substack{x=-1 \\ \text{Substitution}}\right)$ using the result

  $\int_a^b f(x)dx = F(b) - F(a)$ where $F(x)$ is the result of integrating $f(x)$.

Calculation

- Take great care, even with a calculator, as these are not easy calculations!
- Always give exact answers.

see pp 110–111

Q9. (b)

$$A(x) = -\frac{5}{2}x^2 + \frac{15}{2}x + 10$$

$$\Rightarrow A'(x) = -5x + \frac{15}{2}$$

For stationary value set $A'(x) = 0$ ✓

$$\Rightarrow -5x + \frac{15}{2} = 0 \Rightarrow -5x = -\frac{15}{2} \Rightarrow x = \frac{3}{2}.$$

		$\frac{3}{2}$		
x:		→		
$A'(x)$:	$+$		$-$	✓
Shape of graph:	/	$-$	\	
nature:		max		✓

so $x = \frac{3}{2}$ gives a maximum value

$$A\left(\frac{3}{2}\right) = -\frac{5}{2} \times \left(\frac{3}{2}\right)^2 + \frac{15}{2} \times \frac{3}{2} + 10$$

$$= -\frac{45}{8} + \frac{45}{4} + 10 = \frac{-45}{8} + \frac{90}{8} + \frac{80}{8}$$

$$= \frac{-45 + 90 + 80}{8} = \frac{125}{8}$$

Required fraction $= \dfrac{\frac{125}{8}}{\frac{125}{6}} = \dfrac{6}{8} = \dfrac{3}{4}.$ ✓

4 marks

Strategy

- For a maximum value you must find a stationary point where the gradient on the graph is zero, hence set $A'(x) = 0$.

Differentiate and solve

- $-5x(\times 2) = -\frac{15}{2}(\times 2) \Rightarrow -10x = -15$

$$\Rightarrow x = \frac{-15}{-10}$$

So $x = \frac{3}{2}$ is one method of solving the equation

Justify

- You must show $x = \frac{3}{2}$ gives a maximum value hence the need for the 'nature table'.

Interpretation

- You have found that when $x = \frac{3}{2}$ the area of the shaded triangle is at a maximum. The actual shaded area is given by $A\left(\frac{3}{2}\right)$.

  Your calculation should give $\frac{125}{8}$ unit^2 for this area. In part (a) the whole enclosed area was found to be $\frac{125}{6}$ unit^2. To calculate the fraction consider a simpler case:
  What fraction is 2 unit^2 of 6 unit^2? It is $\frac{2}{6}$ or $\frac{1}{3}$
  What fraction is
  $\frac{125}{8}$ unit^2 of $\frac{125}{6}$ unit^2? It is $\dfrac{\frac{125}{8}}{\frac{125}{6}}$

see pp 107–108

Q10.

In triangle PQR

$RQ^2 = PR^2 + PQ^2$

$= 5^2 + 12^2 = 25 + 144 = 169$

So $RQ = \sqrt{169} = 13$

$\Rightarrow \cos 2\theta = \frac{5}{13}$ ✓✓

$\Rightarrow 2\cos^2\theta - 1 = \frac{5}{13}$ ✓

$\Rightarrow 2\cos^2\theta = \frac{5}{13} + 1 = \frac{18}{13}$

$\Rightarrow \cos^2\theta = \frac{9}{13}$

$\Rightarrow \cos\theta = \pm\sqrt{\frac{9}{13}}$

$= \pm\frac{3}{\sqrt{13}}$ but $0 < \theta < \frac{\pi}{2}$ ✓

So $\cos\theta = \frac{3}{\sqrt{13}} = \frac{3\times\sqrt{13}}{\sqrt{13}\times\sqrt{13}}$ ✓

$= \frac{3\sqrt{13}}{13}.$

5 marks

Strategy
- You are using SOHCAHTOA in the large right-angled triangle PQR with angle 2θ.

Value
- Pythagoras' Theorem allows you to calculate the length of the hypotenuse RQ then give the value of $\cos 2\theta$ as $\frac{5}{13}$

Strategy
- The double angle formula allows you to calculate the value of $\cos\theta$.

Value
- 1st quadrant so the value of $\cos\theta$ is positive $\frac{3}{\sqrt{13}}$.

Rationalisation
- Rationalising the denominator – show this clearly.

see pp 62–63

Answers to Exam B

Exam B: Paper 1

Q1.

$\int 2x^2 + 3dx$ ✓

$= \dfrac{2x^3}{3} + 3x + C$ ✓ ✓

3 marks

1st term
- Use of the rule: $\int x^n \, dx = \dfrac{x^n}{n+1} + \ldots$

2nd term
- Knowing how to integrate a constant: $\int a \, dx = ax + \ldots$

Constant of integration
- Don't forget the constant C.

see p 77

Q2. (a)

$f(x) = \dfrac{1}{1+x}$

let $f(a) = \dfrac{1}{1+a} = b$

$\Rightarrow 1 = b(1+a) \Rightarrow \dfrac{1}{b} = 1 + a \Rightarrow \dfrac{1}{b} - 1 = a$ ✓

so $f^{-1}(b) = \dfrac{1}{b} - 1 = a$

$\Rightarrow f^{-1}(x) = \dfrac{1}{x} - 1$ ✓

2 marks

Strategy
- If a goes in to f and b comes out, then doing this in reverse means:
  b goes in to f^{-1} and a comes out.
  In terms of the formulae you have:
  $b = \dfrac{1}{1+a}$ with b the subject (when a goes in)
  $a = \dfrac{1}{b} - 1$ with a now the subject (b goes in)
  The strategy is to change the subject of the formula.

Formula
- Remember to use x for the final formula

Q2. (b)

The non-zero real numbers ✓

1 mark

Domain
- Division by zero in not allowed, so $x \neq 0$

see p 12

Q3. (a)

Here is the sketch of $y = 2\log_3(x + 3)$:

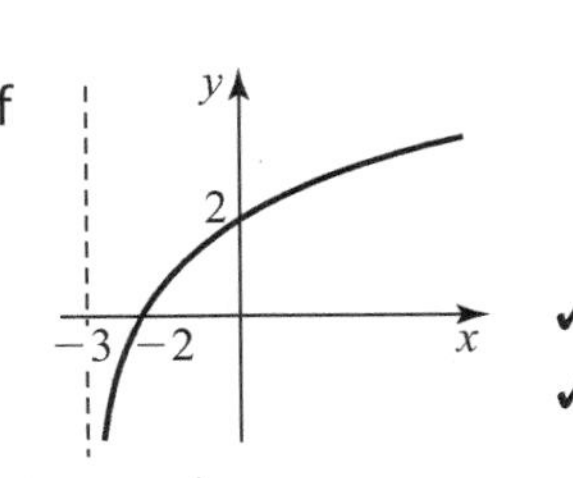

✓ ✓

2 marks

Horizontal translation
- $x + 3$ indicates a left shift of three units so (1,0) ends up at (−2,0)

Vertical scaling
- 2 × at the front of the formula indicates a y-axis scaling by a factor of 2. All heights of points on the graph are doubled, so (0,1) ends up at (0,2) [(3,1) went to (0,1) by the left shift]
- The new graph approaches $x = -3$. The original graph approached the y-axis.

Q3. (b)

$y = \log_3 \dfrac{1}{(x+3)^2}$

$= \log_3(x + 3)^{-2}$

$= -2\log_3(x + 3)$

Here is the graph: ✓ ✓

2 marks

Rearrangement
- You need to rearrange the given formula using the log rules, so that the previous formula $2\log_3(x + 3)$ appears.

Related graph
- The previous formula $2\log_3(x + 3)$ has been multiplied by −1. This indicates you should flip the graph in the x-axis.

see pp 21–23

Q4. (a)

(i) $\boldsymbol{p}.\boldsymbol{p} = |\boldsymbol{p}| \times |\boldsymbol{p}| \times \cos 0 = 1 \times 1 \times 1 = 1$ ✓

(ii) $\boldsymbol{q}.\boldsymbol{q} = |\boldsymbol{q}| \times |\boldsymbol{q}| \times \cos 0 = \sqrt{2} \times \sqrt{2} \times 1 = 2$ ✓

(iii) $\boldsymbol{p}.\boldsymbol{q} = |\boldsymbol{p}| \times |\boldsymbol{q}| \times \cos 45° = 1 \times \sqrt{2} \times \frac{1}{\sqrt{2}} = 1$ ✓

3 marks

Dot product
- You are using the result: $\boldsymbol{a}.\boldsymbol{b} = |\boldsymbol{a}||\boldsymbol{b}| \cos\theta$ where θ is the angle between $\boldsymbol{a}$ and $\boldsymbol{b}$.
- $|\boldsymbol{p}| = 1$ since the edge length of the cube is 1 unit.

Dot product
- The angle between a vector and itself is 0°.
- The diagonal of the square face of the cube is at 45° to the edges of the cube.

Dot product
- You should know the exact value of cos 45°.

Q4. (b)

$$\boldsymbol{r}.\boldsymbol{r} = (2\boldsymbol{p} + \boldsymbol{q}).(2\boldsymbol{p} + \boldsymbol{q}) \quad ✓$$
$$= 4\boldsymbol{p}.\boldsymbol{p} + 4\boldsymbol{p}.\boldsymbol{q} + \boldsymbol{q}.\boldsymbol{q} \quad ✓$$
$$= 4 \times 1 + 4 \times 1 + 2 = 10 \quad ✓$$

3 marks

Strategy
- Evidence of an attempt to expand the brackets: $(2\boldsymbol{p} + \boldsymbol{q}).(2\boldsymbol{p} + \boldsymbol{q})$

Expansion
- The dot product behaves very much like normal multiplication when you expand the brackets. However remember $\boldsymbol{p}^2$ makes no sense, so don't write it!

Calculation
- This calculation uses your calculated values from part (a) of the question.

Q4. (c)

$\boldsymbol{r}.\boldsymbol{r} = |\boldsymbol{r}|^2 = 10 \Rightarrow |\boldsymbol{r}| = \sqrt{10}$ ✓

1 mark

Calculation
- $\boldsymbol{r}.\boldsymbol{r} = |\boldsymbol{r}||\boldsymbol{r}| \cos 0° = |\boldsymbol{r}| \times |\boldsymbol{r}| \times 1 = |\boldsymbol{r}|^2$

see p 45

Q5. (a)

$$\cos(A + \tfrac{5\pi}{6}) - \sin(A + \tfrac{4\pi}{3}) \quad ✓$$
$$= \cos A \cos \tfrac{5\pi}{6} - \sin A \sin \tfrac{5\pi}{6} - (\sin A \cos \tfrac{4\pi}{3} + \cos A \sin \tfrac{4\pi}{3}) \quad ✓$$
$$= \cos A \times (-\tfrac{\sqrt{3}}{2}) - \sin A \times \tfrac{1}{2} - \sin A \times (-\tfrac{1}{2}) - \cos A \times (-\tfrac{\sqrt{3}}{2})$$
$$= -\tfrac{\sqrt{3}}{2} \cos A - \tfrac{1}{2} \sin A + \tfrac{1}{2} \sin A + \tfrac{\sqrt{3}}{2} \cos A = 0 \quad ✓$$

3 marks

Expansions
- Both addition formulae expansions are given to you in your exam:
  $\sin(A \pm B) = \sin A \cos B \pm \cos A \sin B$
  $\cos(A \pm B) = \cos A \cos B \mp \sin A \sin B$
- Be careful to subtract the whole of the sine expansion: $-(\sin A \cos \tfrac{4\pi}{3} + \cos A \sin \tfrac{4\pi}{3})$

Exact values
- $\tfrac{5\pi}{6} = \pi - \tfrac{\pi}{6}$ a 2nd quadrant angle
- $\tfrac{4\pi}{3} = \pi + \tfrac{\pi}{3}$ a 3rd quadrant angle

Simplification
- All the terms cancel. Be very careful with the negative terms and the subtraction.

Q5. (b)

$$\cos(A + \tfrac{5\pi}{6}) - \sin(A + \tfrac{4\pi}{3}) = 0$$
$$\Rightarrow \cos(A + \tfrac{5\pi}{6}) = \sin(A + \tfrac{4\pi}{3})$$

Now let $A = \frac{\pi}{12}$

$$\Rightarrow \cos(\tfrac{\pi}{12} + \tfrac{5\pi}{6}) = \sin(\tfrac{\pi}{12} + \tfrac{4\pi}{3})$$
$$\Rightarrow \cos(\tfrac{\pi}{12} + \tfrac{10\pi}{12}) = \sin(\tfrac{\pi}{12} + \tfrac{16\pi}{12})$$
$$\Rightarrow \cos\tfrac{11\pi}{12} = \sin\tfrac{17\pi}{12} \quad ✓$$

1 mark

Proof
- As with all 'show that …' questions you must be careful to show all the steps clearly.

see p 26, p 32

Q6. (a)

$A(-2, 1)$ and $B(3, 2)$

So $m_{AB} = \frac{2-1}{3-(-2)} = \frac{1}{3+2} = \frac{1}{5}$ ✓

Any line parallel to AB will have the same gradient, i.e. $\frac{1}{5}$ ✓

So the required line has gradient $\frac{1}{5}$ and a point on the line is $C(3, -1)$.

The equation is:

$y-(-1) = \frac{1}{5}(x-3) \Rightarrow y+1 = \frac{1}{5}(x-3)$

$\Rightarrow 5y+5 = x-3 \Rightarrow 5y-x = -8$ ✓

3 marks

Gradient
- You are using the gradient formula: $m_{AB} = \frac{y_2-y_1}{x_2-x_1}$ where $A(x_1, y_1)$ and $B(x_2, y_2)$
- An alternative calculation is $\frac{1-2}{-2-3}$.

Strategy
- Since lines that are parallel have the same gradient, the required line will also have a gradient of m_{AB}, i.e. $\frac{1}{5}$.

Equation
- Here you are using the formula $y-b = m(x-a)$ where (a, b) is a point on the line and m is the gradient of the line.
- Multiply both sides of the equation by 5 to get rid of the fraction $\frac{1}{5}$.
- Alternatives are $5y = x-8$ or $5y-x+8=0$.

see p 86, p 89

Q6. (b)

For x-axis intercept set $y = 0$.

So $5y-x = -8$ becomes $5 \times 0 - x = -8$

giving $-x = -8$ so $x = 8$.

The x-axis intercept is (8, 0) ✓

1 mark

Coordinates
- Substitute $y = 0$ in the equation of the line.
- Since the coordinates of the point are asked for, an answer of $x = -8$ would not gain this mark.

see p 90

Q7.

$f(t) = \frac{2}{3t} = \frac{2}{3}t^{-1}$ ✓

$\Rightarrow f'(t) = -\frac{2}{3}t^{-2} = -\frac{2}{3t^2}$ ✓

$\Rightarrow f'(2) = -\frac{2}{3 \times 2^2} = -\frac{1}{6}$ ✓

3 marks

Preparation for differentiation
- Note that the phrase 'rate of change' indicates that you need to differentiate and 'when $t = 2$' indicates that you need to find $f'(2)$.
- $\frac{2}{3t} = \frac{2}{3} \times \frac{1}{t} = \frac{2}{3} \times t^{-1} = \frac{2}{3}t^{-1}$

Differentiation
- You are using: $f(t) = at^n \Rightarrow f'(t) = nat^{n-1}$ with $a = \frac{2}{3}$ and $n = -1$.

Evaluation
- Substitute $t = 2$ in the derivative formula.
- Cancel down the fraction, i.e. simplify $-\frac{2}{12}$

see pp 66–67

Q8.

$\int 7\cos(3x-1) - 3\sin x \; dx$

$= \int 7\cos(3x-1)dx - \int 3\sin x \, dx$

$= 7 \times \sin\frac{(3x-1)}{3} - 3 \times (-\cos x) + c$ ✓

$= \frac{7}{3}\sin(3x-1) + 3\cos x + c$ ✓ ✓

3 marks

Start 1st integration
- You gain this mark for $\sin(3x-1)$

Complete 1st integration
- Either $\frac{7}{3}\sin(3x-1)$ or $\frac{7\sin(3x-1)}{3}$

Complete 2nd integration
- Don't forget c the constant of integration.

see p 78, p 81

Q9. (a)

$$\boldsymbol{p}+\boldsymbol{q} = \begin{pmatrix} -1 \\ 2 \\ 3 \end{pmatrix} + \begin{pmatrix} 1 \\ 3 \\ 2 \end{pmatrix} = \begin{pmatrix} 0 \\ 5 \\ 5 \end{pmatrix}$$

and $\boldsymbol{p}-\boldsymbol{q} = \begin{pmatrix} -1 \\ 2 \\ 3 \end{pmatrix} - \begin{pmatrix} 1 \\ 3 \\ 2 \end{pmatrix} = \begin{pmatrix} -2 \\ -1 \\ 1 \end{pmatrix}$ ✓

1 mark

Calculations
- Add/subtract the corresponding components.

Q9. (b)

$$(\boldsymbol{p}+\boldsymbol{q}).(\boldsymbol{p}-\boldsymbol{q}) = \begin{pmatrix} 0 \\ 5 \\ 5 \end{pmatrix} . \begin{pmatrix} -2 \\ -1 \\ 1 \end{pmatrix}$$ ✓

$= 0 \times (-2) + 5 \times (-1) + 5 \times 1$

$= 0 + (-5) + 5 = 0$

$\Rightarrow \boldsymbol{p} + \boldsymbol{q}$ is perpendicular to $\boldsymbol{p} - \boldsymbol{q}$ ✓

2 marks

Strategy
- Evidence that you know: dot product $= 0 \Rightarrow$ vectors are perpendicular

Calculations & statement
- Show the calculation clearly – working must be shown to gain this mark.
- Make a clear statement about the vectors.

see p 37, p 43

Q10. (a)

$T_t = T_0 \times 10^{-kt}$

with $T_0 = 3000$, $t = 20$ and $T_t = 300$

$\Rightarrow 300 = 3000 \times 10^{-k\times 20}$ ✓

$\Rightarrow \frac{300}{3000} = 10^{-20k} \Rightarrow \frac{1}{10} = 10^{-20k} \Rightarrow 10^{-1} = 10^{-20k}$

$\Rightarrow -1 = -20k \Rightarrow k = \frac{-1}{-20} = \frac{1}{20}$ ✓

2 marks

Substitution
- Correct substitution of the values into the formula will gain you this mark.

Calculation
- Notice that $a^m = a^n \Rightarrow m = n$. If the base is the same (a) then the indices (m and n) are equal.

Q10. (b)

$T_t = T_0 \times 10^{-\frac{1}{20}t}$

Use $T_0 = 300$ and $t = 20$ ✓

giving: $T_t = 300 \times 10^{-\frac{1}{20}\times 20} = 300 \times 10^{-1} = 30$

The temperature will be 30°C. ✓

2 marks

Substitution
- Alternatively you could have used: $T_0 = 3000$ and $t = 40$

Calculation
- Answer with a clear statement of your result.

see p 19, p 24

Q11. (a)

−4	1	6	0	−32	
		−4	−8	32	✓
	1	2	−8	0	✓

Since the remainder is 0 then $x + 4$ is a factor of $x^3 + 6x^2 - 32$. ✓

3 marks

Strategy
- Use of synthetic division will be expected.

Calculation
- Make sure you line up the numbers in neat columns and rows for easy reading.

Statement
- You must make the statement: 'Since the remainder is 0 then $x + 4$ is a factor'

Q11. (b)

So $x^3 + 6x^2 - 32 = (x + 4)(x^2 + 2x - 8)$ ✓

$= (x + 4)(x + 4)(x - 2)$ ✓

$= (x + 4)^2(x - 2)$

2 marks

1st factorisation
- This mark is for the factor $x^2 + 2x - 8$

Full factorisation
- This is for factorising the quadratic.

see pp 60–61

Q12. (a)

$$\int \sqrt{1+3x}\,dx = \int (1+3x)^{\frac{1}{2}}\,dx$$ ✓

$$= \frac{(1+3x)^{\frac{3}{2}}}{\frac{3}{2}\times 3} + C$$ ✓

$$= \frac{2(1+3x)^{\frac{3}{2}}}{9} + C = \frac{2\left(\sqrt{(1+3x)}\right)^3}{9} + C$$ ✓

3 marks

Preparation
- This involves changing the expression into a form suitable for integration – in this case changing the root to the power of $\frac{1}{2}$

Integration
- Use of $\int (ax+b)^n\,dx = \frac{(ax+b)^{n+1}}{a(n+1)} + \ldots$

Simplification & constant
- To simplify: multiply top and bottom of the fraction by 2.
- The last step is not needed to gain this mark; however, it is needed for the following question.

Q12. (b)

$$\int_0^1 \sqrt{1+3x}\,dx = \left[\frac{2\left(\sqrt{(1+3x)}\right)^3}{9}\right]_0^1$$

$$= \frac{2\left(\sqrt{(1+3\times 1)}\right)^3}{9} - \frac{2\left(\sqrt{(1+3\times 0)}\right)^3}{9}$$ ✓

$$= \frac{2\left(\sqrt{4}\right)^3}{9} - \frac{2\left(\sqrt{1}\right)^3}{9} = \frac{2\times 8}{9} - \frac{2\times 1}{9}$$

$$= \frac{16}{9} - \frac{2}{9} = \frac{14}{9}$$ ✓

2 marks

Substitution
- To proceed with substitution you need to know the meaning or $a^{\frac{m}{n}}$: take the nth root then raise to the power m.

Calculation
- Correct evaluation gains you this mark.

see pp 80–81

Q13. (a)

$u_1 = pu_0 - 1 = p \times 3 - 1 = 3p - 1$ ✓

$u_2 = pu_1 - 1 = p(3p - 1) - 1 = 3p^2 - p - 1$ ✓

2 marks

1st term
- 'in terms of p' so you will not get a numerical value. You require an expression involving p.

2nd term
- u_1 is replaced by the expression $3p - 1$ and you will be expected to simplify the result.

Q13. (b)

$u_2 = 1 \Rightarrow 3p^2 - p - 1 = 1 \Rightarrow 3p^2 - p - 2 = 0$ ✓

$\Rightarrow (3p + 2)(p - 1) = 0$

$\Rightarrow 3p + 2 = 0$ or $p - 1 = 0 \Rightarrow p = -\frac{2}{3}$ or $p = 1$ ✓

Since p is the multiplier and there is only a limit if the multiplier lies between -1 and 1, then $p = -\frac{2}{3}$ is the only value generating a sequence with a limit. ✓

3 marks

Equation
- This is a quadratic equation in the variable p.

Roots
- Solving the quadratic equation and stating the two roots will gain you this mark.

Solution & Reason
- For $u_{n+1} = pu_n - 1$ to have a limit then $-1 < p < 1$. This must appear somewhere in your reasoning when you eliminate $p = 1$.

Q13. (c)

Let the limit $= L$ then since $u_{n+1} = -\frac{2}{3}u_n - 1$

this gives: $L = -\frac{2}{3}L - 1$ ✓

$\Rightarrow 3L = -2L - 3 \Rightarrow 5L = -3 \Rightarrow L = -\frac{3}{5}$

The limit is $-\frac{3}{5}$ ✓

2 marks

Strategy

- If you use $L = \dfrac{c}{1-m}$ then use that method.

Calculation

- Multiply through by 3 to remove the fraction.

see pp 102–104

Q14.

$$y = \frac{1}{16}x^4 - \frac{1}{8}x^2 + x$$ ✓

$$\Rightarrow \frac{dy}{dx} = \frac{1}{4}x^3 - \frac{1}{4}x + 1$$ ✓

The tangent $y = x + c$ has gradient 1

So set

$$\frac{dy}{dx} = 1$$ ✓

$$\Rightarrow \tfrac{1}{4}x^3 - \tfrac{1}{4}x + 1 = 1$$

$$\Rightarrow \tfrac{1}{4}x^3 - \tfrac{1}{4}x = 0 \Rightarrow x^3 - x = 0$$

$$x(x^2 - 1) = 0 \Rightarrow x(x-1)(x+1) = 0$$

$$\Rightarrow x = 0 \text{ or } x = 1 \text{ or } x = -1$$ ✓

For $x = 0$:

$$y = \frac{1}{16} \times 0^4 - \frac{1}{8} \times 0^2 + 0 = 0$$ ✓

so $y = x + c$ gives $0 = 0 + c \Rightarrow c = 0$

The tangent is $y = x$ with contact point (0, 0)

For $x = -1$:

$$y = \frac{1}{16} \times (-1)^4 - \frac{1}{8} \times (-1)^2 + (-1)$$

$$= -\frac{17}{16}$$

So $y = x + c$ gives

$$-\frac{17}{16} = -1 + c \Rightarrow c = -\frac{1}{16}$$

tangent is $y = x - \dfrac{1}{16}$, ✓

contact point is $\left(-1, -\dfrac{17}{16}\right)$

For $x = 1$:

$$y = \frac{1}{16} \times 1^4 - \frac{1}{8} \times 1^2 + 1 = \frac{15}{16}$$

so $y = x + c$ gives

$$\tfrac{15}{16} = 1 + c \Rightarrow c = -\tfrac{1}{16}$$

Tangent is $y = x - \dfrac{1}{16}$ (same as for $x = -1$)

Contact point is $\left(1, \dfrac{15}{16}\right)$ ✓

7 marks

Strategy

- Evidence that you know to differentiate will gain you this mark.

Differentiation

- Be careful with the fractions here: $4 \times \frac{1}{16} = \frac{4}{16} = \frac{1}{4}$ and $2 \times \frac{1}{8} = \frac{2}{8} = \frac{1}{4}$

Gradient

- Compare $y = mx + c$ and $y = x + c$. This leads to $m = 1$ for the gradient of the tangent line.

Strategy

- This second strategy mark is given for knowing to set the gradient formula equal to 1.

Solving

- Remove fractions first by multiplying both sides of the equation by 4. Although this is a cubic, terms are missing, and so it is easily factorised once you realise to remove the common factor x.

Calculation

- This processing mark involves a fair amount of calculation and is gained for clearly stating the possible values of c. These are: $c = 0$ and $c = -\dfrac{1}{16}$

Interpretation

- There are three points of contact: $(0, 0)$, $\left(-1, -\dfrac{17}{16}\right)$ and $\left(1, \dfrac{15}{16}\right)$ but only two equations for the tangent: $y = x$ and $y = x - \dfrac{1}{16}$

  A close examination of the graph shows $y = x - \frac{1}{16}$ is a tangent at two separate points on the curve as is shown in this diagram

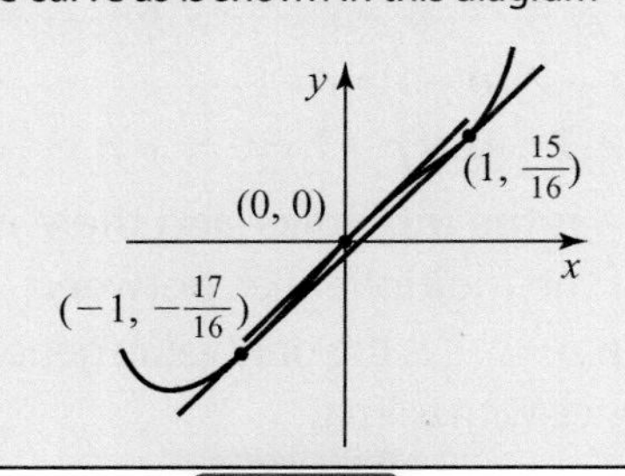

see p 68

Q15. (a)

$f(x) = 2x - 1,\ g(x) = \log_{12}x$

$f(g(x)) + g(f(x)) = 0$ ✓

$\Rightarrow f(\log_{12}x) + g(2x - 1) = 0$ ✓

$\Rightarrow 2\log_{12}x - 1 + \log_{12}(2x - 1) = 0$ ✓

$\Rightarrow \log_{12}x^2 + \log_{12}(2x - 1) = 1$

$\Rightarrow \log_{12}x^2(2x - 1) = 1$ ✓

$\Rightarrow x^2(2x - 1) = 12^1$ ✓

$\Rightarrow 2x^3 - x^2 - 12 = 0$

2	2	−1	0	−12
		4	6	12
	2	3	6	0

✓

So $x = 2$ satisfies the equation and is therefore a solution.

6 marks

Composition
- Combining f and g to get the formula $f(g(x))$ is called the 'composition' of f and g. Showing either $f(\log_{12}x)$ or $g(2x - 1)$ gains you this 1st mark.

Composition
- Reaching either $2\log_{12}x - 1$ or $\log_{12}(2x - 1)$ will gain you this 2nd mark.

Composition
- Correctly finding the 2nd expression $2\log_{12}x - 1$ or $\log_{12}(2x - 1)$ will give you this 3rd mark.

Log Law
- The rule used here is:

$$\log_b m + \log_b n = \log_b mn$$

Exponential form
- The result used is:

$$\log_b a = c \leftrightarrow a = b^c$$
(logarithmic form) (exponential form)

Calculation
- At this stage you must clearly show that $x = 2$ is a solution of $2x^3 - x^2 - 12 = 0$. The table shows that $f(2) = 0$ where $f(x) = 2x^3 - x^2 - 12$.

see p 12, p 22

Q15. (b)

From part(a) equation becomes:

$(x - 2)(2x^2 + 3x + 6) = 0$

Consider $2x^2 + 3x + 6 = 0$ ✓

Discriminant $= 3^2 - 4 \times 2 \times 6$

$= 9 - 48 = -39$

Since Discriminant < 0 there are no Real solutions. ✓

The only Real solution is $x = 2$.

2 marks

Strategy
- One solution, $x = 2$, comes from the factor $x - 2$ equating to zero. Any other solutions will therefore come from equating the other factor to zero, i.e. $2x^2 + 3x + 6 = 0$. If your working shows you knew this you will gain this strategy mark.

Communication
- There must be a clear reason for 'no Real solutions'. In this case, the discriminant of the quadratic equation is negative. Your working should state this fact quite clearly. The result used is:

  Discriminant $< 0 \Rightarrow$ no Real roots.

see p 52

Exam B: Paper 2

Q1. (a)

$f(x) = x^3 + 3x^2 - 4$ ✓

$\Rightarrow f'(x) = 3x^2 + 6x$

For stationary points set

$f'(x) = 0$ ✓

$\Rightarrow 3x^2 + 6x = 0 \Rightarrow 3x(x + 2) = 0$ ✓

$\Rightarrow x = 0$ or $x = -2$

x:	-2		0	
$f'(x) = 3x(x + 2)$:	+	−		+
Shape of graph:	/	\		/
nature:	max		min	

✓

$f(-2) = (-2)^3 + 3 \times (-2)^2 - 4 = 0$

So $(-2, 0)$ is a maximum stationary point

$f(0) = 0^3 + 3 \times 0^2 - 4 = -4$ ✓

So $(0, -4)$ is a minimum stationary point ✓

6 marks

Differentiate
- Correct differentiation will gain this mark.

Strategy
- To find the stationary points set $f'(x) = 0$.

Solutions
- The common factor is $3x$ with two roots: 0 and -2.

Justify
- A 'nature table' is needed to determine maximum and minimum points.

***y*-coordinates**
- Take care with negatives. In general remember: squaring produces positive or zero answers, cubing a negative quantity gives a negative answer. In this case: $(-2)^3 = -8$ and $(-2)^2 = 4$.

Statements
- Clear statements concerning the nature of the points, i.e. maximum or minimum, etc. are expected for this final mark.

see pp 68–69

Q1. (b)

For x-intercepts set $y = 0$

So $x^3 + 3x^2 - 4 = 0$

$\Rightarrow (x - 1)(x + 2)^2 = 0$

$\Rightarrow x = 1$ or $x = -2$

Intercepts are $(-2, 0)$ and $(1, 0)$ ✓

For y-intercept set $x = 0$

So $y = 0^3 + 3 \times 0^2 - 4 = -4$

Intercept is $(0, -4)$ ✓

Sketch: ✓ ✓

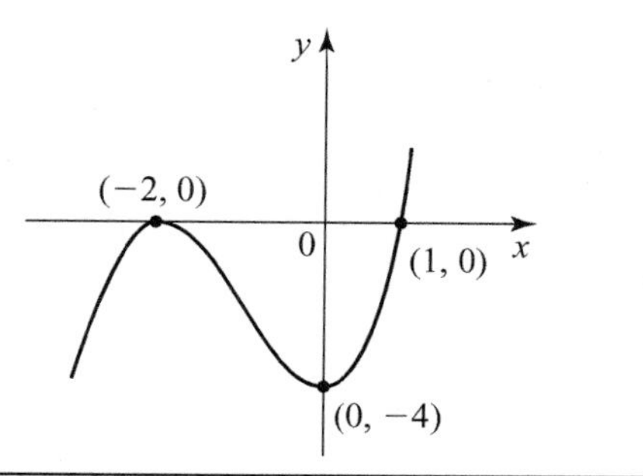

4 marks

***x*-intercepts**
- x-axis intercepts are obtained by setting $y = 0$ or $f(x) = 0$. In this case, a cubic equation is formed, but you have been given the factorisation of the cubic. So you can proceed by setting each factor in turn equal to 0.
- Remember repeated roots, like $x = -2$ in this case, would indicate that the graph touches the x-axis

***y*-intercept**
- 1 mark is allocated for this result.

Sketch
- 1 mark is allocated for your sketch showing the cubic shape correctly, with the two stationary points clearly shown and labelled.
- The 2nd mark is given for your sketch clearly showing the intercepts, i.e. (0,−4) and (1,0), etc.

see p 70

Q2. (a) (i)

$x^2 + y^2 + 4x - 6y + 5 = 0$

Centre is $(-2, 3)$. ✓

1 mark

Centre

- The result used is:

  $x^2 + y^2 + 2gx + 2fy + c = 0$

  Centre: $(-g, -f)$

  This result is on your formulae sheet in the exam.

see p 94

Q2. (a) (ii)

$$\text{Radius} = \sqrt{(-2)^2 + 3^2 - 5}$$

$$= \sqrt{4+9-5} = \sqrt{8} = 2\sqrt{2}$$ ✓

1 mark

Radius

- Notice that:

  $\sqrt{8} = \sqrt{4\times2} = \sqrt{4}\times\sqrt{2} = 2\times\sqrt{2} = 2\sqrt{2}$

- The radius formula: $\sqrt{g^2 + f^2 - c}$ for the circle $x^2 + y^2 + 2gx + 2fy + c = 0$ is given to you on your formulae sheet in the exam.

see p 94

Q2. (b) (i)

Circle B:

$(x - 2)^2 + (y + 1)^2 = 2$

has centre $(2, -1)$ ✓

Distance between $(-2,3)$ and $(2,-1)$ is:

$$\sqrt{(-2-2)^2 + (3-(-1))^2}$$

$$= \sqrt{(-4)^2 + 4^2} = \sqrt{32} = 4\sqrt{2}$$ ✓

2 marks

Centre

The result used is:

- For circle $(x - a)^2 + (y - b)^2 = r^2$ the centre is (a, b) a result also given to you in your exam.

Calculation

- The distance formula is used here.

  If $A\,(x_1,y_1)$ and $B(x_2,y_2)$ then

  $$AB = \sqrt{(x_2 - x_1)^2 + (y_2 - y_1)^2}$$

see p 91, 94

Q2. (b) (ii)

Circle A: radius $= 2\sqrt{2}$

Circle B: radius $= \sqrt{2}$

Distance between centres $= 4\sqrt{2}$

Sum of radii $= 2\sqrt{2} + \sqrt{2} = 3\sqrt{2}$ ✓

less than $4\sqrt{2}$, the distance between centres

$\Rightarrow$ circles do not intersect ✓

2 marks

Strategy

I — Distance between centres is greater than the sum of the radii.

II — Distance between centres is equal to the sum of the two radii.

III — Distance between centres is less than the sum of the two radii.

Communication

- There must be a comparison given. In this case a statement that the sum of the radii is $3\sqrt{2}$, which is less than the centre distance $4\sqrt{2}$, so there is no intersection.

  (This is situation I shown above.)

see p 99

Q2. (c)

Solve $\left.\begin{array}{l} y = x + 5 \\ x^2 + y^2 + 4x - 6y + 5 = 0 \end{array}\right\}$

Substitute $y = x + 5$ in circle equation: ✓

$x^2 + (x + 5)^2 + 4x - 6(x + 5) + 5 = 0$ ✓

$\Rightarrow x^2 + x^2 + 10x + 25 + 4x - 6x - 30 + 5 = 0$

$\Rightarrow 2x^2 + 8x = 0 \Rightarrow 2x(x + 4) = 0$ ✓

$\Rightarrow x = 0$ or $x = -4$ ✓

when $x = -4$ $\quad y = -4 + 5 = 1$

when $x = 0$ $\quad y = 0 + 5 = 5$

So $P(-4, 1)$ and $Q(0, 5)$ ✓

5 marks

Strategy

- In general, to find where graphs $y = f(x)$ and $y = g(x)$ intersect, you equate the formulae i.e. $f(x) = g(x)$ and solve the resulting equation. However in this case, the equation of the circle cannot be written as '$y = ...$' so substitution is used.

Substitution

- Replace all occurrences of y by $x + 5$ in the circle equation.

'Standard form'

- Reducing the equation to $2x^2 + 8x = 0$ gains you this mark.

Solve for x

- The common factor is $2x$ with roots -4 and 0.

Coordinates

- Coordinates are asked for not just values of x and y.

see p 96

Q3. (a)

$A(-5, 2)$ and $B(-3, 8)$ ✓

$$\Rightarrow m_{AB} = \frac{8 - 2}{-3 - (-5)} = \frac{6}{2} = 3$$

$$\Rightarrow m_{\perp} = -\frac{1}{3}$$ ✓

Midpoint of AB is

$$\left(\frac{-5 + (-3)}{2}, \frac{2 + 8}{2}\right) = (-4, 5)$$ ✓

For the perpendicular bisector:

A point on the line is $(-4, 5)$

and the gradient $= -\frac{1}{3}$

So, equation is:

$$y - 5 = -\frac{1}{3}(x - (-4))$$

$\Rightarrow 3y - 15 = -(x + 4)$

$\Rightarrow 3y - 15 = -x - 4$ ✓

$\Rightarrow 3y + x = 11$

4 marks

Strategy

- Here you are using the gradient formula:

  $P(x_1,y_1)$, $Q(x_2,y_2)$ gives $m_{PQ} = \frac{y_2 - y_1}{x_2 - x_1}$

Perpendicular gradient

- If $m = \frac{a}{b}$ then $m_{\perp} = -\frac{b}{a}$. For $m = 3$ you think of 3 as $\frac{3}{1}$. Inverting and changing sign then gives $-\frac{1}{3}$ as shown in the solution.

Strategy

- You have to know that 'bisector' means find the mid point of AB.
- The mid point formula is:

  $P(x_1,y_1)$, $Q(x_2,y_2)$.

  Midpoint is $\left(\frac{x_1 + x_2}{2}, \frac{y_1 + y_2}{2}\right)$

Equation

- Using $y - b = m(x - a)$ with $m = -\frac{1}{3}$ and the point (a,b) is $(-4,5)$.

see p 86, p 89, p 91

Q3. (b)

$A(-5, 2)$ and $D(5, -8)$

Midpoint of AD is:

$$\left(\frac{-5+5}{2}, \frac{2+(-8)}{2}\right) = M(0, -3)$$ ✓

So, using $C(3, 6)$ and $M(0, -3)$

$$m_{CM} = \frac{6-(-3)}{3-0} = \frac{9}{3} = 3$$ ✓

For the median:

A point on the line is $(0, -3)$ and the gradient is 3

So, equation is $y - (-3) = 3(x - 0)$

$\Rightarrow y + 3 = 3x$

$\Rightarrow y - 3x = -3$ ✓

3 marks

Strategy

- The median is the line from C to the midpoint of the opposite side AD. You will therefore need to find the coordinates of the midpoint.

Gradient

- Medians do not involve 'perpendicular' and so when m_{CM} is calculated you use this value, 3, to find the equation of the median.

Equation

- Use $y - b = m(x - a)$ with $m = 3$ and (a,b) being the point $(0,-3)$. Alternatively, spot that $(0,-3)$ is the y-intercept of the line and use $y = mx + c$ with $m = -3$ and $c = -3$ to give $y = 3x - 3$.

see pp 91–92

Q3. (c)

To find the intersection point S: ✓

Solve:

$$\begin{rcases} 3y + x = 11 \\ y - 3x = -3 \end{rcases}_{(\times 3)} \quad \begin{aligned} &\rightarrow 3y + x = 11 \\ &\rightarrow \underline{3y - 9x = -9} \end{aligned}$$

subtract: $10x = 20$

$\Rightarrow x = 2$ ✓

Now substitute $x = 2$ in

$y - 3x = -3$

$\Rightarrow y - 3 \times 2 = -3 \Rightarrow y - 6 = -3$

$\Rightarrow y = 3$ ✓

so S (2,3)

3 marks

Strategy

- To find the point of intersection of two lines you solve the two equations of the lines simultaneously.

Find one variable

- An alternative is to multiply the 1st equation by 3, then add to give $10y = 30 \Rightarrow y = 3$.

Second variable

- Use the 'easier' equation when doing the substitution. If you found $y = 3$ first then the 1st equation is 'easier'.

see p 90

Q4.

$3\cos 2x° + 9\cos x° = \cos^2 x° - 7$

$\Rightarrow 3(2\cos^2 x° - 1) + 9\cos x° = \cos^2 x° - 7$ ✓

$\Rightarrow 6\cos^2 x° - 3 + 9\cos x° = \cos^2 x° - 7$

$\Rightarrow 5\cos^2 x° + 9\cos x° + 4 = 0$ ✓

$\Rightarrow (5\cos x° + 4)(\cos x° + 1) = 0$ ✓

$\Rightarrow 5\cos x° + 4 = 0$ or $\cos x° + 1 = 0$

$\Rightarrow \cos x° = -\frac{4}{5}$ or $\cos x° = -1$ ✓

For $\cos x° = -\frac{4}{5}$

$x°$ is in 2nd or 3rd quadrants

1st quadrant angle is 36·9°

so $x = 180 - 36{\cdot}9$ or

$x = 180 + 36{\cdot}9$

$\Rightarrow x = 143{\cdot}1$ or $x = 216{\cdot}9$

For $\cos x° = -1$

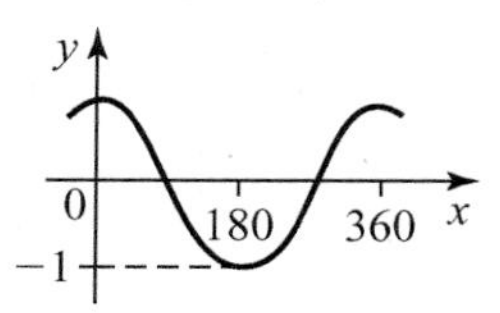

So $x = 180$

Solutions are:

143·1, 180, 216·9 ✓

(to 1 decimal place).

5 marks

Strategy

- There are three expansions of $\cos 2x°$: $\cos 2x° = 2\cos^2 x - 1$ or $\cos^2 x - \sin^2 x$ or $1 - 2\sin^2 x$. Which of these you use is dictated by the surrounding 'landscape' in the equation: There is a '$\cos x$ term' and a $\cos^2 x$ term, but no '$\sin x$ term' or '$\sin^2 x$ term'. So $\cos 2x = 2\cos^2 x - 1$ is used as the other two forms involve $\sin^2 x$.

'Standard form'

- You recognise the equation as a quadratic equation in $\cos x$ and arrange it into the standard form $5\cos^2 x° + 9\cos x° + 4 = 0$.

Factorisation

- Compare $5c^2 + 9c + 4 = (5c + 4)(c + 1)$.
- Remember that you should always check your answer in a factorisation by multiplying out, i.e. working backwards.

Solving for $\cos x$

- From $(5c + 4)(c + 1) = 0$ to $c = -\frac{4}{5}$ or $c = -1$ is no different to what you do at this stage except that the single variable c is replaced by the expression $\cos x°$.

Solutions

- A lot of knowledge and work needed for this final processing mark!
- The quadrant diagram is used:

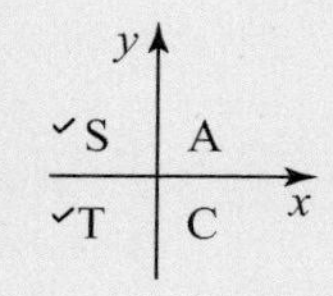

for $\cos x°$ negative.

see p 63

Q5. (a)

$M(0,3,2)$ ✓

$N(5,2,0)$ ✓

2 marks

Point M

- M is the midpoint. Half-way along DG which is 6 units long (y-coordinate) is 3 units.

Point N

- N is $\frac{1}{3}$ of the way along AB so $\frac{1}{3}$ of $6 = 2$ units is the y-coordinate.

see p 36

Q5. (b)

$M(0,3,2)$ and $B(5,6,0)$

$$\overrightarrow{MB} = \boldsymbol{b} - \boldsymbol{m} = \begin{pmatrix} 5 \\ 6 \\ 0 \end{pmatrix} - \begin{pmatrix} 0 \\ 3 \\ 2 \end{pmatrix} = \begin{pmatrix} 5 \\ 3 \\ -2 \end{pmatrix}$$ ✓

also $M(0,3,2)$ and $N(5,2,0)$

$$\overrightarrow{MN} = \boldsymbol{n} - \boldsymbol{m} = \begin{pmatrix} 5 \\ 2 \\ 0 \end{pmatrix} - \begin{pmatrix} 0 \\ 3 \\ 2 \end{pmatrix} = \begin{pmatrix} 5 \\ -1 \\ -2 \end{pmatrix}$$ ✓

2 marks

Components

- The basic result used is:

$P(x_1,y_1,z_1)$ and $Q(x_2,y_2,z_2)$

$$\text{so } \overrightarrow{PQ} = \boldsymbol{q} - \boldsymbol{p} = \begin{pmatrix} x_2 \\ y_2 \\ z_2 \end{pmatrix} - \begin{pmatrix} x_1 \\ y_1 \\ z_1 \end{pmatrix} = \begin{pmatrix} x_2 - x_1 \\ y_2 - y_1 \\ z_2 - z_1 \end{pmatrix}$$

see p 40

Q5. (c)

M, B, N, v, w, $\theta°$

Use $\cos\theta = \dfrac{\boldsymbol{v}.\boldsymbol{w}}{|\boldsymbol{v}|\,|\boldsymbol{w}|}$ ✓

where $\boldsymbol{v} = \begin{pmatrix} 5 \\ 3 \\ -2 \end{pmatrix}$ and $\boldsymbol{w} = \begin{pmatrix} 5 \\ -1 \\ -2 \end{pmatrix}$

$$\boldsymbol{v}.\boldsymbol{w} = \begin{pmatrix} 5 \\ 3 \\ -2 \end{pmatrix} . \begin{pmatrix} 5 \\ -1 \\ -2 \end{pmatrix}$$

$$= 5 \times 5 + 3 \times (-1) + (-2) \times (-2)$$

$$= 25 - 3 + 4 = 26$$ ✓

$$|\boldsymbol{v}| = \sqrt{5^2 + 3^2 + (-2)^2}$$

$$= \sqrt{25 + 9 + 4} = \sqrt{38}$$ ✓

$$|\boldsymbol{w}| = \sqrt{5^2 + (-1)^2 + (-2)^2}$$ ✓

$$= \sqrt{25 + 1 + 4} = \sqrt{30}$$

So, $\cos\theta = \dfrac{26}{\sqrt{38}\sqrt{30}}$

$$\Rightarrow \theta = \cos^{-1}\left(\frac{26}{\sqrt{38}\sqrt{30}}\right)$$

So $\theta = 39{\cdot}64\ldots \doteqdot 39{\cdot}6$ (to 1 dec pl.) ✓

5 marks

Strategy

- This strategy mark is for the use of the 'scalar' or 'dot product' formula.

Calculation

- The 'dot product' formula is:

$$\begin{pmatrix} x_1 \\ y_1 \\ z_1 \end{pmatrix} . \begin{pmatrix} x_2 \\ y_2 \\ z_2 \end{pmatrix} = x_1x_2 + y_1y_2 + z_1z_2$$

Magnitudes

- The magnitude formula is:

$$\left\| \begin{pmatrix} x_1 \\ y_1 \\ z_1 \end{pmatrix} \right\| = \sqrt{x_1^2 + y_1^2 + z_1^2}$$

Angle

- Be careful with your calculator calculation

INV cos (2 6 ÷ (√ 3 8 × √ 3 0)) EXE

The brackets are vital: $\cos^{-1}(\ldots)$ and $(\sqrt{38} \times \sqrt{30})$.

see pp 42–43

Q6. (a)

$7 + 4x - x^2$

$= -x^2 + 4x + 7$

$= -(x^2 - 4x) + 7$ ✓

$= -[(x - 2)(x - 2) - 4] + 7$

$= -(x - 2)^2 + 4 + 7$

$= -(x - 2)^2 + 11$ ✓

$p(x + q)^2 + r$ This comparison gives:

$p = -1$, $q = -2$ and $r = 11$ ✓

3 marks

Common factor

- It is much easier to deal with $x^2 - 4x\ldots$ than to deal with $-x^2 + 4x\ldots$ hence the need to take out the common factor -1.

Complete the square

- Compare $x^2 - 4x$ with $(x - 2)^2$. Multiplying out $(x - 2)(x - 2)$ gives $x^2 - 4x + 4$. A constant term $+4$ has appeared. You need to remove it, hence -4 in the square brackets.

Values

- The question specifically asks for the values of the p, q and r. If you don't state these values then you will not gain this mark.

see p 14

Q6. (b)

Given: $2x^2 - 3x + (1 - k) = 0$

Discriminant $= (-3)^2 - 4 \times 2 \times (1 - k)$ ✓

$= 9 - 8(1 - k)$

$= 9 - 8 + 8k$

$= 1 + 8k$ ✓

So if $1 + 8k < 0$ there will be no real roots

$1 + 8k < 0 \Rightarrow 8k < -1 \Rightarrow k < -\frac{1}{8}$ ✓

3 marks

Use the discriminant

- For $ax^2 + bx + c = 0$ the discriminant is $b^2 - 4ac$. In this case $a = 2$, $b = -3$ and $c = 1 - k$.

Simplify and impose condition

- Note that $\leq$ is wrong since this includes the possibility that the discriminant is zero and in that case there would be one root (equal roots).

State range

- Solve the inequation correctly for this mark.

see p 55

Q7.

$h(x) = \int h'(x)dx = \int \frac{2\sqrt{x} + x^4}{4x}\, dx$

so, $h(x) = \int \frac{2x^{\frac{1}{2}}}{4x} + \frac{x^4}{4x} dx$

$= \int \frac{x^{-\frac{1}{2}}}{2} + \frac{x^3}{4} dx$ ✓

$= \frac{x^{\frac{1}{2}}}{2 \times \frac{1}{2}} + \frac{x^4}{4 \times 4} + c$

$= \sqrt{x} + \frac{x^4}{16} + c$ ✓

$h(4) = 17 \Rightarrow \sqrt{4} + \frac{4^4}{16} + c = 17$ ✓

$\Rightarrow 2 + 16 + c = 17$

$\Rightarrow c = -1$

so $h(x) = \sqrt{x} + \frac{x^4}{16} - 1$ ✓

4 marks

Preparation for integration

- Your 1st step is to 'split the fraction' the pattern is: $\frac{p+q}{r} = \frac{p}{r} + \frac{q}{r}$. In this case each of the terms $2\sqrt{x}$ and x^4 is divided by $4x$
- Change $\sqrt{x}$ to $x^{\frac{1}{2}}$
- You are using: $\int ax^n dx = \int \frac{an^{n+1}}{n+1} dx$

Integrate 1st term

- Notice that the index $-\frac{1}{2}$ is increased by 1 giving $-\frac{1}{2} + 1 = \frac{1}{2}$

Complete the integration

- Notice that $\frac{x^3}{4}$ can also be written $\frac{1}{4}x^3$ so alternatively: $\int \frac{1}{4}x^3 dx = \frac{\frac{1}{4}x^4}{4} + c$. Now multiply top and bottom of fraction by 4 to give $\frac{x^4}{16} + c$.

Correct expression

- Finding $c = -1$ is not enough. You must state the expression for $h(x)$.

see p 79

Q8. (a)

$m = 1$ and $n = \sqrt{3}$ ✓ ✓

2 marks

Interpret graphs

- The 'normal' sine graph has amplitude $= 1$. This gives $m = 1$. The amplitude of the cosine graph shown is $\sqrt{3}$ so $n = \sqrt{3}$.

see pp 28–29

Q8. (b)

$f(x) = \sin x$ and $g(x) = \sqrt{3}\cos x$

So $f(x) - g(x) = \sin x - \sqrt{3}\cos x$

Let $\sin x - \sqrt{3}\cos x$

$= k\sin(x - a),\ k > 0$

$\Rightarrow \sin x - \sqrt{3}\cos x$

$= k\sin x\cos a - k\cos x\sin a$ ✓

Now equate coefficients of $\sin x$ and $\cos x$:

$\left.\begin{array}{l} k\cos a = 1 \\ k\sin a = \sqrt{3} \end{array}\right\}$ since both $\sin a$ and $\cos a$ are positive a is in 1st quadrant ✓

Divide: $\dfrac{k\sin a}{k\cos a} = \dfrac{\sqrt{3}}{1}$

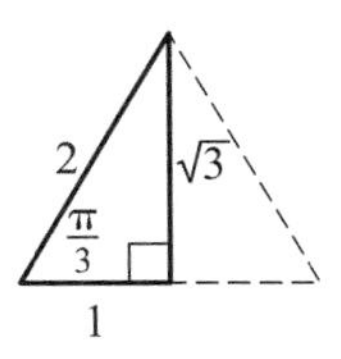

$\Rightarrow \tan a = \sqrt{3}$

$\Rightarrow a = \dfrac{\pi}{3}$ ✓

Square and add:

$(k\cos a)^2 + (k\sin a)^2 = 1^2 + \left(\sqrt{3}\right)^2$

$\Rightarrow k^2\cos^2 a + k^2\sin^2 a = 1 + 3$

$\Rightarrow k^2(\cos^2 a + \sin^2 a) = 4$

$\Rightarrow k^2 \times 1 = 4 \Rightarrow k = 2\ (k > 0)$ ✓

So, $f(x) - g(x) = \sin x - \sqrt{3}\cos x$

$= 2\sin\left(x - \dfrac{\pi}{3}\right)$

4 marks

Strategy

- You must clearly show the expansion of $k\sin(x - a)$. To do this you use the addition formulae: $\sin(A \pm B) = \sin A\cos B \pm \cos A\sin B$ which is given on your formulae sheet.

Compare coefficients

compare $(1)\sin x \quad -(\sqrt{3})\cos x$

$(k\sin x\cos a) \quad -(k\cos x\sin a)$

giving $k\cos a = 1$ and $k\sin a = \sqrt{3}$

Find a

- You should recognise $\sqrt{3}$ as an exact value leading to the angle $\frac{\pi}{3}$.

Find k

- For finding a you used $\frac{\sin a}{\cos a} = \tan a$ and for finding k you use $\sin^2 a + \cos^2 a = 1$. If you try to apply a learnt 'formula' for finding k sometimes mistakes creep in. It is better to understand that squaring and adding both sides of the two equations leads to the value of k, and just 'do the mathematics' at the time, $k^2 = 4$, $k = -2$ is not an allowable value since $k > 0$.

see pp 34–35

Q8. (c)

$$y = 2\sin\left(x - \frac{\pi}{3}\right)$$

$$\Rightarrow \frac{dy}{dx} = 2\cos\left(x - \frac{\pi}{3}\right) \quad ✓$$

For a gradient of 2, set $\frac{dy}{dx} = 2$

$$\Rightarrow 2\cos\left(x - \frac{\pi}{3}\right) = 2$$

$$\Rightarrow \cos\left(x - \frac{\pi}{3}\right) = 1$$

$$\Rightarrow x - \frac{\pi}{3} = 0 \text{ (or } 2\pi)$$

$$\Rightarrow x = \frac{\pi}{3}\left(\text{or } 2\pi + \frac{\pi}{3}\right)$$

But $0 \le x \le \pi$ so $x = \frac{\pi}{3}$ is the only solution. ✓

2 marks

Strategy

- You are told the gradient so you have to differentiate and find the x-value that satisfies $\frac{dy}{dx} = 2$. The word 'hence' is very important. It is telling you to use the previous result. You will lose marks if you do not do this.

  Differentiation involves the chain rule:

  $y = 2\sin(g(x)) \Rightarrow \frac{dy}{dx} = 2\cos(g(x)) \times g'(x)$

  In this case $g(x) = x - \frac{\pi}{3}$ so $g'(x) = 1$ so

  $\frac{dy}{dx} = 2\cos\left(x - \frac{\pi}{3}\right) \times 1 = 2\cos\left(x - \frac{\pi}{3}\right)$

Equation and solution

- For $\cos\theta = 1$

  $\theta = \ldots -2\pi, 0, 2\pi \ldots$

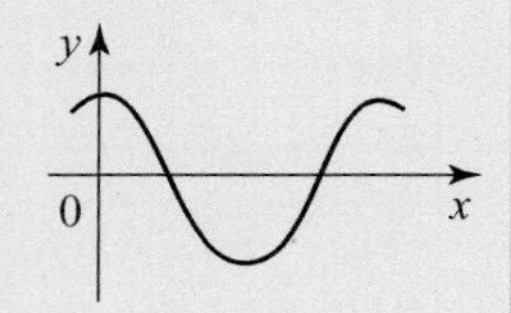

- In this question, the angle is $x - \frac{\pi}{3}$ so to finally find x you must add $\frac{\pi}{3}$.

see p 62, pp 72–73

Q9.

Intersection with $y = 15$ solve:

$$\left.\begin{array}{l} y = 15 \\ y = x^4 - 1 \end{array}\right\} \quad \text{so } x^4 - 1 = 15 \Rightarrow x^4 = 16 \Rightarrow x = 2 \text{ or } -2$$

For x-intercept set $y = 0$ ✓
so $x^4 - 1 = 0 \Rightarrow x^4 = 1$
$\Rightarrow x = 1$ or -1 ✓

Here is the diagram:

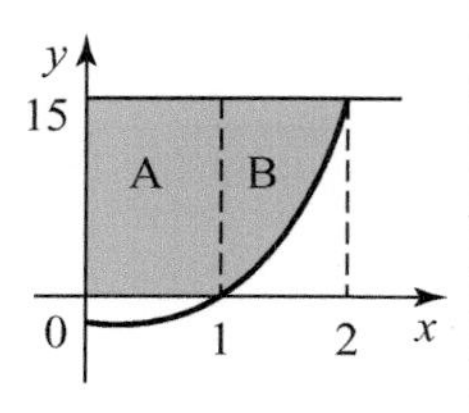

Area A is a rectangle:

Area $A = 1 \times 15 = 15$ unit^2

(continued to next page)

Limits

- It is essential to find the x values of the intersection of the lines $y = 15$ and x-axis with the curve. These values will be used later in the integral for finding the area between two graphs.

x-values

- The relevant values are $x = 1$ and $x = 2$. Notice that all the work can be done in the 1st quadrant and then doubled at the end, since the y-axis is an axis of symmetry for the diagram.

Q9. Continued

Area $B =$

$= \int_1^2 15-(x^4-1)\,dx$ ✓

$= \int_1^2 16-x^4\,dx = \left[16x - \frac{x^5}{5}\right]_1^2$ ✓

$= \left(16\times 2 - \frac{2^5}{5}\right) - \left(16\times 1 - \frac{1^5}{5}\right)$ ✓

$= 32 - \frac{32}{5} - 16 + \frac{1}{5} = 16 - \frac{31}{5} = \frac{49}{5}$ unit^2 ✓

Area A + Area B

$= 15 + \frac{49}{5} = \frac{75}{5} + \frac{49}{5} = \frac{124}{5}$ unit^2 ✓

By symmetry, the required area

$= 2 \times \frac{124}{5} = \frac{248}{5}$ units^2 ✓

8 marks

Strategy
- You should know to use integration to find the area between $y = 15$ and the curve (area B) and add the area of Rectangle A. There are other methods possible.

Integration
- Notice

$\int a\,dx = ax + c$ and $\int x^n = \frac{x^{n+1}}{n+1} + c$

(a is a constant)
- The constant c is not needed when there are limits.

Limits
- A mark is allocated for correct use of limits 1 and 2.

Evaluate
- Using $\int_a^b f(x)dx = [F(x)]_a^b = F(b) - F(a)$

Where $F(x)$ is the result of integrating $f(x)$.

Strategy
- Knowing what to add together!

Calculation
- Final answer is gained by doubling as you have only found the area in the 1st quadrant.

see pp 110–111

Q10. (a)

(0,9) lies on the curve

So $y = 9$ when $x = 0$

$y = -x^2 + a$ gives $9 = -0^2 + a$

$\Rightarrow a = 9$. ✓

1 mark

Calculation
- If (p,q) lies on a graph with equation $y = f(x)$ then $q = f(p)$. In other words, the coordinates of the point can be substituted into the equation. In this case, this gives the value of a.

see p 11

Q10. (b)

$f(x) = -x^2 + 9 = 9 - x^2$

To find coordinates of P set $x = m$

So $f(m) = 9 - m^2 \Rightarrow P(m, 9 - m^2)$

$AP = 9 - m^2$ ✓

1 mark

Calculation of y-coordinate
- As can be seen from this diagram, the length AP is the y-coordinate of P. The x-coordinate of P you know is m, i.e. $x = m$.

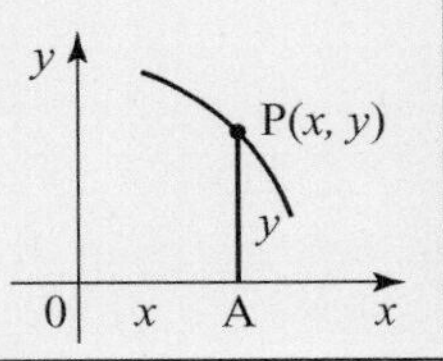

see p 11

Q10. (c)

The area of the rectangle, $A(m)$ is given by:

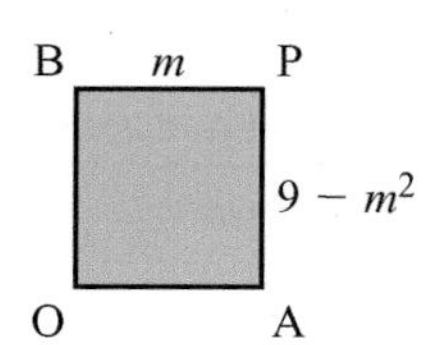

$A(m) = m(9 - m^2)$ ✓

$= 9m - m^3 \Rightarrow A'(m) = 9 - 3m^2$ ✓

For stationary value set $A'(m) = 0$ ✓

$\Rightarrow 9 - 3m^2 = 0 \Rightarrow 3m^2 = 9$

$\Rightarrow m^2 = 3$

So $m = \sqrt{3},\ m \neq -\sqrt{3}$ since $m > 0$ ✓

m:		$\sqrt{3}$	
$A'(m) = 9 - 3m^2$:	$+$		$-$
Shape of graph:			
nature:	/	max	\

✓

So $m = \sqrt{3}$ gives a maximum value for the area of rectangle $OAPB$. ✓

6 marks

Communication

- Area of a rectangle is given by length × breadth, and in this case the dimensions are m units × $(9 - m^2)$ units.

Differentiate

- The function $A(m)$ gives the area of the rectangle, whereas $A'(m)$ is the gradient function and determines the slope of the graph $y = A(m)$ showing the various areas as m changes.

Strategy

- Setting $A'(m)$ to zero determines which values of m give stationary points on the Area graph.

Solve

- Here there is a positive and a negative value of m. Only the positive value makes sense, as you are told that $0 < m < 3$.

Justify

- The 'nature' table is required to show that $m = \sqrt{3}$ does give a maximum value for the Area.

Communication

- A clear statement summarising your findings is needed for this final mark.

see pp 107–108

Q10. (d)

$A(\sqrt{3}) = \sqrt{3}\left(9 - (\sqrt{3})^2\right)$

$= \sqrt{3}(9 - 3)$

$= \sqrt{3} \times 6$ ✓

So $6\sqrt{3}$ unit² is the maximum area.

1 mark

Calculation

- When you read the word 'exact' you know to steer clear of decimal approximations. Your answer will be an integer value, a fraction (rational number) or a surd (involving roots) or perhaps an expression involving π or e, … but never approximate decimals.

see p 108